Atmospheric Research in Antarctica

Atmospheric Research in Antarctica

Present Status and Thrust Areas in Climate Change

Edited by
Neloy Khare
Ministry of Earth Sciences, India

CRC Press
Taylor & Francis Group
Boca Raton London New York

CRC Press is an imprint of the
Taylor & Francis Group, an **informa** business

CRC Press
Taylor & Francis Group
6000 Broken Sound Parkway NW, Suite 300
Boca Raton, FL 33487-2742

Printed on acid-free paper

International Standard Book Number-13: 978-0-367-40880-0 (Hardback)

Library of Congress Control Number: 2019952862

Visit the Taylor & Francis Web site at
http://www.taylorandfrancis.com

and the CRC Press Web site at
http://www.crcpress.com

Contents

Foreword

The vast potential of the icy continent Antarctica for conducting basic research relevant to atmospheric sciences is deeply acknowledged by the scientific community at large, as it holds several unique features and answers to many scientific challenges. The unique location of the earth's aurora region, which lies typically between the 65° and 75° geomagnetic latitudes in both hemispheres, represents a region where there is a direct transfer of solar wind energy and energetic particles into the earth's near atmosphere environment and offers an opportunity for the study of solar terrestrial energy coupling.

With the establishment of a second Indian research base, Maitri, during 1987–1989, India captured the scientifically significant location, which is just equatorwards of the auroral oval for magnetically quiet conditions but is well within the oval under geomagnetically disturbed conditions. This makes Maitri unique for the study of the response of the auroral oval to changing solar and geomagnetic conditions.

We have completed 39 years of the Indian Antarctic Programme. During this time of almost four decades, research in Antarctic meteorology and atmospheric science has emerged as an exciting opportunity and reward for scientists and society. It is therefore imperative to discuss, collate, and evaluate the data, results, and significant achievements of the past.

The present book, *Atmospheric Research in Antarctica: Present Status and Thrust Areas in Climate Change,* has ably covered most of the scientific aspects of Antarctic meteorology and atmospheric sciences. I am sure this book will act as a catalyst for budding researchers to take up atmospheric science as a challenging career.

Dr. K. J. Ramesh

Preface

The icy continent Antarctica holds the answers to many scientific riddles and thus offers tremendous scope for carrying out research in various scientific fields. Besides other scientific disciplines, this region provides challenging opportunities for atmospheric scientists and meteorologists.

Maitri is an ideal site to study the response of the auroral oval to changing solar and geomagnetic conditions. Additionally, the study of the ozone depletion phenomenon over the Antarctic region has been an area of major excitement.

The present volume represents a panoramic view of the developments in the field of Antarctic atmospheric sciences and meteorology. Each participating organization has reviewed the progress broadly in the following areas:

Geomagnetism and aeronomy

Middle atmospheric studies

Global and climate change studies

Tremendous progress has been made in the front-ranking areas of solar terrestrial research, ozone depletion and climate change, with a multi-instrumental and multi-institutional (India Meteorological Department [IMD], Indian Institute of Geomagnetism [IIG], National Physical Laboratory [NPL], Physical Research Laboratory [PRL], Indian Institute of Tropical Meteorology [IITM], Space Applications Centre [SAC], Barkatullah University, and the K. Banerjee Centre for Atmospheric and Ocean Studies, University of Allahabad [KBCAOS]) approach, during the initial two decades of Indian Antarctic program and the same is continued with renewed vigour and zeal.

The importance of the location of Maitri in the study of solar terrestrial research of magnetic disturbances has been illustrated by several measurements made around Maitri. In middle atmospheric studies, measurements reveal that the recovery of ozone depletion does not take place in Antarctica as opposed to the Arctic due to frequent breakup of the polar vortex by disturbances, the mechanism for which lies in the chlorine reservoir in the polar stratosphere. Although regular operation is being continued for geomagnetic measurements, ionospheric absorption, very low frequency (VLF) emission, and optical aeronomy, the setting up of the facility, such as digital ionosonde, high frequency (HF) Doppler, and for visual auroral imaging, has not been achieved in the initial two decades of Indian Antarctic program. A secular variation in horizontal components of the total field at Maitri is observed as compared to other stations, and systematic study is to be continued.

Meteorological and boundary layer parameters are being regularly monitored by India Meteorological Department (IMD) and National Physical

Laboratory (NPL) for which adequate arrangements have been made that include surface as well as radiosonde and satellite data, tower-based boundary layer measurements, and so on . However, there is a need for mean, daily, seasonal, and inter-annual climatology that is to be derived, and its usage must be ensured, as the same may be a useful data set for various applications. Daytime auroral measurements, although weak compared to background, are now possible with equipment developed at Physical Research Laboratory (PRL). During the 13th and 14th expeditions, unique results were obtained with meridional scanning mode, which gives evidence for wave generation due to joule heating of atmospheric electrojet current. It is now possible to locate polar cap boundaries by monitoring 6300 Å oxygen emission, and efforts shall be made to have such a facility at Maitri, especially given that the instrument is simple and elegant.

Monitoring of ozone and UV radiation, total and vertical profiles, trace gases, aerosols, and atmospheric electricity is being done by NPL, but successful operationalization of a microwave spectrophotometer and laser heterodyne spectrometer has not been achieved, and light detection and ranging (LIDAR) was not installed at the site in the initial two decades of Indian Antarctic Program. Although IMD regularly observes ozone and trace gases that match other comparable findings, efforts should be made for the detailed study of (a) the evolution and decay of the ozone hole in different years using profile data, its interannual variability, and the role of dynamics vs chemistry; (b) ozone distribution during the dark period and impact on the development of the ozone hole, if any; and (c) potentiality of microwave-based instruments for the study of chemistry. Antarctica is pristine; no seasonal variations are observed in trace gases such as CO_2, CH_4, CO, and SO_2 in the absence of local sources and with anthropogenic activities and aerosol being very low at Maitri as compared to Southern Ocean and Northern Ocean regions and coastal areas. The station should be exploited for long-term monitoring programs. Systemic studies have been undertaken with NPL's multi-wavelength radiometer and PRL's Microtop and Sun photometer in a regular manner for aerosol studies, and these should be continued.

The comparative study of discrete and continuous VLF emissions showed the evidence of their presence at Antarctica. Such studies cover diurnal variations, propagation mechanisms, generation and origin of whistlers, and mapping of electric fields from low to high latitudes. A triangular loop antenna has been installed at Maitri by the Barkatullah University, and diurnal variations were observed during the 20th and 22nd expeditions. In the next phase, mapping is in progress with the help of direction finding of VLF emissions in general and whistlers.

KBCAOS is embarking upon modeling and simulation studies. Similarly, Indian Institute of Geomagnetism (IIG) plays a critical role in observations for studies in geomagnetism and allied fields. The location of Maitri, as a sub-auroral station, has been reported to be important in predicting space weather such as nowcasting, cryosphere weather, and interplanetary weather and in the study of global electric circuits and conjugate area. The regular observation recorded at Antarctica by IIG and other historical records shows a rapid decrease in the total geomagnetic field intensity at and around Maitri; the rate of decrease drops as one moves away from Maitri to the poleward and equator directions, as seen from the data after 1940.

India was the first country to take up the magnetometer triangulation experiment to determine the presence and movement of small-scale auroral current systems over Maitri. The efforts are on to explore the storm–sub-storm relationship and to determine the velocity of the overhead auroral current system, variations in air-earth currents, and GPS measurements for crustal deformations and glacial movement studies in collaboration with the Survey of India.

IMD has generated a rich data set by ongoing efforts since the very first expeditions.

On the other hand, NPL's contribution in monitoring aerosols in Antarctica and the results obtained are encouraging to show that stability has been achieved after the Montreal Protocol. There has been some failure as well, such as the implementation of the laser heterodyne system. It is advisable to install micropulse LIDAR at Antarctica for regular measurements.

It is heartening to mention that India also tapped the opportunity of observing shadow bands during the unique total solar eclipse on November 23, 2003, in the first-ever opportunity from the icy continent. The observations have been analyzed for the study of shadow bands and their relationship with the total solar eclipse, Antarctic lower atmosphere boundary layer, solar corona, and other features of solar activity during the declining phase of the sunspot cycle. The effort has been laudable, especially in view of spontaneous and coherent efforts of the scientific community and logistics team. Efforts have also been made for the calibration of the AWIFS sensor onboard Resourcesat-1 at Antarctica.

Having completed almost four decades of the Indian Antarctic Programme, it is most appropriate now to collate and consolidate the results and achievements obtained during the initial two decades of Indian Antarctic Program of various participating organizations in the field of Antarctic meteorology and atmospheric sciences.

Accordingly, a review of Antarctic aeronomy and atmospheric sciences has been incorporated in this book, which covers an overview of Antarctic aeronomy and atmospheric sciences by Subbaraya and contains articles on greenhouse gases and ozone monitoring by Jain et al., as well as VLF phenomena by Gwal et al., Antarctic meteorology by Lal, and mathematical modeling of atmosphere and ocean processes around Antarctica by Pandey et al. Similarly, the significance of the data obtained from the permanent geomagnetic observatory at Antarctica is documented by Sinha et al., atmospheric electricity and aerosol investigations over Antarctica by Kamra and Deshpande, total solar eclipse–related studies by Bhandari, calibration of the AWIFS sensor by Manjul et al., and space weather on GPS performance at high-latitude regions and its comparison with low-latitude regions by Gwal and Choudhary.

It is hoped that this book will be a reference for researchers in the field of Antarctic meteorology and atmospheric sciences.

Neloy Khare
New Delhi

Acknowledgments

Dr. M. N. Rajeevan, Secretary to the Government of India, Ministry of Earth Sciences, who has shown keen interest in and provided valuable guidance and encouragement for addressing the issues of Antarctic meteorology and atmospheric sciences in this book, is gratefully acknowledged.

Deep gratitude is expressed to all former Secretaries of Ministry of Earth Sciences (erstwhile Department of Ocean Development) especially Dr. Harsh K. Gupta, Dr. Prem Shankar. Goel, Dr. Shailesh R. Nayak for their academic and moral support to bring out the present book on the most relevant topic

All contributing authors are deeply acknowledged for the significant scientific inputs that they provided to this book. Without their support and cooperation, this book would never have materialized in time.

We gratefully appreciate the valuable technical expertise and help provided by Dr. P. C. Joshi (SAC), Dr. P. C. S. Devera (IITM), Dr. Vikram Kumar (NPL), Dr. M. K. Tiwari (NPL), Dr. K. J. Ramesh (IMD), Dr. Akhilesh Gupta (DST), Dr. O. P. Mishra (NCS), Dr. Rajiv Nigam (NIO), Dr. A. Arunachalam (ICAR), Prof. Archana Bhattacharya (IIG), Shri B. Lal (IMD), and Dr. M. R. Ramesh Kumar (NIO).

The officers and staff of Ministry of Earth Sciences and the National Centre for Polar & Ocean Research, Goa, deserve appreciation for their direct or indirect involvement in the preparation of this book.

Special thanks are due to Dr. Debishree Khan (NEERI), Dr. Shabnam Choudhary (MoES), Shri Hari Dass Sharma (MoES) and Shri Hemant Kumar (MoES) for their assistance and support rendered in the preparation of this book.

Editor

Dr. Neloy Khare, presently Adviser/Scientist G to the government of India at MoES, has very distinctive acumen, not only of administration but also of quality science and research in his areas of expertise covering a large spectrum of geographically distinct locations like the Antarctic, Arctic, Southern Ocean, Bay of Bengal, Arabian Sea, Indian Ocean, and so on. Dr. Khare has almost 30 years of experience in the field of paleoclimate research using paleobiology (paleontology)/teaching/science management/administration/ coordination for scientific programs (including the Indian Polar Programme) and so on. Having completed his doctorate (PhD) on tropical marine regions and Doctor of Science (DSc) on southern high-latitude marine regions and environmental/climatic implications using various proxies, including foraminifera (micro-fossils), he has made significant contributions in the field of the paleoclimatology of southern high-latitude regions (Antarctic and Southern Ocean) using micropaleontology as a tool. These studies, coupled with his paleoclimatic reconstructions from tropical regions, help explain causal linkages and tele-connections between the processes taking place in southern high latitudes with that of climate variability occurring in tropical regions.

Dr. Khare has been awarded Honorary Professor and Adjunct Professor by many Indian universities. He has a very impressive list of publications to his credit (121 research articles in national and international scientific journals, 3 special issues of national scientific journals as guest editor; he edited a special issue of *Polar Science* [Elsevier] as its managing editor and has authored/edited many books, contributed 130 abstracts to various seminars, and written 23 popular science articles and 5 technical reports). The government of India and many professional bodies have given him many prestigious awards for his humble scientific contributions to past climate changes/oceanography/ polar science and southern oceanography. The most coveted award is the Rajiv Gandhi National Award-2013, conferred by the Honourable President of India. Others include the ISCA Young Scientist Award, Boyscast Fellowship, CIES French Fellowship, Krishnan Gold Medal, Best Scientist Award, Eminent Scientist Award, ISCA Platinum Jubilee Lecture, and IGU Fellowship, in addition to many others.

Dr. Khare has made tremendous efforts to popularize ocean science and polar science across the country by way of delivering many invited lectures and radio talks and publishing popular science articles. Dr. Khare sailed in the Arctic Ocean as a part of "Science PUB" in 2008 during the International Polar Year campaign for scientific exploration in the Arctic Ocean and became the first Indian to sail in the Arctic Ocean.

Contributors

B. C. Arya has worked as a senior scientist at the National Physical Laboratory, New Delhi.

S. M. Bhandari has worked as a senior scientist at the Space Applications Centre, Ahmedabad. He led the Special Expedition to Antarctica during the total solar eclipse event November 23, 2003.

S. Choudhary is a researcher at Department of Physics, Rabindra Nath Tagore University, Raisen.

C. G. Deshpande is a scientist at the Indian Institute of Tropical Meteorology, Pune, with research interest in the fields of aerosol physics and atmospheric electricity.

A. Dhar has worked as a technical officer at Indian Institute of Geomagnetism, Navi Mumbai, and has been the Leader of 18th IAE. He has wintered in Antarctica thrice and stayed many times during summer period. His main field of work is Antarctic geomagnetism.

H. N. Dutta has worked as a senior scientist at the National Physical Laboratory New Delhi and has been associated with Antarctic research for more than two decades.

S. Dwivedi is a Professor in the Physics Department at Allahabad University.

S. D. Ghude is a researcher in the Indian Institute of Tropical Meteorology (IITM), Pune.

S. Gurubaran is a resercher at Indian Institute of Geomagnetism (IIG), Mumbai.

A. K. Gwal has worked as a professor and head of the Department of Physics and Electronics at Barkatullaha University, Bhopal. He has extensive experience in the field of VLF propagation in the Antarctic region. Presently, he is the Vice chancellor of Rabindranath Tagore University, Raisen.

A. Hanchinal has worked as a scientist at Indian Institute of Geomagnetism, Navimumbai, and has been the Leader of 22nd IAE. He has wintered in

Antarctica twice and stayed many times during the summer period. He is also Technical Officer–III working in observatory and data analysis (ODA), with his main field working in Antarctic geomagnetism.

S. L. Jain has worked as a senior scientist at the National Physical Laboratory, New Delhi, and has been associated with Antarctic research for more than two decades.

A. K. Kamra has worked as an emeritus scientist at the Indian Institute of Tropical Meteorology, Pune, with research interest in the fields of aerosol physics and atmospheric electricity.

P. S. Kulkarni is a researcher at the Radio and Atmospheric Science Division of the National Physical Laboratory.

R. P. Lal served as Director at the India Meteorological Department. He has wintered twice in Antarctica and led the 21st IAE.

S. S. Manjul has worked as a senior scientist at the Space Applications Centre, Ahmedabad.

A. P. Mishra has completed his Master's degree in geophysics from Banaras Hindu University and submitted his D. Phil. dissertation at the University of Allahabad. He has performed modeling experiments to study air-sea interaction processes.

A. Mitra has worked as senior research fellow at K. Banerjee Centre of Atmospheric and Ocean Studies at Allahabad University.

P. Narayanbabu is a researcher at the Space Applications Centre, Ahmedabad.

S. Nath is a researcher at the Radio and Atmospheric Science Division of the National Physical Laboratory.

A. C. Pandey earned three Master's degrees, M.Sc. (physics, 1984), M.B.A. (marketing, 1993), and M.Sc. (mathematics, 1996), from University of Allahabad. He earned his D. Phil. from University of Allahabad in 1995. Presently he is the Director at Inter University Accelerator Centre (IUAC).

V. K. Pandey is a faculty in the Physics Department at Universityof Allahabad.

B. M. Pathan has served as senior scientist and head of the Antarctic Division at Indian Institute of Geomagnetism, Navi Mumbai. He has extensive experience in the field as Associate Professor Head of Geomagnetic Observatories.

S. Rai is a faculty in the Physics Department at University of Allahabad.

D. R. M. Samudraiah has worked as a researcher at the Space Applications Centre, Ahmedabad.

R. Singh has worked as a senior scientist at the National Physical Laboratory, New Delhi, and was associated with Antarctic research for more than two decades.

S. K. Singh is a faculty at Department of Physics,Laxmi Narayan College, Bhopal.

A. K. Sinha is a Scientist at Indian Institute of Geomagnetism, Navi Mumbai.

B. H. Subbaraya was an eminent scientist of Indian Space Research Organization (ISRO) at Bangalore. He has extensive experience in the field of atmospheric science.

R. S. Tanwar is a scientist at the National Physical Laboratory, New Delhi.

K. C. Tripathi has worked as a senior research fellow at K. Banerjee Centre of Atmospheric and Ocean Studies at Allahabad University.

Abbreviations

ACC	Antarctic Circumpolar Current
AOD	Aerosol Optical Depth
AOS	Acousto-Optic Spectrometer
APT	Automatic Picture Transmission
CAN	Cosmic Radio Noise absorption
CCD	Charged Couple Device
CGA	Cup Generator Anemometer
CPS	Central Plasma Sheet
DIWE	Distant Indicating Wind Equipment
DMS	Di Methyl Sulfide
EAE	Eastward Auroral Electrojet
ELF	Extremely Low Frequency
EOF	Empirical Orthogonal Function
FAC	Field Aligned Currents
FFT	Fast Fourier Transform
GEC	Global Electric Circuit
GFDL	Geophysical Fluid Dynamics Laboratory
GPS	Global Positioning System
GTR	Ground Truth Radiometers
GTS	Global Telecommunication System
HFA	Heat Flux Anomaly
IMF	Interplanetary Magnetic Field
MOM	Modular Ocean Model
MWR	Microwave Radiometer
OGCM	Ocean General Circulation Models
PACIO	Pacific and Indian Oceans in the Indonesian Archipelago
POM	Princeton Ocean Model
PTFE	Polytetrafluoroethylene
SIO	Southern Indian Ocean
SSH	Sea Surface Height
SST	Sea Surface Temperature
TEC	Total Electron Content
TSE	Total Solar Eclipse
UV-B	Ultra Violet-B
VLF	Very Low Frequency

1

Indian Antarctic Programme—Aeronomy and Atmospheric Sciences: An Overview

B. H. Subbaraya*

CONTENTS

Introduction

The Indian Antarctic Programme was initiated in 1981. The major effort in the first decade was in establishing the Indian station at Dakshin Gangotri and later at Maitri and the operationalization of these stations. After this phase, in the next decade, there was a major facelift to scientific programs. After an exhaustive review by an INSA committee under the chairmanship of Dr. S. Varadarajan, an expert committee headed by Prof U.R. Rao worked toward a long-term plan of research in Antarctica and formulated a 10-year work profile in the fields of earth sciences, atmospheric sciences and aeronomy, biological and medical sciences, and environmental sciences, as well as engineering and technical services.

In atmospheric sciences and aeronomy, the committee pointed out the high potential of Antarctica in general and the location of the Indian station

* Deceased

for frontier research in diverse areas such as solar-terrestrial research, middle atmospheric sciences, meteorology, and boundary layer studies, as well as global change studies. Further, it was recognized that atmospheric sciences, unlike other disciplines, require the collection of continuous data over an extended period, and certain aspects of polar science need correlated measurements of several related parameters. A long-term plan of an operational nature with the participation of several institutions was formulated.

The emphasis of the long-term plan has been on solar-terrestrial research, the impact of charged particle emission from the sun on terrestrial electric and magnetic field parameters and ionospheric behavior, ozone hole and related chemistry, trace gases, and aerosols of interest in global change studies. It provided a goal of setting up a permanent laboratory with a few instruments continuously running on an operational (routine) basis both for upper and middle atmospheric studies as well as meteorology. During the last few years, a good portion of this long-term plan has been realized, though not fully. The undertaken programs and studies could be covered under the following major headings: geomagnetism and aeronomy solar terrestrial physics, middle atmospheric studies, ozone and trace gases, meteorology and boundary layer studies, and global change studies.

Geomagnetism and Aeronomy

The program consisted of geomagnetic field measurements—x, y, and z components and total field micropulsation riometers and ionospheric absorption optical aeronomy—daytime auroral studies, auroral imaging, visual auroral imaging, digital ionosonde, and HF Doppler (not yet done).

Middle Atmospheric Studies

Ozone and UV radiation, Dobson spectrophotometer, Brewer spectrophotometer, and balloon ozonesonde UV-B measurements.

Trace Gas Measurements: CO_2, CH_4, NOx

Aerosols and radiation

Atmospheric electricity

Laser heterodyne spectrometer and millimeter (GHz) spectrometer were attempted, but did not materialize into an operational program

Meteorology and Boundary Layer Studies

Routine meteorological parameters, radiosonde

Boundary-layer—Tower based as well as slow-rising balloons

The following sections will provide selected highlights of the results from the program over the last 10 years or so, more an illustration of studies conducted rather than an exhaustive summary of the scientific results obtained.

Daytime Auroral Emission Measurements

Auroral emissions are the outcome of the complex interaction between energetic charged particles (of solar wind and magnetospheric origin) and the constituents of the earth's atmosphere. Most of the visible emissions emanate from different altitudes in the lower thermosphere, 100 to about 150 km. The auroral zone, the region of visibility of the auroral, is a circular zone of approximately 20° radius around the geomagnetic pole. The locus of maximum occurrence—the "auroral oval"—is a region typically 75° to ±85° in the sunward and 65° to ±75° latitude band in the anti-sunward direction. The precipitating particles are of three types, low-energy (soft) electrons, high-energy (hard) electrons, and protons (Basu et al., 1993). The emissions are mostly from oxygen and nitrogen atoms and molecules and arise both from neutral and ionized species. They are both in the UV and visible regions of the electromagnetic spectrum.

The auroral oval is different in the southern hemisphere from the northern hemisphere in shape, size, and location. The shape and size of the oval change with magnetic activity. At high magnetic activity levels, the oval expands.

The Indian Antarctic station Maitri is at the fringe of the oval, just outside during quiet conditions and within it under disturbed/storm conditions. Hence it is ideally situated for monitoring changes due to magnetic activity—the so-called space weather phenomena.

It is well known based on rocket and satellite experiments (Sivjee, 1983) that the characteristics of daytime auroral emissions are quite different from night-time emissions. Atomic emissions dominate during sunlit hours, whereas molecular bands dominate during nighttime. The study of daytime auroral characteristics has been handicapped for a long time due to lack of suitable instrumentation—difficulties associated with detection and measurement of weak emission features in the presence of a bright sunlit background.

A unique multiwavelength photometer capable of detecting faint emission features that constitute 0.05% of the sky and background was developed at PRL in the 80s (Narayanan et al., 1989) for rocket experiments and ground-based dayglow measurements. Using this development, a program of monitoring daytime auroral emissions by ground-based multiwavelength photometry was undertaken at Maitri (Sridharan et al., 1995) in the summers of 1994 and 1995—the 13th and 14th Indian Antarctic expeditions.

The photometer was operated in meridional scanning mode (changing the elevation in steps of 10° from 10° to 50°), as well as zenith observations mode pointing toward the south. Around-the-clock observations were made on 13 clear days during 1994. The 1995 expedition experienced mostly cloudy skies, and very few observations could be made.

Measurement Findings

1. A significant day-to-day variability in emission intensities—a typical characteristic of charged particle fluxes. Very high on some days, almost saturating the detectors.
2. Broad similarities between the characteristics of different emissions on any given day about the diurnal and latitudinal variations (Figure 1.1).
3. The diurnal variation shows an overall low level of intensities during most of the day, with a sharp peak around noon hours. Sometimes the peak is quite narrow, almost a spike. The enhanced intensities are confined to a narrow latitude zone around 64°S mag lat. All wavelengths show this feature. But sometimes there is a slight shift in the peak position between high-energy electron excited emissions and low-energy electron excited emissions by about 1°.
4. Very often a large wave of activity is associated with enhanced emissions (Figure 1.2), resulting in secondary peaks on some days. Wave activity is, however, limited to about 68° inv lat (Sridharan et al., 1995)

   The sharp peak is attributed to the fact that the emissions are geomagnetically controlled and have a fixed orientation with respect to the sun. Due to the separation between the geomagnetic and geographic poles and consequent precession of the magnetic pole with respect to the axis of rotation, the observational site cuts through the region of deposition around a specific time, in this case (Pallam Raju and Sridharan, 1998).
5. As expected, the overall intensities are much higher during geomagnetically disturbed periods. For example, February 6, 1994, $\sum kp = 43$, when compared to February 19, $\sum kp = 31$—almost a factor of two variability. Low-energy emissions, however, 5577 Å, do not show a large dependence on solar activity. This can be explained by atomic processes.

These results are unique. The experiment has not been repeated at Antarctica. Similar experiments were, however, conducted in the Arctic in later years. More extensive observations as well as a coordinated multi-technique, both ground-based and satellite-borne, is needed to enable a detailed study of space weather phenomena.

Locating the Polar Cap Boundary from 6300 Å Auroral Emission

This is a simple experiment with high scientific potential for a long-term/regular operational program. Canadian observations in the Arctic (Blanchard et al., 1965) have shown that a meridional scan of the 6300 Å oxygen emission

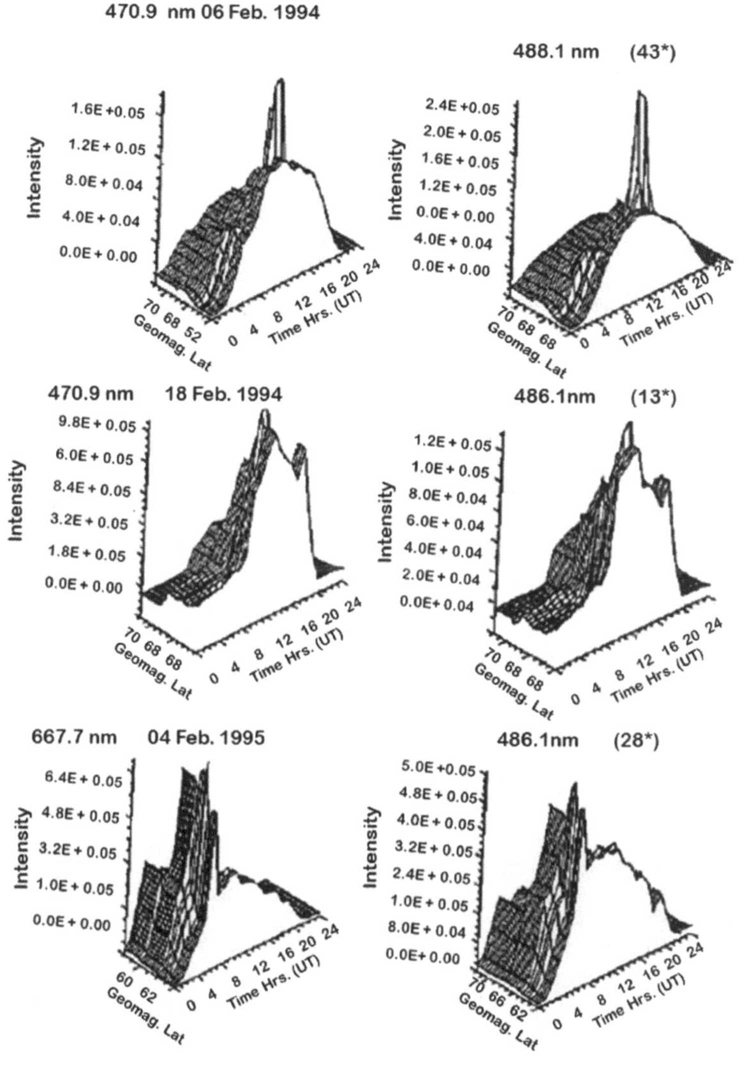

FIGURE 1.1
Emission patterns of the high-energy electron excited, 4861 Å; low-energy electron excited, 5577 Å; and the proton, 4709 Å, measured at the Indian Antarctic Station, Maitri (Sridharan et al., 1995).

will show a characteristic step function in emission intensity while crossing the polar cap boundary. The polar cap is the boundary between the open and closed field lines and hence its movement would track the changes in regions of particle precipitation. The boundary, as detected from the emissions, is found to correlate with that obtained from satellite-borne charged particle data. Hence, this is a simple and cheap method of monitoring the polar cap boundary to within a precision of 1° inv latitudes. A scanning photometer or

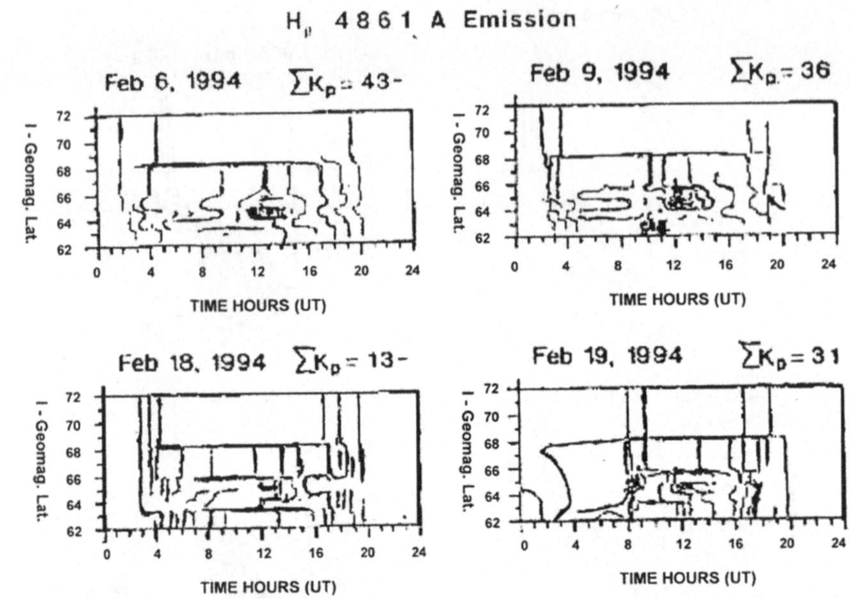

FIGURE 1.2
Ground projections of the HP 4861 Å emission intensities for magnetically quiet and disturbed periods depicting wave activity in the emission contours (Sridharan et al., 1995).

an all-sky imaging photometer can be used. Several groups in the country have the necessary instrument capability.

Geomagnetic Measurements

This was one of the components of the Indian Antarctic Programme right from the beginning. But in recent years, a systematic program of coordinated geomagnetic and ionospheric (riometer) measurements from at least two stations (if not three) has been undertaken. The goal of three stations has not yet been successful due to logistical problems. The main elements of the program have been three component measurements, x, y, and z; total field F occasionally; and riometer absorption and micropulsation on occasions. In recent years, new magnetometers with high sensitivity (0.1 nT) are being used instead of the old ones. The program has been handled mostly by Indian Institute of Geomagnetism scientists. The data is being studied along with satellite data on the interplanetary magnetic field.

Several interesting solar-terrestrial events have been studied in detail. Good correlation between riometer absorption and the signatures in the magnetic field is always found. An attempt has been made to determine the velocities associated with the currents under both quiet and disturbed conditions. Figures 1.3 and 1.4 illustrate some typical results. Figure 1.3 shows the data for May 11, 1999, a very quiet day with kp = 2, the day on

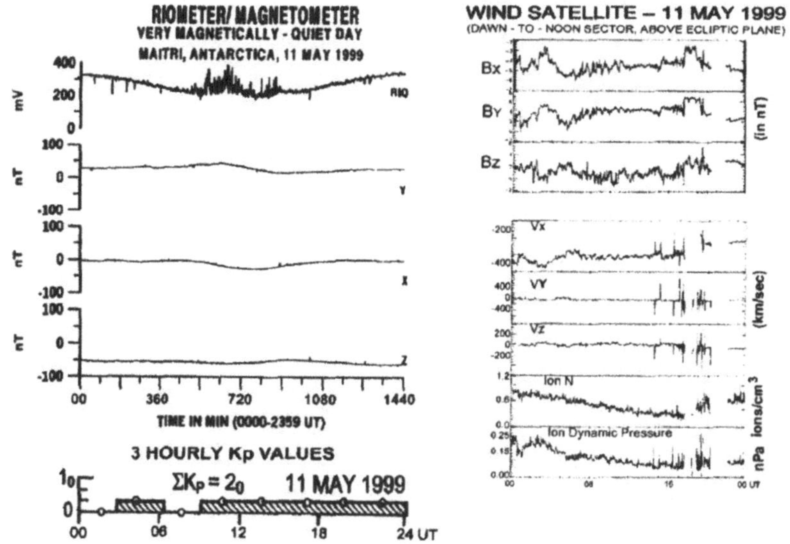

FIGURE 1.3
Geomagnetic and riometer measurements at Maitri on May 11, 1999, a very quiet day on which the solar wind almost disappeared (Rajaram et al., 2002).

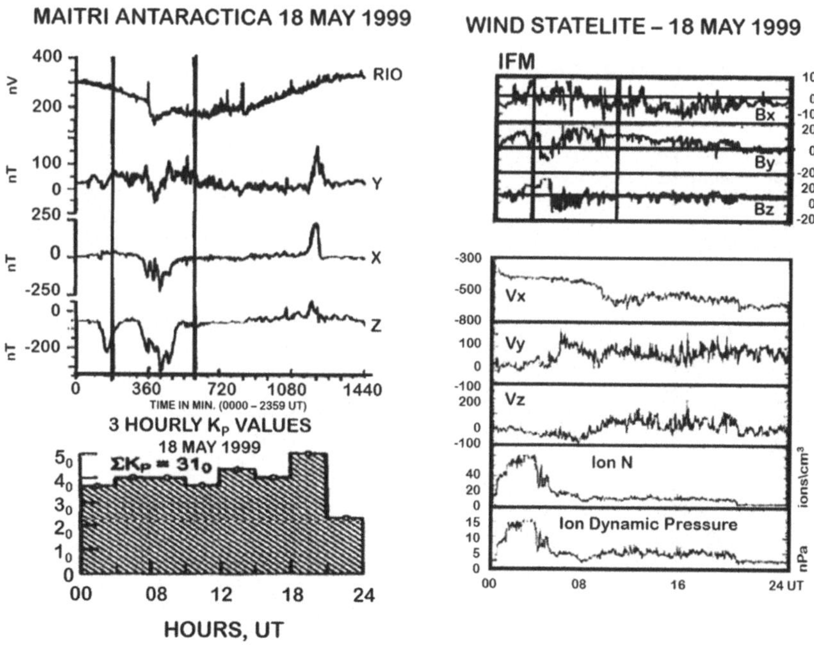

FIGURE 1.4
Geomagnetic and riometer measurement at Maitri on a disturbed day, May 18, 1999 (Rajaram et al., 2002).

which solar wind almost disappeared. All field components show very smooth plots with very little diurnal variation. Riometer data shows no charged particle effect. May 18, 1999, a week later, was a very disturbing day kp = 31 (Figure 1.4). Significant D region absorption on the riometer is seen, associated with signatures in magnetic field data. The day was characterized by high solar wind speeds of 500–700 km/sec, and Bz was negative during the disturbance.

Several studies have been made of the characteristics during quiet and disturbed time periods: secular variation in F and rapid decline in magnetic field intensity around Maitri in recent decades. The total magnetic field has been measured occasionally using flux gate and proton precession magnetometers at the Indian stations DG and Maitri. Measurements revealed a large decline (more than 900 nT) in the total magnetic field intensity from 1986 to 1996. Several other stations in the neighborhood of Maitri, ex-Russian station Novolazarevskaya and German station George Foster, also showed a similar large decline. This prompted a more detailed study of the problem using IGRF data (Rajaram et al., 2002). The analysis showed a decrease of about 9000 nT in the last 75 years (since 1920) in the region around Maitri (Figure 1.5). The zone of maximum decline is confined to a narrow horseshoe-shaped belt in the Antarctic region encompassing Maitri.

Outside this belt, the decrease is much smaller (Figure 1.6).

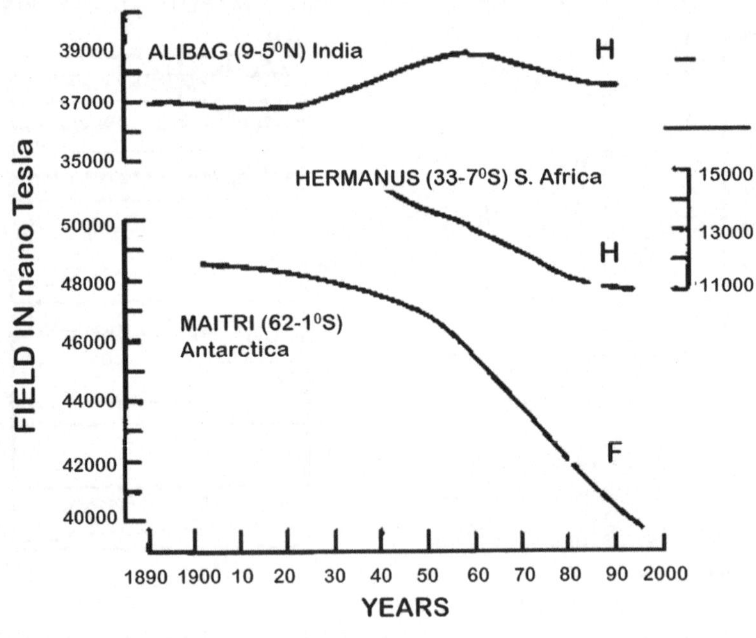

FIGURE 1.5
Rapid decline in the magnetic field intensity at the Indian Antarctic Station, Maitri, in recent decades (Rajaram et al., 2002).

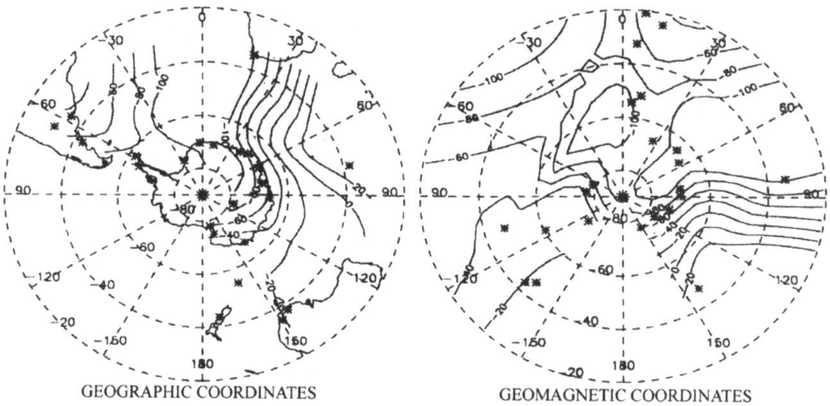

FIGURE 1.6
Total magnetic field variation in the Antarctic region during recent decades (Rajaram et al., 2002).

Middle Atmospheric Studies: Ozone and Trace Gases

The Indian Meteorological Department has been making regular measurements of total ozone using a Dobson spectrophotometer (which has been replaced by the more sophisticated Brewer spectrophotometer in recent years) and the vertical distribution of ozone by balloon ozonesonde. During September–November, the ozone hole period, the frequency of balloon ozonesonde ascents has been enhanced to make a detailed study of the evolution of the ozone hole and its recovery. There have been other attempts off and on, mostly by NPL scientists, using different instruments to study this problem; Figure 1.7 shows a typical variation of total ozone at Maitri for the year 1999 during August–December (Lal and Krishnamurthy, 2002). During the entire September–October month, total ozone values hovered around 150 DU, reaching minimum values of 130–140 DU for about 10 days during the peak of the ozone hole period. In November total, ozone recovered to pre-ozone hole values of 200–250 DU, reaching values close to 300 in December.

Figure 1.8 shows the ozone vertical distribution for October 4, 1999, when the ozone hole was at its severest. Comparing with the vertical distribution on April 23, it is seen that almost the whole of the ozone content in the region of the main ozone layer—12 to about 18 km—is eaten away during the ozone hole period. These are typical results. IMD has collected similar data for about 8 years now. This data can be used for a detailed study of the ozone hole phenomenon in the Maitri region. Several scientifically interesting studies connected with the role of dynamics vs chemistry can be addressed with this data. For example, the interannual variability with regular alternate year maximum and minimum in the ozone hole intensity, which was noticed in earlier years and vanished after the strong perturbation produced by the Pinatubo volcanic eruption in 1991, can be investigated. Whether the role

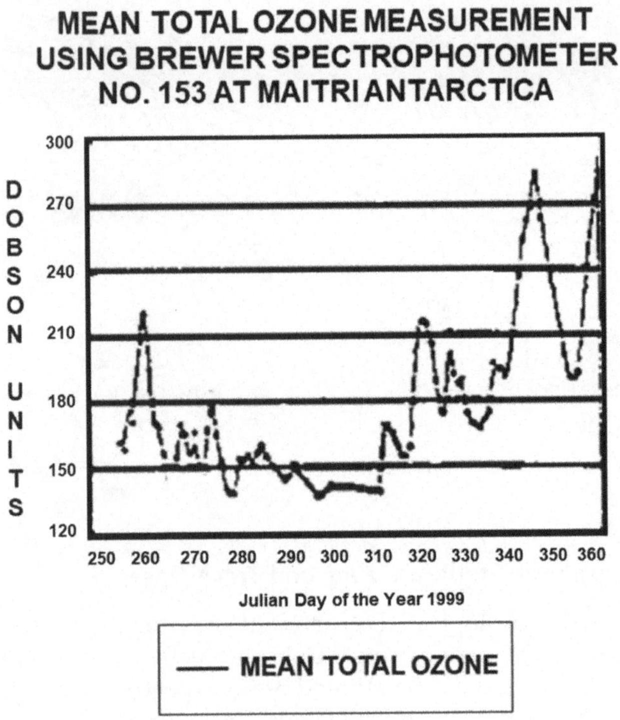

FIGURE 1.7
Total ozone measurement during the period August–December 1999 made by the IMD Brewer spectrophotometer at Maitri indicating the springtime ozone hole (Lal and Krishnamurthy, 2002).

of chemistry has finally taken over dynamics due to the very high chlorine levels needs to be ascertained.

NPL sun photometer measurements on total ozone (Jain, personal communication) show broadly similar features regarding the ozone hole, and its recovery shows a rapid rise from the low ozone hole values to values of the order of 300–350 DU in the recovery phase, which takes less than 7 days. Also interesting are the anomalously larger values at peak during September. This is attributed by the authors to the breaking of the polar vortex by waves. Such features are common in the Arctic but are not generally expected in the Antarctic. The dynamics associated with this need to be studied in greater detail. In Figure 1.7, the Brewer spectrometer data also shows the signature of wave activity on ozone distribution. Notice the low value (<200 DU) in November after the recovery of the ozone hole.

Another study that needs to be conducted is a detailed study of the vertical distribution of ozone during the dark period (May, June, July months). Ground-based instruments like the Dobson or Brewer that depend on the sun are not effective. Moon-based measurements are plagued by

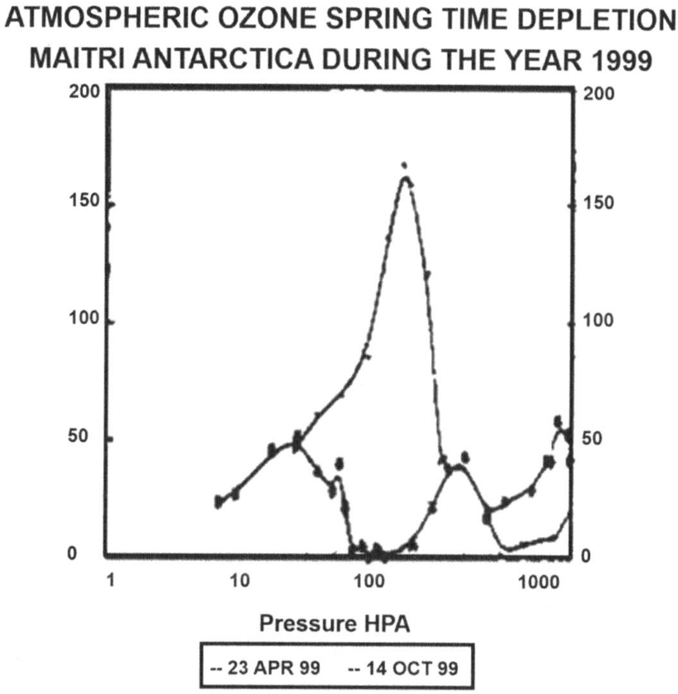

FIGURE 1.8
Comparison of the vertical distribution of ozone at Maitri in April and October 1999 to illustrate the catalytic destruction of ozone in the 12–18 km region (Lal and Krishnamurthy, 2002).

effects of scattered light, and their reliability has very often been doubted. The millimeter wave spectrometer (100 GHz) that was used by NPL is not affected by these problems, but the instrument could not collect enough data. Similarly, the laser heterodyne system suffered from logistics problems. Both these techniques are highly sophisticated and can be tuned to other trace gases such as ClO and HOCl that play prime roles in the heterogeneous chemistry responsible for the ozone hole phenomenon. There are very few measurements of these species made by any of the groups in the world. The Brewer spectrophotometer gives, in addition to ozone information, NO_2, SO_2, and ground-reaching UV B. Data collected over the past few years can be used for comparison of the features of O_3 and NO_2 since both these oxidizing agents have several common features with regard to their chemistry, the correlation between O_3 and ground-reaching UV especially since the instrument can give the spectrum in the UV B range, calibration of the UV biometer by the Brewer spectrophotometer, and so on. For NO_2 and SO_2 observations, NO_2 is of interest in the ozone hole problem. The release of NO_2 in the springtime, as well as ClO, from inert $ClONO_2$ is responsible for the ozone hole. SO_2 is a by-product of the Brewer measurement protocol and is an indicator of atmospheric pollution in urban areas. However, at a site like Antarctica, SO_2

measurements would give global background values and are an indicator of sulfate aerosols in the stratosphere. Figure 1.9 is an illustration of the type of results that are being obtained. Note the steady increase of NO_2 from spring to December–January (Figure 1.9a) as the solar elevation increases. Whether this is genuine or a measurement artifact needs to be confirmed by a separate

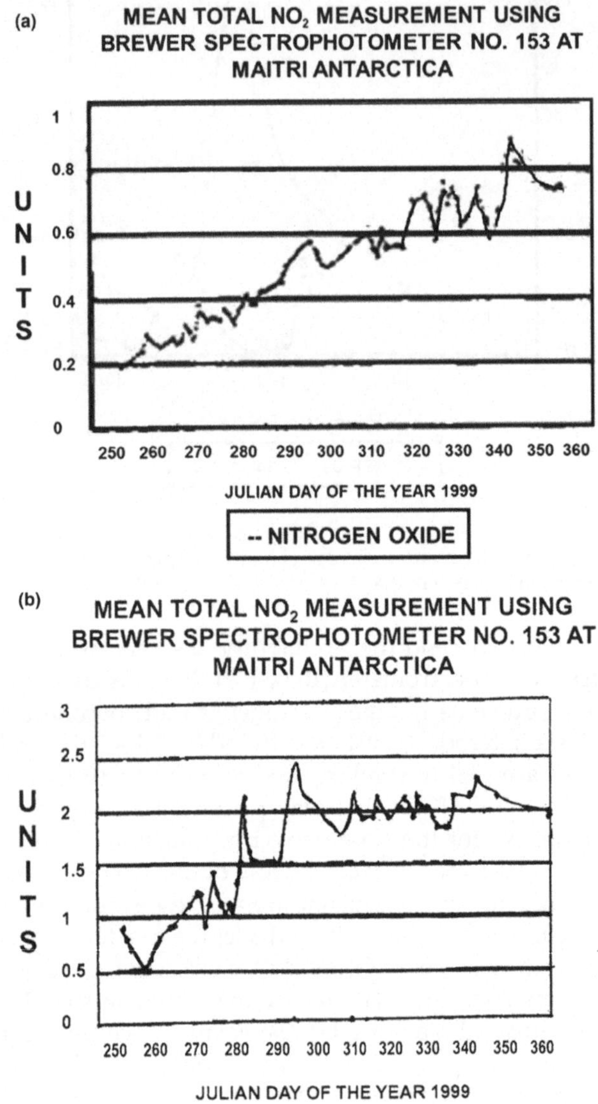

FIGURE 1.9
(a) steady increase of NO_2 from spring to December/January recorded at Maitri (Lal and Krishnamurthy, 2002). (b) SO_2 measurements at Maitri during August–December 1999 made by the IMD Brewer spectrophotometer (Lal and Krishnamurthy, 2002).

measurement, especially since SO_2 also shows similar behavior (Figure 1.9b). This is possible but needs to be confirmed because NO_2 is produced by solar radiation, photolysis of OClONO, and oxidation of N_2O. But NO_2 is also photolyzed by solar radiation. What is the equilibrium situation? Hence the need for confirmation.

The program has just begun and already showed some interesting results. The SO_2 values are comparable to those obtained at other remote sites. Measurements are indicative of the impact of long-range transport.

Trace Gases: CO_2, CH_4, and CO

Recently, NPL (Jain and co-workers) has taken up a program to monitor trace gases of interest in tropospheric chemistry and climate change problems—CO_2, CH_4, and CO at the Indian Antarctic Station, Maitri. Sample data for 2002–2003 shows that for both CO_2 and CH_4 (Figure 1.10), there is no seasonal variation at Maitri, unlike most northern hemisphere stations. This indicates the absence of any local source for these gases and that Maitri is indeed a pristine site. Mean methane concentrations are on the order of 1.7 ppm ± 0.1. While there is significant day-to-day variability, most of the observations lie between 1.685 and 1.715 ppm. One-year data (2002–2003) shows no increasing trend; while the data length is small, the observations are in agreement with observations made by other groups elsewhere the past few years showing a

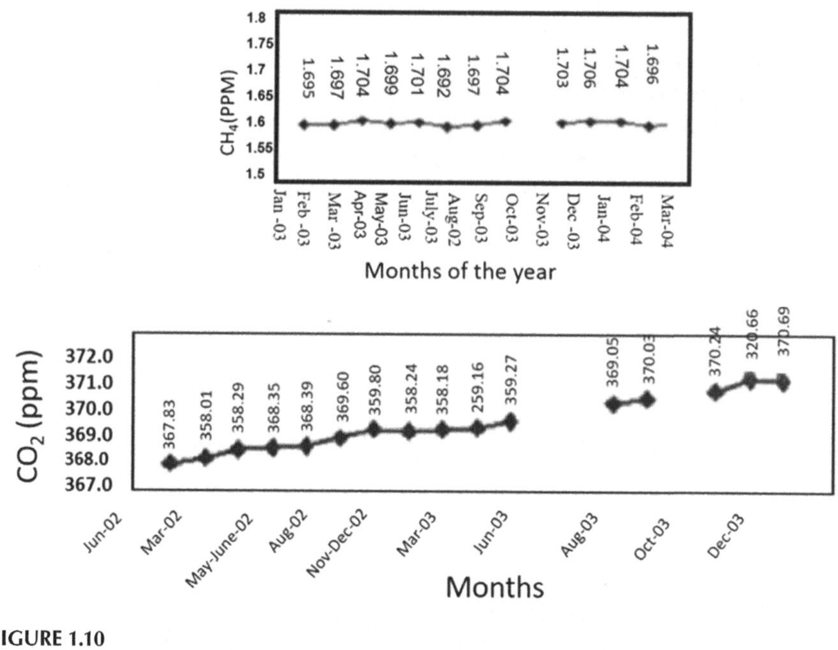

FIGURE 1.10
CO_2 and CH_4 measurements during 2002–2003 made at Maitri by NPL.

leveling of the atmospheric methane indicative of an increase in the oxidizing capacity of the atmosphere. Whether this state will continue or CH_4 growth rates will recover needs to be monitored. CO_2 observations also show significant day-to-day variability. Mean concentrations were 368.4 ppm in 2002 and 369.7 ppm in 2003, an increase of 1.3 ppm in 1 year, amounting to a growth rate of 0.35% per year. This compares favorably with the mean growth rate of 0.4% per year observed during the last decade. Observations should continue and should be augmented with surface ozone measurements, either the IMD instrument or an independent optical measurement, to study the troposphere chemistry at Maitri and compare it with the tropical tropospheric chemistry features obtained from mainland stations in India.

Aerosols

Aerosols were the subject of investigations even in the earliest Indian Antarctic expeditions. Dr. C. R. Sridharan of IMD had made AOD measurements along the sea cruise to Antarctica and noted the very low values in the southern Indian oceans. During the last decade, NPL had an MWR program on several expeditions that confirmed the low AOD values in the Maitri region and made some comparative studies with data over the Indian urban polluted (New Delhi) characteristics.

More recently, PRL used a hand-held sun photometer and a Microtop instrument for multiwavelength observations of AOD and water vapor. Mean aerosol optical depths in the Maitri region were around 0.02 when compared to the mean values of 0.04–0.08 over the Southern Indian Ocean, the Northern-hemisphere values of 0.1–0.3 over the ocean, and 0.3–0.5 or more over the land mass of coastal India (Figure 1.11). The lowest observed AODs are encountered over Maitri. However, over and above the mean low values, episodes of larger values due to long-range transport have been observed despite the pristine environment. So far there is no evidence of local sources. This feature needs to be watched in the coming years to monitor the possible impact of human-made local regional pollution in Antarctica.

Currently (2002–2003), surface level aerosol concentrations are in the region of 7.0 $\mu g\ m^3$. Further, most of the aerosols (more than 60%) are coarse mode particles, giving rise to a nearly flat AOD spectrum. Observed AODs are used to calculate the top of atmospheric aerosol radiative forcing, which is needed for climate impact studies.

An interesting by-product of the study is the effect of surface characteristics (surface albedo) on aerosol radioactive forcing. Another interesting feature is the observation that, because of the strongly varying sun-earth geometry over polar regions, radiative forcing exhibits strong seasonal dependence, varying from 0.7 in December–January to 0.15 in April–May—negative in dark periods and positive in summer periods (Figure 1.12).

Figure 1.13 shows that, for the same aerosol composition and optical depth, radiative forcing is strongly dependent on the nature of the surface. A large

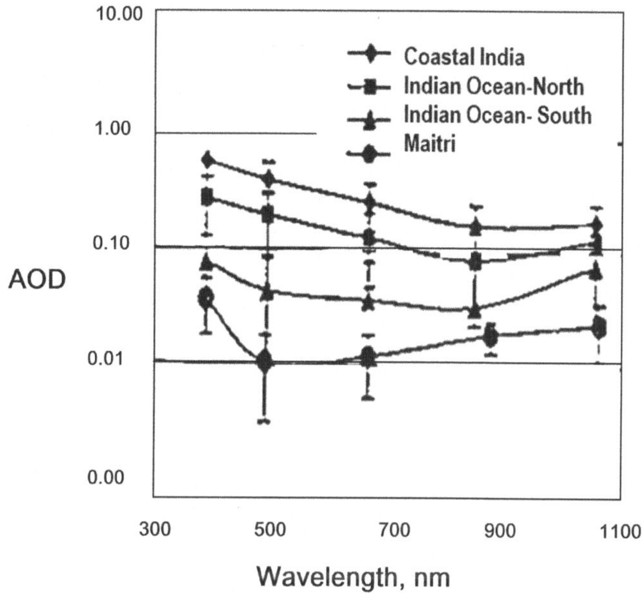

FIGURE 1.11
Aerosol optical depths measurements made by PRL at Maitri compared with the values of southern hemisphere oceans, northern hemisphere oceans, and land regions in northern hemisphere (Gadhvi and Jayaraman, 2004).

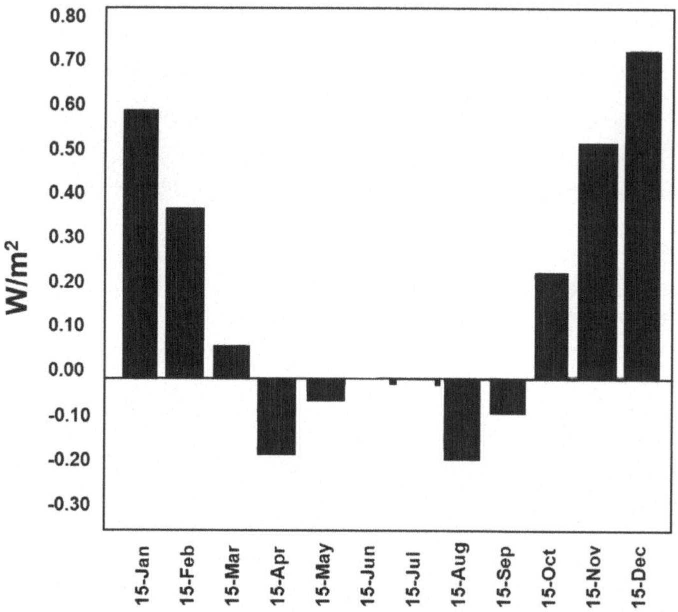

FIGURE 1.12
Annual variation of aerosol radiative forcing at Maitri (Gadhvi and Jayaraman, 2004).

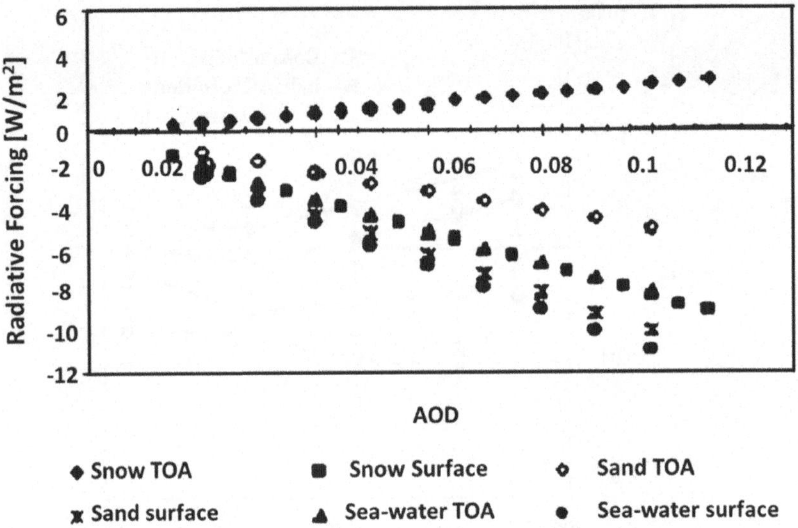

FIGURE 1.13
Dependence of aerosol radiative forcing on the nature of the earth's surface (Gadhvi and Jayaraman, 2004).

change even in the sign (negative to positive) is seen from sand and sea water surface to snow-covered and other highly reflecting surfaces. This is because aerosols over a bright surface have an overall effect of reducing the surface albedo, while over dark surfaces they have an overall effect of enhancing the albedo. These results are significant for climate change problems.

References

Basu B., Jasperse J. R., Strickland D. J. and Daniell Jr. J. R. E. 1993. *J. Geophysic. Res.* v. 98, p. 21517.

Blanchard G. T., Lyons L. R., Sanson J. C. and Rich I. J. 1965. *J. Geophy. Res.* v. 100, p. 7855.

Gadhvi H. and Jayaraman A. 2004. *Curr Sci.* v. 86, p. 296.

Lal R. P. and Krishnamurthy K. 2002. *Scientific Report on the 18th Indian Expedition to Antarctica 2002.* DOD Tech Pub No 16 P 17.

Narayanan R., Desai I. N., Modi N. K., Raghavarao R. and Sridharan R. 1989. *Appl Optics*, v. 28, p. 2138.

Pallam R. D. and Sridharan R. 1998. *Proc. Indian Acad. Sci (Earth Planet Sci)* v. 107, p. 203.

Rajaram G., Emperumal S., Singh K., Hanchinal A. N. and Dhar A. 2002. *Scientific Report on 18th Indian Antarctic Expedition.* DOD Tech Report No. 16 P 01.

Sivjee G. G. 1983. *J. Geophys Res.* v. 88, p. 435.

Sridharan R. D., Pallam R., Narayan R., Modi N. K., Subbaraya B. H. and Raghavarao R. 1995. *Curr Sci.* v. 68, p. 830.

2

Geomagnetic Research at Indian Research Base, Antarctica: Initial Efforts and Salient Achievements

A. K. Sinha, S. Gurubaran, A. Dhar, A. Hanchinal, and B. M. Pathan

CONTENTS

Introduction

Geomagnetic research has been one of the major thrust areas of investigation of the Indian Antarctic Program ever since the first Indian scientific expedition was launched back in 1981 and the Indian research bases Dakshin Gangotri and Maitri were made operational. Due to the scientific relevance of the geomagnetic investigations and the unique location of the Indian research base Maitri in Antarctica, it was considered imperative to undertake intensive research in the contemporary area of geomagnetism at the Indian Antarctic research base.

The remarkable efforts of the Indian Institute of Geomagnetism (IIG) in the initiation of geomagnetic and allied research in the Antarctic region since 1985 have been well documented in diverse research publications available in scattered form. It is, therefore, important to assess the initial meaningful contributions of the Indian scientific community and highlight the salient achievements made in the field of geomagnetism and allied aspects in the Antarctic region from those efforts. Sinha et al. (2017) have provided an exhaustive account of the scientific findings in the contemporary areas of geomagnetic research over the Antarctic region, with special reference to the significant contributions made by the IIG.

In this chapter, an attempt has been made to review the efforts made by Indian scientists pertaining to the initial two decades. A brief account of the significant breakthroughs achieved in the field of Antarctic geomagnetism is also provided in this chapter.

Initial Efforts in Antarctic Geomagnetic Research

As mentioned earlier, the Indian Antarctic Geomagnetic program commenced with the launching of the first Indian Antarctic Expedition (IAE) in 1981–1982. The Indian base was set up at the geographical latitude of 70.8°S and longitude of 12°E, in the Queen Maud region of the east Antarctic ice shelf. Nearly 10 days of magnetic data were collected with a proton precession magnetometer (PPM) in January 1982. Since then, the stay and observational period kept increasing, with the setting up of permanent stations, Dakshin Gangotri (DG) and Maitri, nearly 76 km from DG. The geomagnetic field has been continuously monitored, except for a few years' gap. The geographic locations of DG and Maitri have proved to be unique and interesting from a geomagnetic point of view, as will be elaborated upon in the results and discussion. With barely 2% of the Antarctic continent free from ice, Maitri is located on one such ice-free rocky area, Schirmacher Oasis, with a number of freshwater lakes available within this area.

Scientific Importance

The central plasma sheet (CPS) represents the region of the magnetosphere wherein a large amount of solar-terrestrial energy exchange takes place, and this energy ultimately finds its way into the auroral ovals of the Earth. Thus, the magnetic variations recorded on the ground in the auroral regions provide clues to physical processes occurring in the distant magnetosphere and hence to those in the interplanetary medium and on the sun.

The IIG has been taking part in IAEs since the beginning and has carried out geomagnetic, ionospheric, air-Earth current, and global positioning system (GPS) measurements in Antarctica. The following instruments were operated for various studies over different periods of time.

i. PPM for recording of total magnetic field (F) intensity.
ii. Fluxgate magnetometer to record variations in the three orthogonal components X, Y, and Z.
iii. A 30-MHz riometer to measure the strength of the cosmic radio noise that impinges on the Earth's atmosphere, causing secondary ionization in the D region.
iv. An imaging riometer, commissioned during 2009–2010.
v. Triangulation experiment comprising fluxgate magnetometers deployed at the vertices of a triangle to determine the velocity of small-scale auroral current systems over Maitri.
vi. A Maxwell current experiment to determine variations in the air-Earth current system accompanied by electric field mills and Gerdien conductivity meters for monitoring vertical electric potential gradient and bipolar conductivity.
vii. GPS observations.

The IIG has been investigating the phenomenon of geomagnetic substorms over the last few decades (Roy, 1992; Roy and Lakhina, 2002, for example). It has already been shown by numerical simulations that, prior to the onset of magnetic reconnection in the near-Earth geotail region, there is a plasma flow toward the Earth. Therefore, this is a mechanism of entry of plasma sheet particles to the interior of the magnetosphere to form ring currents and therefore drive geomagnetic storm processes. Magnetic data recorded at Maitri have been utilized (Rajaram et al., 2001, 2002) to investigate some aspects of the auroral electrojet currents and the substorm phenomenon. The ground geomagnetic data recorded at the chain of magnetic observatories operated by the IIG in India can be used to monitor the dynamics and solar-interplanetary origin of geomagnetic storms. Thus, the combined data from Antarctic station together with the data from the Indian magnetic observatories are valuable for any investigation of the storm-substorm relationship and its solar-interplanetary origins. Sinha et al. (2017) have categorically highlighted the recent progress of geomagnetic research in Antarctica.

Results and Discussion

Quiet-day Variations: A Comparison between Dakshin Gangotri and Novolazarevskaya

With Novolazarevskaya, the Russian research base at Antarctica, being in the close vicinity of DG and Maitri, a comparative study of the quiet-day features of the geomagnetic field at these two nearby Antarctic stations was carried out. The mean diurnal variation pattern in the three magnetic elements at the two stations showed that the two stations were in close conformity with the anticipated diurnal patterns for such geographical locations during the local summer months close to the solar minimum epoch (Figure 2.1). The range of variations at both stations was nearly the same. However, in the H component, the diurnal variation curve at Novolazarevskaya does not coincide with that of Dakshin Gangotri. Both the maximum and minimum fields appear elevated. In the D component, the opposite seems to be true in the sense that maximum and minimum at Dakshin Gangotri seem to be greater in magnitude.

With the change in geospace disturbance, sub-auroral locations are ideal to sense rapid changes in space weather (refer to Figure 2.2). DG and Maitri

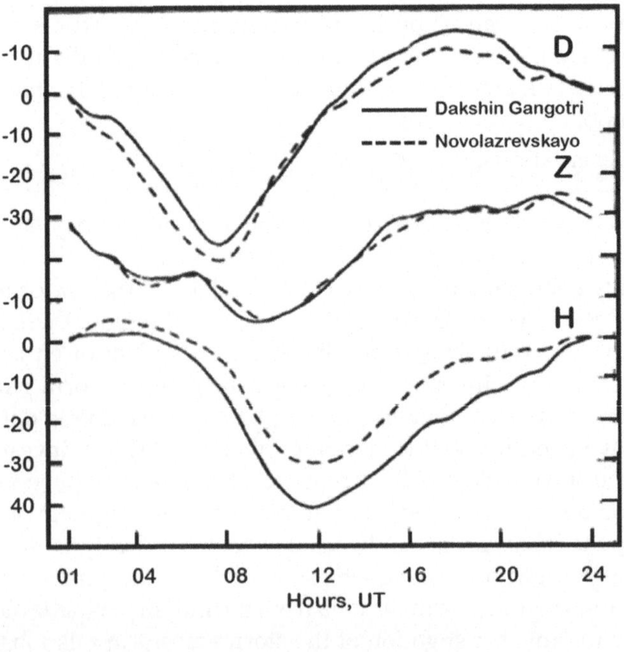

FIGURE 2.1
Mean diurnal variations in H, Z, and D derived from records on quiet days at Dakshin Gangotri and Novolazarevskaya (Sinha et al., 2017).

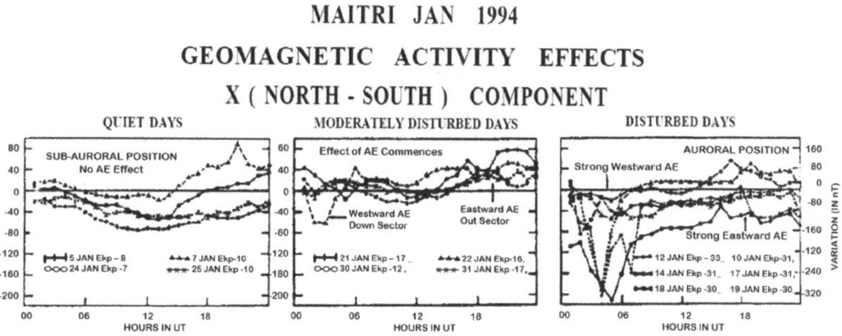

FIGURE 2.2
Response of the X (north-south) geomagnetic field component at Maitri to increasing disturbance in the geospace environment (Sinha et al., 2017).

occupy such an important location and shift in and out of the auroral oval with rapidly changing magnetic activity (Banola et al., 1997).

Changes in HF Communication between India and Antarctica associated with Geospace disturbance

In the initial phase of setting up experiments in Antarctica (1982–1995), the radio communication link on the high frequency (HF) band between India and Antarctica was relied upon. Later, due to the advancement of technologies and availability of direct satellite connectivity in the Antarctic region, the HF communication link between India and Antarctica was discontinued. Nevertheless, the efforts by the IIG in those initial years are briefly outlined below only for historical perspectives.

Earlier studies showed how HF communication can follow a single-hop or double-hop propagation path. While being reflected from the ionosphere, the radio waves would be absorbed and weakened. Much of the absorption occurs in the lower auroral ionosphere through which the HF wave propagates. During magnetic disturbance, considerable precipitation of energetic particles from deep space takes place along geomagnetic field lines. They cause secondary ionization in the D region (60–85 km) of the ionosphere. The secondary electrons draw energy from the radio wave as it passes through the lower ionosphere, causing weakening of radio signal received on the ground—this would manifest as absence of the HF signal at the receiver end (Jeeva et al., 1995, Sridharan et al., 1997). The above-mentioned observations clearly suggest that the quality of HF communication between India and Antarctica depended on:

a. The magnetic activity prevailing on the day:
   i. The power of the Tx-Rx system
   ii. The prevailing season
   iii. The ionospheric conditions at the reflection points as well as the number of hops taken by the radio wave in traversing the India–Antarctica distance

Conjugate Area Studies

In view of the fact that the field line originating from Maitri falls into the ocean in the northern hemisphere, the Indian research base does not have a conjugate location in the opposite hemisphere. Therefore, magnetic data from two nearby stations, Leirvogur and Narsarsuaq (which fall on either side of the conjugate location of Maitri), was obtained and analyzed. Leirvogur and Narsarsuaq are auroral zone stations, show the variations reflected by the auroral electrojet, and thus differ from the signatures recorded by Maitri during quiet magnetic conditions, except during disturbed conditions, which thus confirms the location of Maitri (Figures 2.3–2.6) as sub-auroral (Dhar et al., 1993; Kalra et al., 1995; Sinha et al., 2017). Figure 2.6 shows the magnetic field variations in the three components X, Y, and Z at Maitri and near the conjugate locations of Narsarsuaq and Leirvogur. The variations at the three locations match during magnetically disturbed conditions.

Three Station Fluxgate Magnetometer Experiments around Maitri

This experiment was started in January 1992 and has been carried out six times since then. Three stations were located at the vertices of a triangle around Maitri, with sides of 100–250 km. The experiment was carried out to determine the presence and velocity of mobile, small-scale auroral current systems over Maitri. These currents, while flowing over Maitri and surroundings, leave signatures over the ground-based magnetometers. From the time lags in these pulsations, the velocities can be determined. The velocities of these mobile auroral current systems typically lie between 0.5 and 3.0 km/sec, and they tally well with the velocities obtained by various experimental groups in the northern auroral oval. The only other such campaign in the southern hemisphere was carried out by Australia (Neudeg et al., 1995), which was a one-day campaign. During a particularly disturbed period of January 28–31,

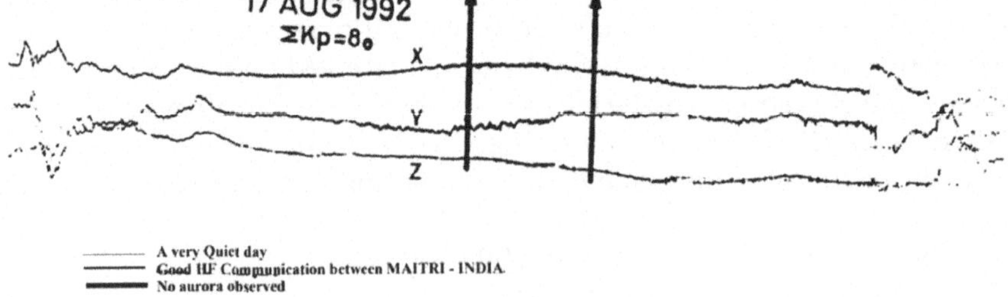

FIGURE 2.3
Traces of X, Y, and Z components on a magnetically quiet day (August 17, 1992) at Maitri (Jeeva et al., 1995; Sridharan et al., 1997).

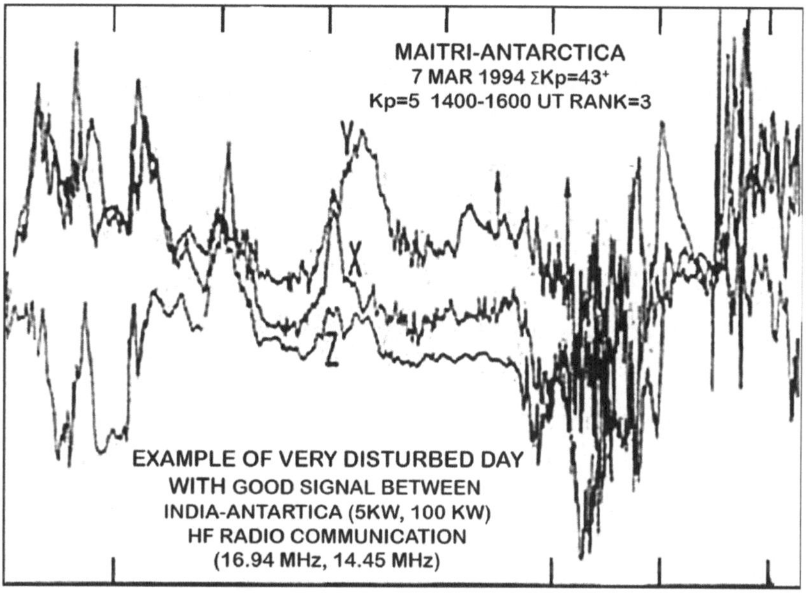

FIGURE 2.4
Traces of X, Y, and Z components recorded at Maitri on March 7, 1994 on a magnetically disturbed day ($\sum Kp = 44$) (Dhar et al., 1993; Kalra et al., 1995).

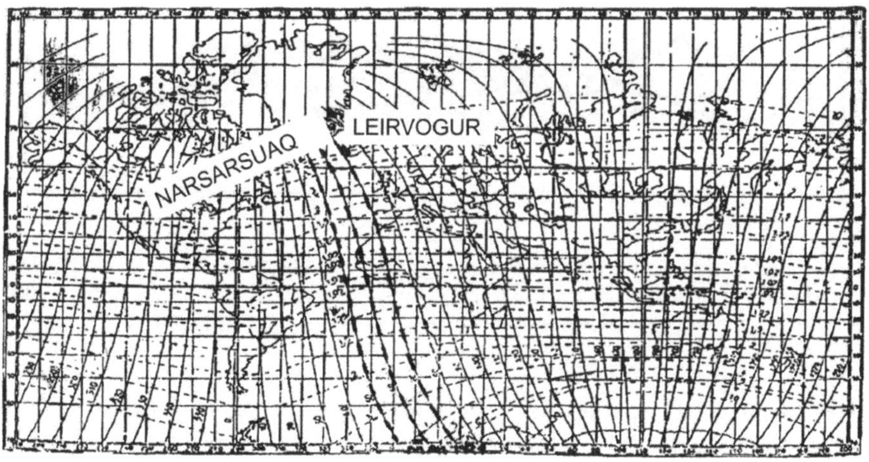

FIGURE 2.5
In the absence of conjugate location for Maitri, magnetic data from two nearby stations, Leirvogur and Narsarsuaq, was obtained and analyzed. These two stations fall on either side of the conjugate location of Maitri (Sinha et al., 2017).

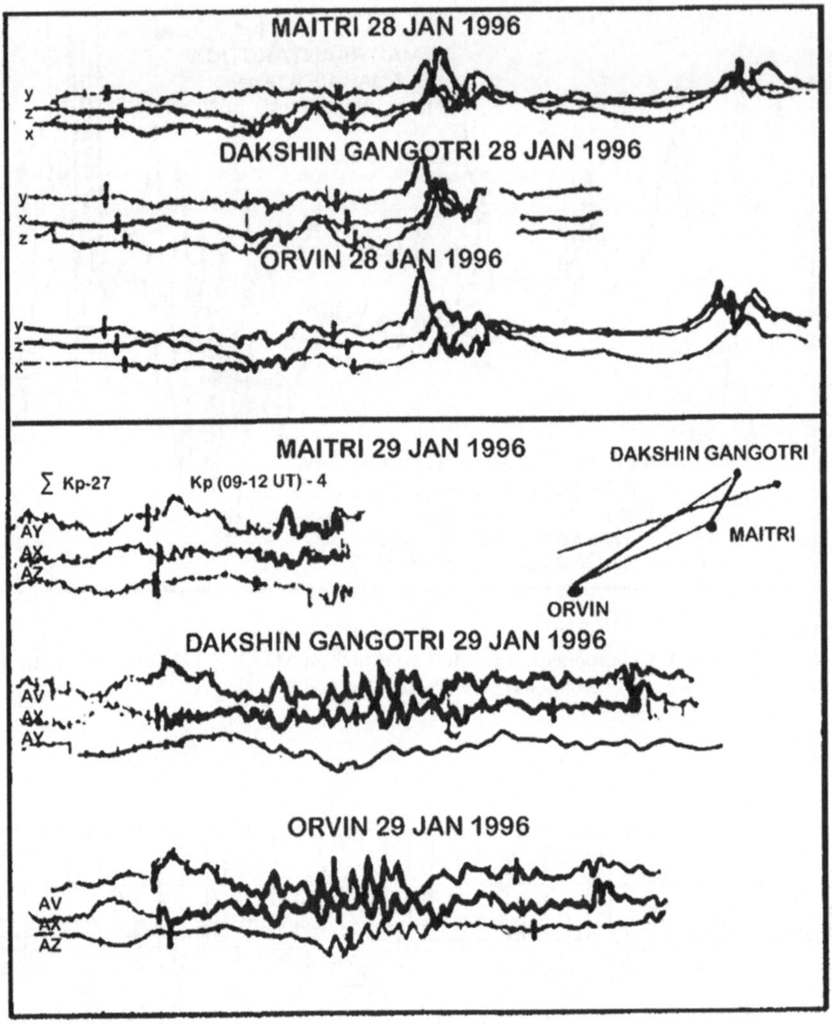

FIGURE 2.6
Long-period pulsation activity with a shorter period overriding them recorded at the three locations Maitri, Dakshin Gangotri, and Orvin during January 28, 20 UT and 29, 09 UT 1996. Delay in times could be used to calculate the velocity of overhead auroral currents.

1996, very large drift speeds of 3 to 18 km/sec were obtained from the time lags at the three stations. This could be due to the presence of the eastward drifting omega bands and westward traveling surges (Kalra et al., 1998; Dhar et al., 1999). Figure 2.7 shows the magnetic variations at three locations, Maitri, Dakshin Gangotri, and the Orvin Mountains, during 1996. The velocities of the overhead mobile current systems were calculated from the time lags in pulsations recorded at the three stations.

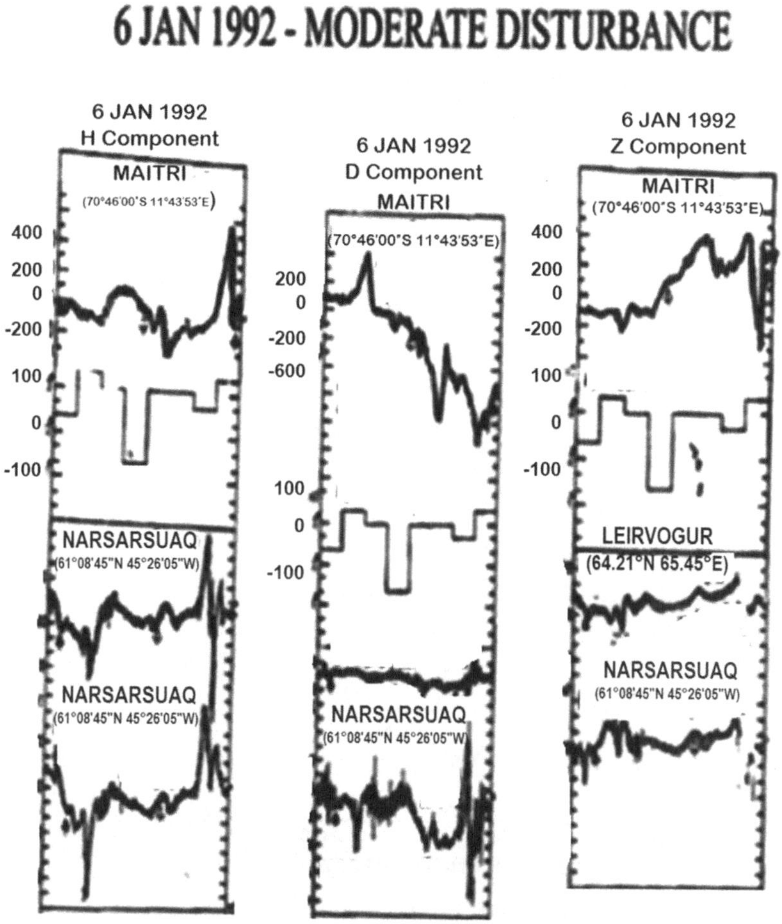

FIGURE 2.7
Variations of H, D, and Z components at Maitri and its northern hemisphere conjugate points on January 6, 1992, a moderately disturbed day.

Operating the Riometer (30 MHz) for Determining the Occurrence of Auroral Particle Precipitation over Maitri

The 30-MHz riometer in operation in the initial years (this was replaced by a more sophisticated state-of-the-art 38-MHz imaging riometer during 2009–2010) sensed the omnidirectional cosmic radio noise and recorded the intensity of this signal. At auroral latitudes, the riometer records the absorption of this radio signal caused by energetic particle precipitation along magnetic field lines from outer space into the auroral ionosphere. A large positive variation in the X-component (positive downward) indicates the presence of strong eastward auroral electrojet (EAE) current in the dusk-side ionosphere over Maitri. This implies that Maitri has moved from its quiet-time sub-auroral position into the auroral oval. The positive variation in the Z-component implies that the EAE was located equatorward of Maitri. The variations in the Y-component reflect field-aligned currents associated with energetic particle precipitation from deep space into the auroral ionosphere over Maitri (Figure 2.8).

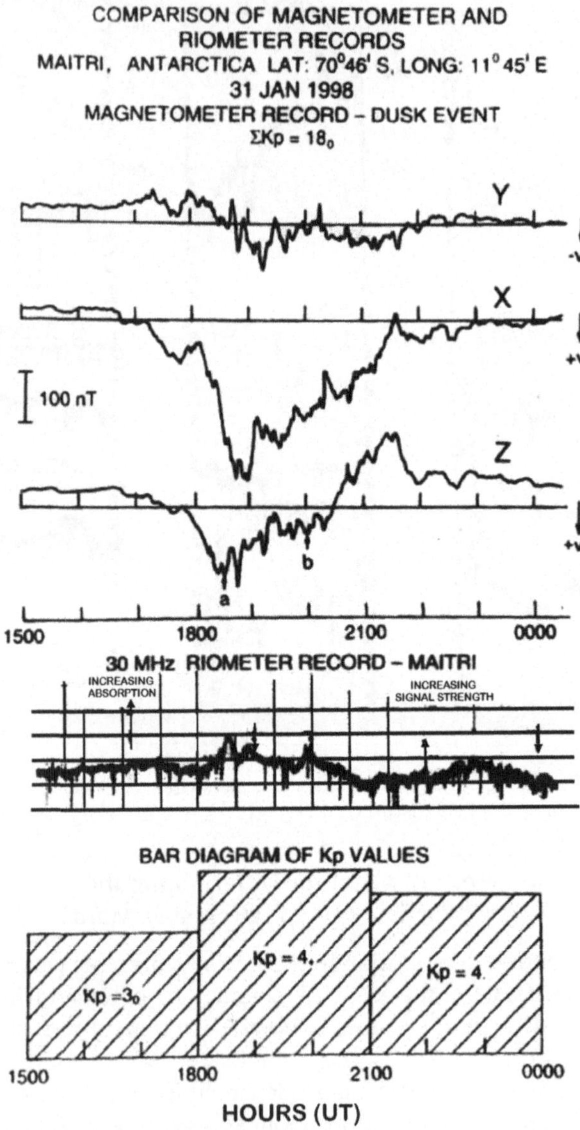

FIGURE 2.8
Variations in magnetometer and riometer records and 3-hourly Kp magnetic indices during magnetically disturbed conditions on January 31, 1998 (15–24 UT).

Rapid Decrease in Total Magnetic Field Intensity F at Maitri and Its Surrounding Area

The total geomagnetic field intensity F was measured at DG during 1982 and 1986. The measurements were repeated during 1996 at Maitri. A large drop was noticed between 1987–1996 in the F values at DG and Maitri. When compared with the values of F for the geographic location of Maitri obtained from the international geomagnetic reference field (IGRF) for 1990, it showed a markedly decreasing trend for the years 1982, 1996, 1990, and 1996. In examination of the F values for 1922, 1951, and 1960 from earlier magnetic charts for this location, the resulting curve suggested a drop of about 8000 nT in the past 75 years. A decrease of nearly 17% was observed since 1922, which is a considerable magnitude within 75 years (Arun et al., 2000; Rajaram et al., 2002a,b).

An examination of F data from other (Figures 2.9 and 2.10) stations in the northern and southern hemisphere suggested that this decrease was occurring only in the southern hemisphere. Bormann et al. (1995) reported a decrease of about 100 nT/yr at Novolazarevskaya (Russian) and George Foster (German) stations, both stations located near Maitri. The rate of decrease observed at southern hemisphere high-latitude stations is as follows: Novolazarevskaya:

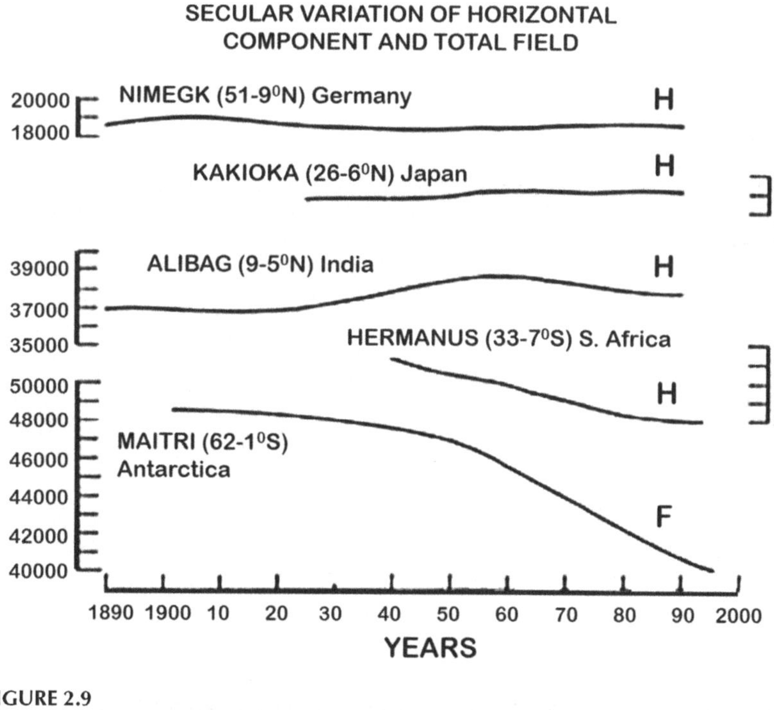

FIGURE 2.9
Variation of the principal component of the geomagnetic field over the past century (1890–1996) at northern hemisphere stations (Nemegk, Kakioka, and Alibag) and at southern hemisphere stations (Harmanus and Maitri) (Sinha et al., 2017).

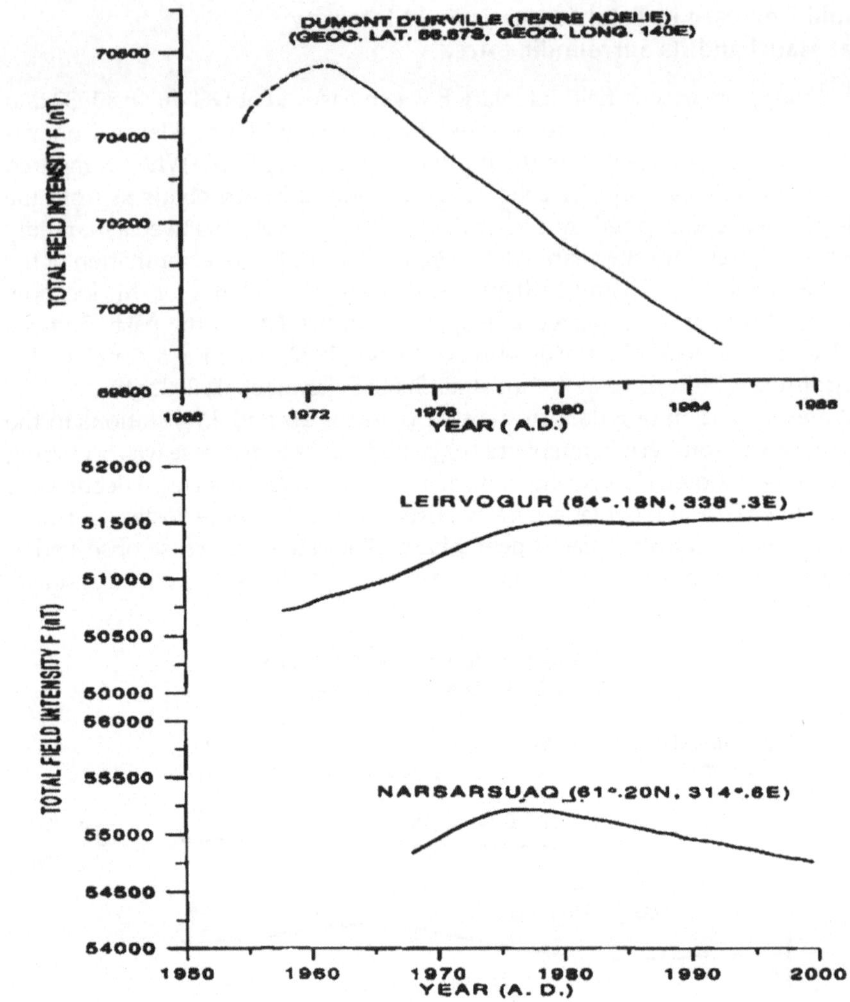

FIGURE 2.10
Geomagnetic total field (F) variation at two northern hemisphere stations, Narsarsuaq and Leirvogur, and southern hemisphere station Dumont D'Urville. (Sinha et al., 2017)

108 nT/yr, Sanae: 108 nT/yr, Hermanus: 105 nT/yr, Argentine Island: 103 nT/yr, Syowa: 94 nT/yr, Halley Bay: 91 nT/yr, Davis: 73 nT/yr, Mawson: 72 nT/yr. Dumont d'Urville showed a drop of 50 nT/yr until 1994, after which the drop was only ~16 nT/yr, possibly due to a new absolute pier adopted in 1995. Further analysis of magnetic data from high-latitude northern and southern stations has shown that this rapid decline is confined to a narrow horseshoe-shaped belt in the Antarctic region, encompassing Maitri. The rate of decline falls as one moves away in the poleward or equator ward direction. Gubbins (1989) defined these areas as regions of reverse magnetic flux, and

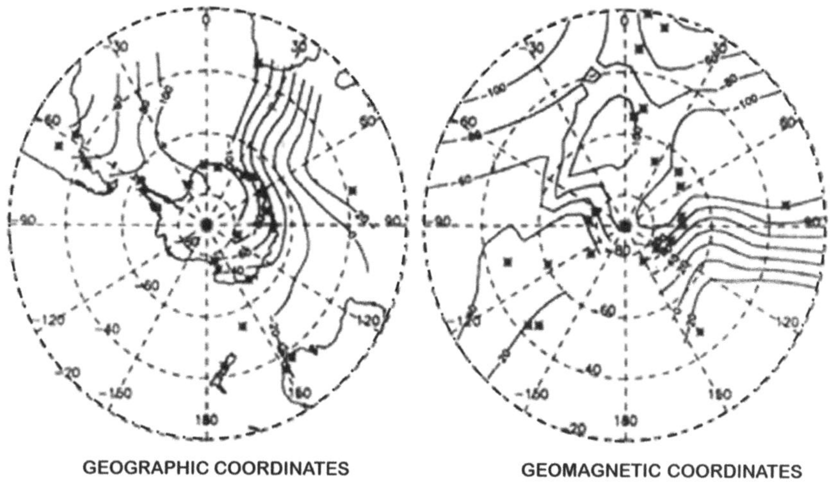

FIGURE 2.11
Conditions of total magnetic field F variation in nT/yr. for the period 1950–2000, using data from stations located in the 30–90°S latitude region (Sinha et al., 2017).

continuous monitoring of the F values as well as efforts to model the decline will be useful in the future. Figure 2.11 shows the contours of a rate of average decrease in F (nT/yr) over the past five decades (1950–1999) for Antarctic and sub-Antarctic stations between 30°–90° S in geographic and geomagnetic coordinates. The locations of the stations considered are shown using star symbols.

Maitri as an Ideal Location for Predicting Geospace Weather

The first two decades of experimental geomagnetism at the Indian Antarctic station Maitri, using magnetometers, riometers, and observations of optical aurora, has led us to the following conclusion:

Maitri is ideally suited for nowcasting geospace weather and perhaps interplanetary weather. The former refers to the degree of disturbance in electric and magnetic fields and particle population in various regimes of the Earth's magnetosphere. The latter refers to the magnitude and direction of solar magnetic field components and the velocity, density, and dynamic pressure of solar wind particles in interplanetary space.

Comparison of Y, X, and Z variations and the riometer absorption patterns at Maitri with the interplanetary magnetic field (IMF) variations and the interplanetary solar wind parameters measured by the WIND satellite during 1999 (Figure 2.12) show good agreement between the magnetometer and riometer variations and the southward component of the IMF, its intensity, the speed of the solar wind, and the ion dynamic pressure of the solar wind. Work was continued to place these correlations on a quantitative footing (Rajaram

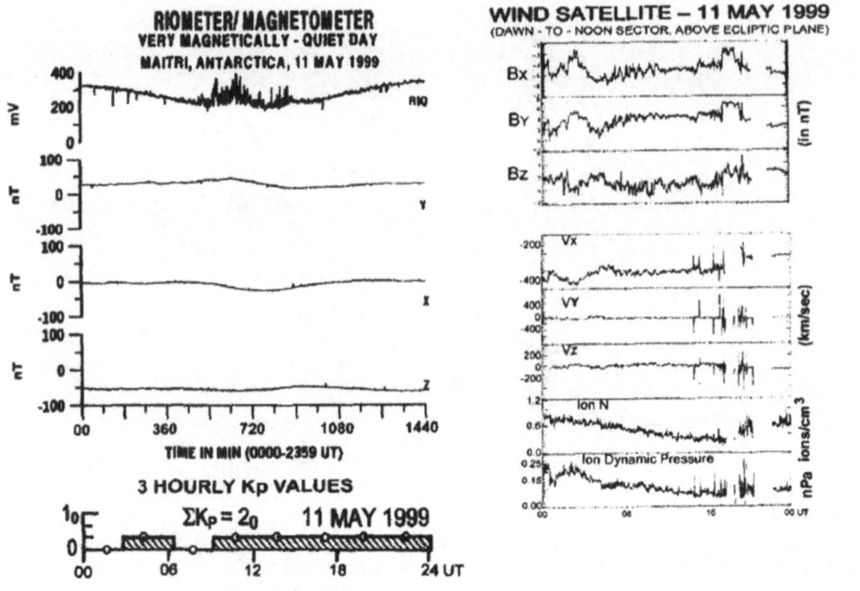

FIGURE 2.12
30 MHz riometer and geomagnetic Y, X, and Z variations at Maitri for the disturbed day of May 11, 1999. Below are 3-hourly Kp values. This is the day the solar wind almost disappeared. IMF parameters for the same periods are shown on the right-hand side for comparison (Sinha et al., 2017).

et al., 2001). A splendid display of aurora australis was observed over Maitri on the night of March 4/5, 1999 (Figure 2.13). The changes recorded by a fluxgate magnetometer and a 30-MHz riometer operating at Maitri were interpreted in terms of then the understood auroral physics. Interplanetary magnetic field and plasma parameters recorded by the WIND satellite indicate very high solar wind velocities (N600 km/sec) and prolonged southward Bz of average value (N5nT) during the auroral event. Ion density was however, an average 3–4/cm^3 and ion pressure only 2–3 nanoPascals during the event (Rajaram et al., 2002a,b).

Programs Subsequently Pursued following the Initial Two Decades of Geomagnetic Research from Antarctica

i. *Storm–substorm relationship*: Solar wind energy enters the Earth's upper atmosphere through the geomagnetic storm and substorm. Near the high-latitude regions, which are closed through the auroral electrojet, field aligned currents (FACs) are generated and can be picked up for monitoring purposes through operating magnetometers and riometers at high latitudes. Charged particles are pushed inside the magnetosphere, originating substorms in high latitudes, a nighttime phenomenon.

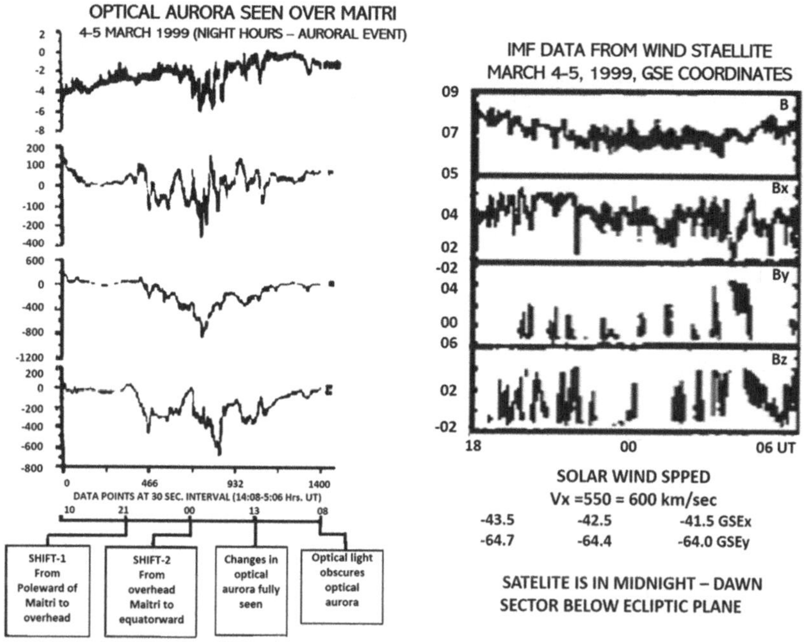

FIGURE 2.13
Shift and change in the movement of optical aurora over Maitri and its ground geomagnetic signature from March 4 18 UT to March 5 06 UT 1999. Variations in IMF parameters are shown on right-hand side for the same period (Sinha et al., 2017).

In order to identify various stages of substorms, the magnetic data recorded at Maitri were utilized. For understanding the storm-substorm relationship, the ground magnetic data recorded at various magnetic observatories in India operated and maintained by the IIG were proposed to be used.

ii. *Variations in air-Earth currents through the Maxwell current experiment*: In an attempt to understand the solar-terrestrial relationship, a study of the global electric circuit (GEC) was undertaken to assess its associated changes in surface weather with solar output.

Atmospheric electricity parameters, namely the air-Earth current, the surface vertical electric field, and electrical conductivity, have been continuously measured to help understand the near-Earth electrical environment and the electric field.

The variation of the atmospheric current in different geomagnetic conditions is important to understand the electrodynamic coupling between the upper and lower atmosphere. Maitri, due to its unique location, offers an opportunity to monitor changes in the atmospheric electric parameters with the context of varying upper atmospheric conditions. Accordingly, during the austral summer of the 19th Indian Antarctic Expedition, measurement of the Maxwell current was carried out at Maitri station. Figure 2.14 shows the diurnal variation of the observed Maxwell current on a few fair-weather days. The maximum is around 1900 UT, and the minimum is around 0300 UT. Similarly, Figure 2.15 shows the comparison of the current density with that of the electric field, and both have a significant resemblance to the famous Carnegie curve. It also clearly reveals the electric environment over Maitri is not masked by any other artificial electrical disturbances and the site is suitable for long-term measurement.

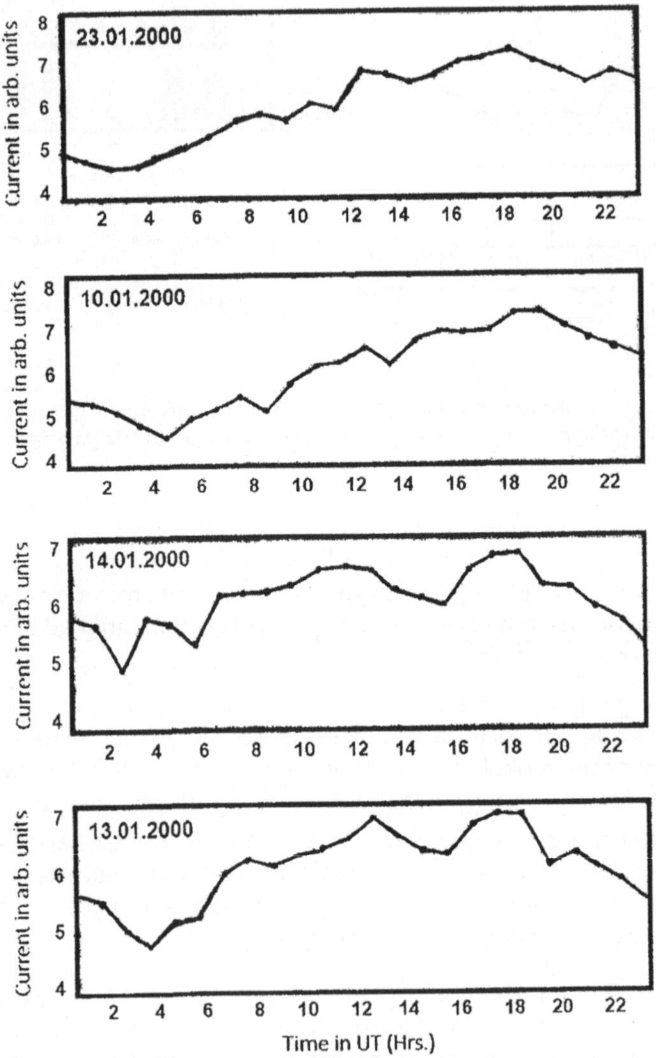

FIGURE 2.14
Maxwell current for the selected fair-weather days during January 2000.

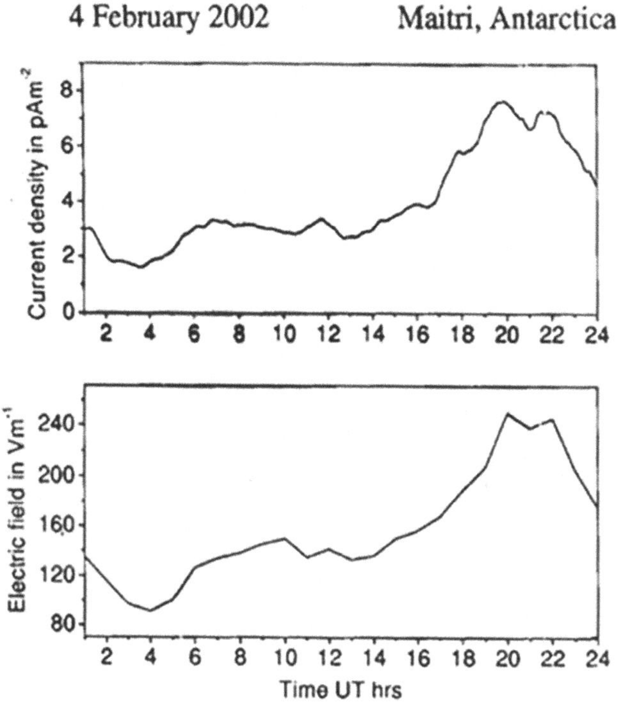

FIGURE 2.15
Comparison of current density and electric field on February 4, 2002 (Sinha et al., 2017).

Long-term monitoring of the atmospheric current will be studied in the context of varying geomagnetic conditions and the possible electrodynamic coupling between the upper and lower atmosphere.

Some Recent Developments

Imaging Riometer Studies

In February 2010, an imaging riometer was successfully installed at Maitri. Results showed that the diurnal pattern experienced a sideral shift of 2 hours, which validates the data. This is attributed to the solar ionization of the D region of the ionosphere, leading to high electron density where the frequency of collision was already high. This imaging riometer was installed at Maitri with the objective of studying the magnetospheric-ionospheric coupling during substorm processes. Many events of disturbed time cosmic radio noise absorption (CAN) associated with storm-time and non-storm-type substorms were analyzed, and it was found that CNA was more conspicuous during storm substorms as compared to non-storm-time substorms (Sinha et al., 2017).

Crustal Deformation, Dynamics of Ice, Mass Balance, and Tropospheric Studies through Global Positioning System

Global Positioning System (GPS) observations are useful to measure the displacement rate of geodetic monuments, which can later be compared with the observed values and those predicted from the ice sheet and response models of the polar region (Fukuzaki et al., 2005). At present, there are nine continuous GPS sites on the Antarctic continent as a part of a global monitoring network. Some initial GPS observations have resulted in an estimation of the Antarctic plate motion and were also compared with contemporary global plate motion models (Yamada et al., 1998; Bouin and Vigny, 2000; Dietrich et al., 2001, 2004; Prawirodirdjo and Bock, 2004) and glacial isostatic adjustment (GIA) models (Raymond et al., 2004). On par with the international scientific efforts in the Antarctic region, the IIG has also recently set up GPS for observations and data collection with an aim to understand crustal deformation, dynamics of ice, mass balance, and tropospheric processes occurring over the Antarctic region. Since the Antarctic plate is a divergent plate boundary with broken structures and ice mass, this precise geodetic measurement can shed light on the deformations and dynamic processes taking place in the Antarctic continent.

Acknowledgments

The authors are grateful to the Indian Institute of Geomagnetism for encouragement and support to the Antarctic Geomagnetism project. We acknowledge the hard work and initiative of all the members from IIG who participated in Indian Antarctic Expeditions and collected good-quality geomagnetic data on a regular basis. We are grateful to National Center for Polar & Ocean Research (NCPOR)/MoES, Government of India, for arranging annual Antarctic expeditions and providing full logistic support to enable participating scientific organizations to carry out scientific experiments.

References

Arun, T., Patil, A. G., Dhar, A. and Rajaram, G. 2000. Rapid decrease in total magnetic field intensity at Indian Antarctic Station, Maitri. *J. Indian Geophysical Union*, v. 4, pp. 119–128.

Banola, S., Dhar, A. and Rajaram, G. 1997. *Movement of Indian Antarctic station Maitri into auroral oval during geomagnetic activity, Technical Report of Thirteenth Indian Expedition to Antarctica,* Technical Publication No. 11, Dept. of Ocean Development, New Delhi, pp. 91–105.

Bouin, N. M. and Vigny, C. 2000. New constraints on Antarctic plate motion and deformation from GPS data. *J. Geophys. Res.*, v. 105, pp. 28279–28293.

De Santis, A., Barraclough, D.R. and Tozzi, R. 2003. Spatial and temporal spectra of the geomagnetic field and their scaling properties. *Phys. Earth. Planet. Inter.*, v. 135, pp. 125–134.

Dhar, A., Patil, A., Sankaran, S. and Rajaram, G. 1999. *Three-station magnetometer experiment at Antarctica during Jan-1996 to determine the velocity of disturbed-time overhead auroral current system. In Scientific Report of Fifteenth Indian Expedition to Antarctica,* Technical Publication No. 13, Dept. of Ocean Development, New Delhi, pp. 1–20.

Dhar, A., Rajesh, K., Jeeva, K., Unnikrishnan, K. Daga, D. M. and Rajaram, G. 1993. Changes in position and intensity of auroral electrojet inferred from conjugate area studies. *Presented at Indo-Russian Symposium on "Nature and Variations of the Geomagnetic Field"*, New Delhi.

Dietrich, R., Dach, R., Engelhardt, G., Ihde, J., Korth, W., Kutterer, H.-J., Lindner, K., Mayer, M., Menge, F. and Miller, H. et al. 2001. ITRF coordinates and plate velocities from repeated GPS campaigns in Antarctica—an analysis based on different individual solutions. *J. Geodyn.*, v. 74, pp. 756–766.

Dietrich, R., Rulke, A., Ihde, J., Linder, K., Miller, H., Niemeier, W., Schenke, H.-W. and Seeber, G. 2004. Plate kinematics and deformation status of the Antarctic Peninsula based on GPS. *Global Planet. Change.* v. 42, pp. 313–321.

Fukuzaki, Y., Shibuya, K., Doi, K., Ozawa, T., Nothnagel, A., Jike, T., Iwano, S., Jauncey, D. L., Nicolson, G. D. and McCulloch, P. M. 2005. Results of the VLBI experiments conducted with Syowa Station, Antarctica. *J. Geodyn.*, v. 79, pp. 379–388.

Gubbins, D., Jones, A. L. and Finlay, C. 2006. Fall in Earth's magnetic field is erratic. *Science*, v. 321, pp. 900–903.

Jeeva, K., Hanchinal, A., Dhar, A., Daga, D. M. and Rajaram, G. 1995. *HF communication between India-Antarctica during 1992 and its dependence on a) geomagnetic activity and b) seasons, Scientific Report of Eleventh Indian Expedition to Antarctica,* Technical Publication No. 9, Dept. of Ocean Development, New Delhi, pp. 103–113.

Kalra, R., Dhar, A., Unnikrishnan, K., Jeeva, K., Daga, D. M. and Girija, R. 1995. *Changes in auroral electrojet currents inferred from geomagnetic field variations at Maitri and Northern Conjugate stations. Scientific Report of Eleventh Indian Expedition to Antarctica,* Technical Publication No. 9, Dept. of Ocean Development, New Delhi, pp. 87–101.

Kalra, R., Hanchinal, A. N., Sridharan, M. and Rajaram G. 1998. *Three station fluxgate magnetometer experiment to determine auroral current velocities over Indian Antarctic stations.* Scientific Report of Fourteenth Indian Expedition to Antarctica.

Neudeg, G. A., Borns, G. B., Fraser, B. J., Menk, F. W., Morris, R. J., Hansen, H. J. and Underwood, M. J. 1995. *Determination of Pc 1-2 ULF wave velocity and direction using a closely-spaced Antarctic magnetometer array.* Australian National Antarctic Research Expedition Report, ANARE, pp. 126–151.

Prawirodirdjo, L. and Bock, Y. 2004. Instantaneous global plate motion model from 12 years of continuous GPS observations. *J. Geophys. Res.*, v. 109, p. B08405.

Rajaram, G., Arun, T. and Dhar, A. 2002a. Diagnostics of magnetosphere-ionosphere coupling over Indian Antarctic Station Maitri, from magnetometer and riometer observations during the optical auroral event of 4–5 March 1999. *Adv. Space Res.*, v. 30 (10), pp. 2195–2201.

Rajaram, G., Arun, T., Dhar, A. and Patil, A. G. 2002b. Rapid decline in total magnetic field F at Antarctic stations– its relationship to core–mantle features. *Antarct Sci.*, v. 14 (1), pp. 61–68.

Rajaram, G., Dhar, A. and Kumar, S. 2001. Response of geomagnetic variations and 30 MHz riometer absorption, at Indian Antarctic Station Maitri, to conditions of "zero" and "high" solar wind. *Adv. Space Res.*, v. 28 (11), pp. 1661–1667.

Raymond, C. A., Ivins, E. R., Heflin, M. B. and James, T. S. 2004. Quasi-continuous Global Positioning System measurements of glacial isostatic adjustment in the Northern Transantarctic Mountains. *Global Planet. Change*, v. 42, pp. 295–303.

Roy, M. and Lakhina, G.S. 2002. Numerical simulation of magnetic reconnection and plasmoid dynamics in the geotail. *Indian journal of Radio and space physics.*, v. 31, pp. 183–189.

Roy, M. and Tajima, T. 1992. The effect of hot inhomogeneous plasma on geomagnetic micropulsations. *J. atm. terrestrial phy.*, v. 54 (2), pp. 209–215.

Sridharan, M., Dhar, A. and Rajaram, G. 1997. *Dependence of HF radio communication between India-Antarctica on seasonal and geomagnetic conditions and some anomalies.* Scientific Report of Thirteenth Indian Expedition to Antarctica, Technical Publication No. 11, Department of Ocean Development, New Delhi, pp. 106–122.

Sinha, A. K., Dhar, A., Singh, A. K., Behera, J. K. and Gurubaran, S. 2017. India's contribution to geomagnetism and allied studies in antarctica–a review. *Proceedings of the Indian National Science Academy*, v. 83 (2), pp. 299–326.

Yamada, A., Maruyama, K., Ootaki, O., Itabashi, A., Hatanaka, Y., Miyazaki, S. I., Negishi, H., Higashi, T., Nogi, Y., Kanao, M. and Doi, K. 1998. Analysis of GPS data at Syowa Station and IGS tracking stations. *Polar Geosci.*, v. 11, pp. 1–8.

3

Study of Antarctic Meteorology and Atmospheric Ozone: A Review

R. P. Lal

CONTENTS

Introduction

The India Meteorological Department (IMD) has participated in the Indian Antarctic Expedition since 1981. A permanent scientific station, known as Dakshin Gangotri (69° 59′ 23″ S, 11° 56′ 26″ E), was established in 1983 during the third expedition. IMD took observation of various meteorological parameters until 1989. A second permanent Indian Antarctic Station, Maitri, was set up at Schirmacher Oasis during the eighth expedition in 1988–1989. Since 1990, all meteorological observations such as pressure, temperature, wind speed and direction, total global solar radiation, surface ozone, radiometersonde, ozonesonde, and so on have been taken at Maitri. So far, the India Meteorological Department has participated in 26 Indian Antarctic Expeditions and contributed towards creation of databases of meteorological parameters for the studies in the fields of meteorology, ozone, and radiation. Regular ozone measurements at these stations are being taken to augment international efforts for the study of the ozone-hole phenomenon over Antarctica and the effect of depletion of ozone on global climate change.

Antarctica has attracted the attention of the scientific community all over the world since it is one of the most important climate regimes, with the potential to influence long-term global climatic patterns. To study these, IMD's meteorological program is specifically aimed at preparing a data archive of meteorological parameters over Antarctica. This may also enable us to understand the global aspects of Indian weather in general. The daily synoptic observations recorded at Maitri, Antarctica, are entered into the global telecommunication system (GTS) and contribute to international efforts for weather forecasting. Time series analysis of multiyear weather records from Antarctic stations gives valuable clues for long-range forecasts of monsoon (Datta, 1989). Meteorological studies during various Indian Scientific Expeditions to Antarctica have been published in the Technical Publications of the Department of Ocean Development. These reports contain annual weather summaries of Dakshin Gangotri (Rizvi, 1986; Lal, 1987; Rao, 1988, 1997), Maitri (Singh and Kulandaivlu, 1994; Gulhane and Katariya, 1995; Koppar, 1995; Venkateswarlu and Lal, 1996; Rao and Gaonkar, 1997; Hosalikar and Machnurkar, 1998; Rasal and Mohar, 1999; Bishnoi and Singh, 2000; Venkateswarlu and Singh, 2000; Lal and Krishnamurthy 2002; Kulandaivelu and Rao, 2005; Ram and De, 2006; Lal and Machnurkar, 2007), and other features of vertical profiles of atmospheric ozone. Reports for the years 2003 to 2005 are yet to be published. Complete data for the years 1984 to 2003 is held in the computer database of the National Data Centre of India Meteorological Department at Pune, India. Daily weather summaries of the station are available in manuscript form at IMD headquarters, New Delhi. Data from all sources has been used in this study.

Indian meteorologists have published several scientific research papers in various national and international scientific journals. Meteorological studies during various Indian Scientific Expeditions to Antarctica have been published in the Technical Publications of the Department of Ocean Development (now Ministry of Earth Sciences). In this chapter, an attempt has been made to provide an overview of the contributions made by the India Meteorological Department to Antarctic meteorology right from the establishment of scientific stations Dakshin Gangotri and Maitri. The climatology of both stations has been reported along with salient features and the comparison of some meteorological parameters recorded at both stations.

Scientific Program

Antarctica is one of the most important climate regimes, with the potential to influence long-term global climatic patterns. Therefore, India's meteorological program is specially aimed at preparing a data archive of meteorological parameters over Antarctica in this area. This may also enable us to understand the global aspects of Indian weather in general. The meteorological program has been an integral part of all the scientific expeditions to Antarctica. The main objectives of the meteorological program are follows:

- To build up a data set for climatology of Antarctica.
- To conduct research on various aspects of the weather at Antarctica and its influence on global weather in general and weather over the Indian subcontinent.
- To organize observational programs for the study of radiation budget, ozone hole phenomena, and so on.

The following were the main tasks for the meteorological program during the Indian Antarctic Expeditions: continuous recording of various surface meteorological parameters, including synoptic hour observations and real-time transmission of main synoptic messages (four times a day) over GTS for use by weather forecasters all over the world.

Results and Discussion

i. Recording of surface ozone and measurement of vertical distribution of ozone in the atmosphere through weekly ozonesonde ascents since 1987.
ii. Radiation budget studies—Direct and diffuse radiation observations are supplemented by taking radiometersonde ascents.

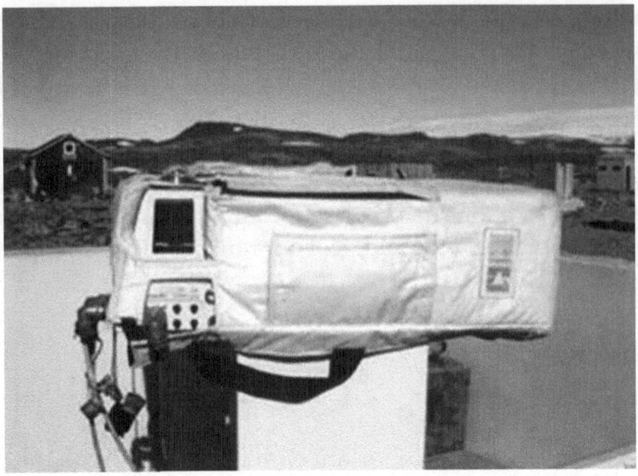

FIGURE 3.1
Brewer spectrophotometer.

iii. Turbidity measurement for aerosol study.
iv. Reception of satellite cloud imageries from polar orbiting satellites and weather chart broadcast over radio-facsimile as part of logistical support to the expedition.
v. Recording of rain/snow.
vi. Recording of surface observations during onward and return cruise of the expedition and transmission of main hour synoptic data over GTS.
vii. Measurement of total atmospheric total ozone by Dobson spectrophotometer installed at Dakshin Gangotri during 1987–1989.
viii. Measurement of total atmospheric total ozone, ground-reaching ultra violet-B (UV-B), and concentration of SO_2 and NO_2 and vertical profile of Ozone by Brewer spectrophotometer.
ix. Spectrophotometer (Figure 3.1) installed at Maitri during 18th Expedition in 1999 (Lal and Krishnamurthy, 2002). Regular ozone measurements at Maitri augment the international efforts for study of the ozone-hole phenomenon over Antarctica and the effect of depletion of ozone on global climate change.

Meteorological and Climatological Studies

A good database is essential to study the climate of the region. Thus, Dakshin Gangotri and Maitri were equipped with various instruments to monitor weather systems around the clock. The field observatory at Maitri is shown in Figure 3.2. Similar instruments were installed at Dakshin Gangotri and on the shelf ice. At the Dakshin Gangotri anemometer, a wind vane and Stevenson screen were mounted on the tower at appropriate heights. The entire instrumentation for the expedition was designed and developed by India Meteorological Department except the Brewer spectrophotometer and automatic picture transmission (APT) instrument.

Along with synoptic observations, the meteorological program also includes the study of radiation, surface ozone, total ozone and atmospheric turbidity, and so on. The regular reception of gridded APT satellite pictures in infrared and visible channels from polar orbiting satellites and analyzed weather charts from Pretoria (South Africa) help in monitoring and forecasting of weather systems approaching and affecting the stations. The auto recorder data of temperature, wind speed, wind direction, pressure, global, solar radiation, and surface ozone is regularly tabulated and computed.

For the study of climatology of Dakshin Gangotri and Maitri, daily synoptic data was used for computation of daily and monthly mean of temperature, wind, pressure, and clouding. The daily mean values were derived by averaging eight synoptic observations. Thereafter, monthly and annual values were calculated. A total of 17,410 and 46,720 observations recorded at Dakshin Gangotri and Maitri, respectively, were used to compute the climatological statistics of the stations given in Tables 3.1 and 3.2.

FIGURE 3.2
Meteorological observatory at Maitri.

TABLE 3.1

Climatological statistics of Dakshin Gangotri (69° 59′ 23″ S, 11° 56′ 26″ E), Antarctica, for the period 1984–1989

	Temperature (°C)					Pressure (HPA)			Wind Speed (Kt)		AVG/ Max	Number of Days with						
Months	Daily Mean	Mean Max.	Mean Min.	EXT. Max.	EXT. Min.	Mean Pr.	Pr. Max.	Pr. Min.	Mean	Max/ Gust	W. Speed >23 KT	Snow Fall	Fog	Blizzard	Clear Sky	Overcast Sky	Average Cloud OCTA	Number of Blizzards
January	−2.5	0.1	−5.9	9.9	−14.0	989.8	1009.3	976.0	13.2	39.2	7.0	9.5	2.4	5.2	5.0	11.4	5.4	1.8
February	−6.6	−4.3	−10.4	0.2	−19.4	983.6	998.7	967.2	18.1	49.4	10.6	4.3	0.4	9.6	3.0	12.0	5.7	4.2
March	−12.1	−9.4	−15.5	1.0	−28.0	984.6	1007.0	961.7	19.4	73.3	11.5	5.5	0.4	11.5	3.0	12.4	5.0	4.5
April	−17.9	−14.3	−21.3	−3.5	−32.5	984.7	1011.6	955.8	18.8	66.8	9.3	8.8	0.0	10.7	3.3	11.6	5.0	5.0
May	−22.1	−19.3	−26.5	0.0	−43.5	988.8	1008.4	953.4	18.8	74.3	10.8	8.5	0.2	12.0	5.5	11.2	4.2	4.7
June	−22.1	−18.3	−25.6	−2.4	−44.0	988.9	1015.0	956.3	20.1	67.0	10.2	7.5	0.4	12.8	6.0	14.0	4.4	4.5
July	−26.6	−22.9	−30.8	−10.0	−46.0	982.5	1016.8	935.0	16.8	68.0	8.5	8.3	0.4	10.2	6.3	11.0	4.0	5.5
August	−28.1	−23.7	−31.0	−8.0	−52.0	985.2	1023.0	948.1	14.8	65.2	6.3	5.3	0.6	8.0	8.8	7.2	3.2	4.3
September	−25.5	−22.4	−30.4	−8.0	−46.5	982.1	1014.9	946.0	17.5	65.8	8.3	9.0	0.8	9.3	7.0	9.0	3.8	3.7
October	−19.5	−15.2	−24.2	−3.6	−37.5	980.6	1006.6	951.5	18.3	67.0	9.8	7.3	1.2	10.5	8.5	12.6	4.3	5.2
November	−11.2	−7.5	−15.9	3.0	−30.0	981.6	1001.2	946.6	17.3	53.7	8.3	6.8	0.2	9.5	5.0	12.0	4.8	5.2
December	−4.7	−1.7	−9.2	6.0	−17.5	987.2	1003.5	965.0	12.6	42.7	4.2	7.6	3.7	5.3	5.5	11.2	4.7	3.0
Year	−16.6	−13.2	−20.6	9.9	−52.0	985.0	1023.0	935.0	17.1	61.0	104.9	88.1	10.6	114.6	66.8	135.6	4.5	51.5

TABLE 3.2

Climatological statistics of Maitri, Antarctica (Lat 70° 45′ 53″S, Long 11° 45′ 03″E) for the period 1990–2005

	Temperature (°C)					Pressure (HPA)			Wind Speed (Kt)		AVG/ Max	Number of Days with							
Months	Daily Mean	Mean Max.	Mean Min.	EXT. Max	EXT. Min	Mean Pr.	Pr. Max.	Pr. Min	Mean	Max/ Gust	W. Speed >23 KT	Snow Fall	Fog	Bliz-zard	Clear Sky	Overcast Sky	Average Cloud OCTA	Number of Blizzards	Snow-Fall in mm
January	0.6	3.4	−2.7	9.4	−9.7	987.9	1007.6	964.7	13.2	51/80	8.3	4.9	0.8	0.3	5.6	9.3	4.8	0.1	4.9
February	−2.4	0.3	−5.6	12.2	−13.8	985.9	1006.0	963.4	16.1	46/84	12.3	3.6	0.0	0.6	4.6	10.9	4.6	0.5	2.2
March	−7.2	−4.5	−10.3	3.5	−19.7	984.0	1006.8	948.4	17.9	53/91	16.3	4.9	0.0	2.8	6.3	9.1	4.5	1.6	4.5
April	−11.2	−8.5	−14.3	2.5	−28.1	984.2	1030.5	945.1	19.4	60/77	16.8	6.1	0.1	5.4	5.5	12.3	4.3	3.2	17.6
May	−13.1	−10.2	−16.4	−0.5	−30.8	986.0	1018.6	955.0	20.8	70/>100	17.9	5.4	0.3	6.2	7.1	10.2	4.3	3.3	16.8
June	−13.8	−11.0	−17.0	−0.1	−32.3	990.0	1015.9	954.3	20.1	66/90	17.1	5.1	0.0	5.1	8.1	9.9	4.1	2.3	27.1
July	−16.4	−13.2	−19.8	−0.6	−33.6	985.8	1012.4	948.9	19.2	78/100	17.3	5.8	0.1	6.1	7.4	11.6	3.8	3.1	22.7
August	−17.4	−14.0	−21.0	0.2	−34.5	983.4	1018.6	930.0	17.9	80/106	15.8	5.4	0.0	7.1	6.9	8.5	4.6	3.5	43.3
September	−16.2	−13.0	−20.0	−0.8	−32.6	985.0	1024.2	943.2	16.9	94/110	13.5	4.9	0.0	5.7	7.4	8.7	4.5	2.9	26.8
October	−11.3	−8.4	−14.8	1.0	−27.2	983.5	1011.9	951.1	17.7	66/82	15.6	4.6	0.0	4.0	6.5	10.7	4.5	2.6	12.1
November	−5.3	−2.6	−8.8	6.2	−21.6	986.4	1020.4	958.1	16.5	58/81	12.7	3.4	0.1	2.8	6.5	7.1	3.8	1.3	2.0
December	−0.2	2.4	−3.4	9.2	−11.6	987.6	1009.1	963.0	13.1	56/86	8.1	6.1	2.2	0.9	4.5	6.4	5.1	0.5	2.1
Year	−9.5	−6.6	−12.9	12.2	−34.5	985.8	1030.5	930.0	17.4	94/110	171.6	60.1	3.7	46.9	76.5	114.7	4.4	24.9	182.1

Surface Temperature

For measurement of surface air temperature, maximum temperature, and minimum temperature, a standard Stevenson screen is installed in the field and well-calibrated thermometers are installed therein. For around-the-clock recording of surface air temperature, a thermolinear thermistor is installed in the same screen at the height of the dry bulb thermometer, and the temperature output is displayed and recorded in the IMD lab in the main station. The temperature value from the autographic record is computed manually.

The basic mean and extreme long-term temperature data for Dakshin Gangotri and Maitri are shown in Figures 3.3 and 3.4 based on the data available for the period of study.

The annual average air temperature of Dakshin Gangotri is –16.6°C, whereas it is –9.9°C for Maitri. Comparison of monthly means of both stations (Figure 3.5) shows that temperatures recorded at Dakshin Gangotri are always lower than those at Maitri and, as the season advances towards midwinter, the difference reaches up to 10°C. The lowest mean temperature and extreme daily minimum temperature are recorded in August, and highest mean and extreme maximum temperatures in January and December, when readings well above zero are recorded. A highest temperature of 9.9°C was recorded

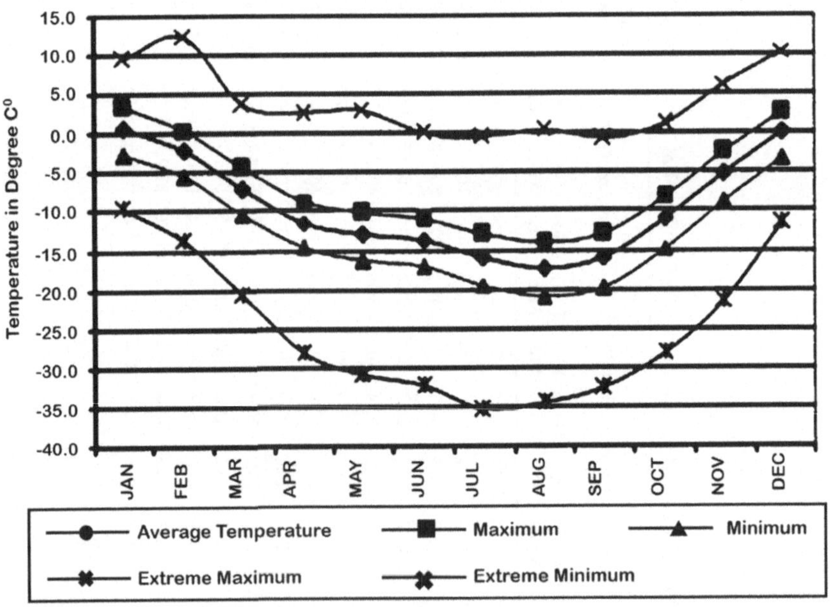

FIGURE 3.3
Mean monthly temperatures for Dakshin Gangotri, Antarctica, for the period 1984–1989.

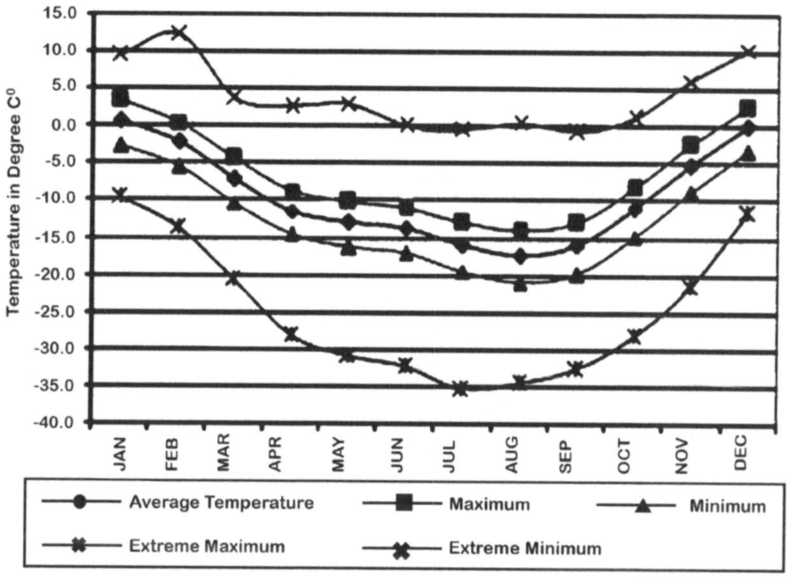

FIGURE 3.4
Mean monthly temperature for Maitri, Antarctica, for the period 1990–2005.

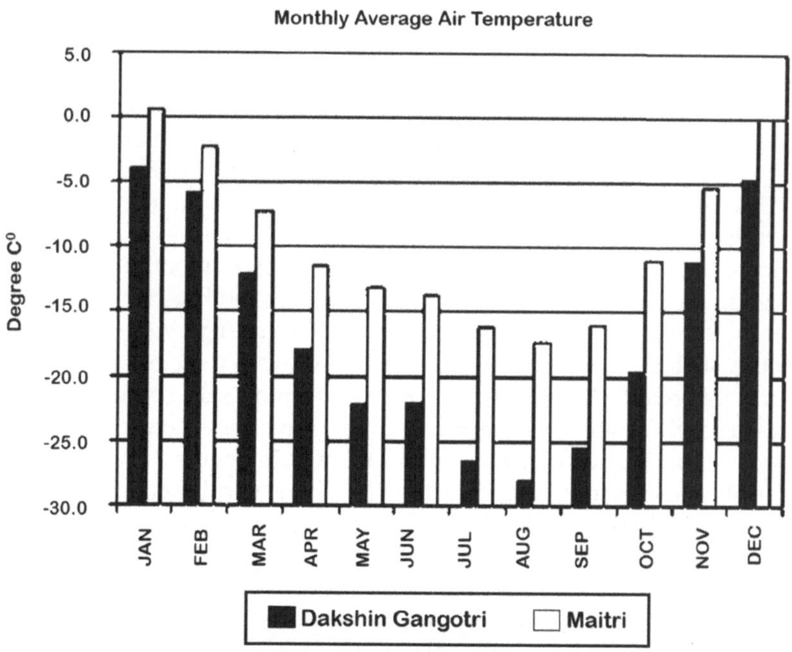

FIGURE 3.5
Comparison of average air temperature of Dakshin Gangotri and Maitri.

at Dakshin Gangotri on January 10, 1988, whereas February 1996 was an exceptional month when on February 3, the maximum temperature rose to 12.2°C at Maitri. The lowest temperature of −52.2°C was recorded on August 16, 1988, at Dakshin Gangotri, and −34.5°C was recorded on August 28, 1993, at Maitri.

In the colder winter months, abnormally warm situations are possible, usually associated with rapid translation of warm oceanic air from the north, disturbance of low-level inversion, or substantial low-level subsidence and warming in high-pressure centers or ridges extending to high latitudes. An example of such an extreme maximum of 0.2°C was recorded on August 25, 1996, at Maitri under a similar synoptic situation. On this date, the highest MSL pressure of 1011.3 hPa was recorded at the station.

Wind

For continuous recording of wind speed and direction, a distant indicating wind equipment (DIWE) composed of one cup generator anemometer (CGA) and Selsyn windvane was installed on the roof of Dakshin Gangotri and Maitri stations. Output from this system was displayed and recorded in the IMD lab. Wind speed and direction recorded on the recorder are used for synoptic observation, and hourly values are also tabulated from these records throughout the year.

The surface wind regime at both the stations is characterized by alternating spells of strong wind and light wind or calm. The duration of these spells varies considerably from a few hours to several days. The mean daily wind speed varies between 5 and 10 ms^{-1} (~10 and 20 kt). Wind maxima recorded in each month varies between 28 and 45 ms^{-1} (~55 and 90 kt). Wind speed exceeding 100 knots has been recorded during winter months.

Wind direction is a significant feature of weather at both stations. It serves as a sure indication of the type of ensuing weather. The most predominant direction is southeast (28.5%). Sectorwise, it is the east–southeast sector (090°–135°) that accounts for 56% at Maitri. The next important sector is the south (160°–200°).

Katabatic winds are characterized by high directional constancy but show large variations in speed: their direction at Maitri is invariably southerly, since the elevation increases in that direction. Katabatic winds are more frequently experienced at Maitri, as it is located at the margin of the polar ice cap. The frequency of katabatic winds at Maitri is about 28%. Although katabatic winds occur throughout the year, their frequency is maximum during the winter season from April to August, followed by spring and autumn. Their frequency is lowest in summer (November–January). During autumn and spring, katabatic winds set in by late evening (1800 hrs) and continue up to morning (0600 hrs). But during winter, these winds occur at any time of the

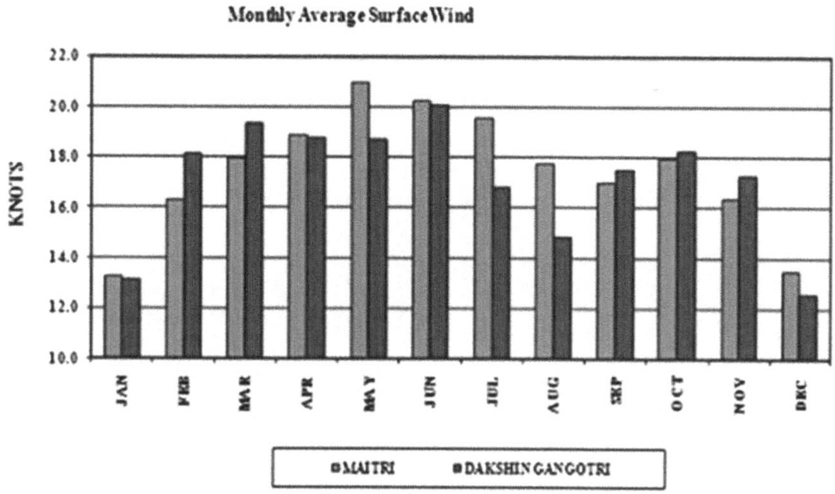

FIGURE 3.6
Monthly variation of average surface wind speed at Dakshin Gangotri and Maitri, Antarctica.

day. The speed of katabatic winds may vary from 5 to 15 ms^{-1} (10 to 30 knots) and shows large variations frequently over short time intervals. Sudden onsets with the wind speed jumping from almost calm to 12–15 ms^{-1} (25 to 30 knots) and equally sudden cessation (lull) are common.

The month of May is the windiest month of the year at Maitri, whereas it is June at Dakshin Gangotri. Comparison of average surface wind speed at both stations is shown in Figure 3.6, and they do not differ substantially in the mean from those for the shorter periods. Annual average surface wind speed at Dakshin Gangotri was 17.1 knots, and at Maitri it was 17.5 knots. On average, there are about 104 days in a year where the daily average wind speed exceeds more than 23 knots at Dakshin Gangotri, and this value is 178 days at Maitri. Monthly distributions of average number of days with daily average wind speed exceeding more than 23 knots are shown in Figure 3.7. It is evident from the above figures that Maitri is windier than Dakshin Gangotri; this is because of the geographical location of the stations. Maitri is more influenced by katabatic winds than Dakshin Gangotri.

Surface Pressure

Dakshin Gangotri and Maitri stations lie between the high-pressure region centered around the South Pole and the circumpolar trough of low pressure roughly along 63°S. Therefore, atmospheric pressure at these stations is influenced by the relative position and strength of these features. At both

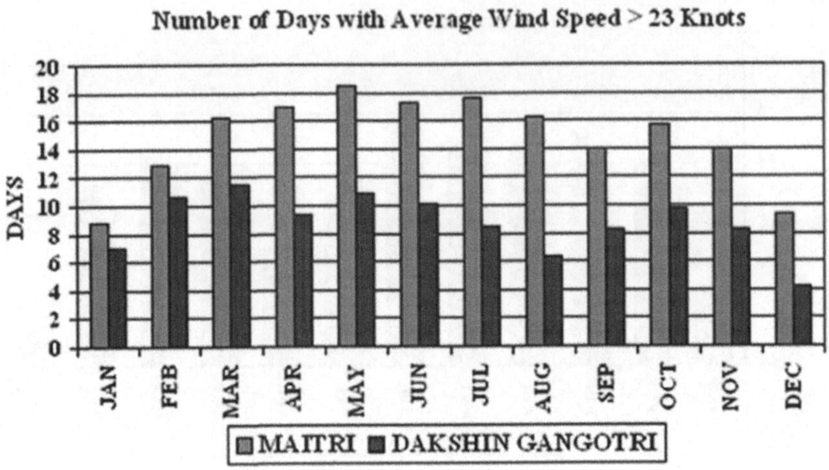

FIGURE 3.7
Monthly variation of number of days with wind speed >23 knots at Dakshin Gangotri and Maitri, Antarctica.

stations, seasonal variation in pressure shows a trend of alternate rise and fall with the highest values in summer and winter and lowest values in autumn and spring. Day-to-day variation in pressure can be very dramatic.

The station level pressure is recorded using a precision aneroid barometer, which is calibrated with reference to a regional standard kept at the office of the DDGM (SI) Pune and converted to mean sea level pressure using height and temperature correction tables specially prepared for this station. A Belfort marine weekly barograph is used for continuous recording of station level pressure. Hourly values are tabulated from the barogram.

The mean sea level (MSL) pressure at the station depicted in Figure 3.8 exhibits a half yearly cycle, with the lowest pressure being recorded in the spring season and a secondary minimum in March. Similarly, maximum pressure is recorded in June and January. A mean monthly variation in the latitude of the axis of the Antarctic trough also indicates similar fluctuations throughout the year. The Antarctic trough is a feature of mean circulation of the southern hemisphere, which controls the weather in the region.

Mean annual mean sea level pressure at Dakshin Gangotri is 985.0 hPa, and at Maitri it is 985.9 hPa. During the period 1984–1989, the lowest MSL pressure of 935.0 hPa was recorded at Dakshin Gangotri on July 30, 1985, and the highest MSL pressure 1023.0 hPa on August 29, 1989. For the period 1990–2005 at Maitri, the highest MSL pressure of 1030.5 hPa was recorded on April 29, 1990, and the lowest MSL pressure of 930.0 hPa was recorded on August 9, 2005, during the passage of a low-pressure area. This was the lowest mean sea level pressure recorded at this station since 1990.

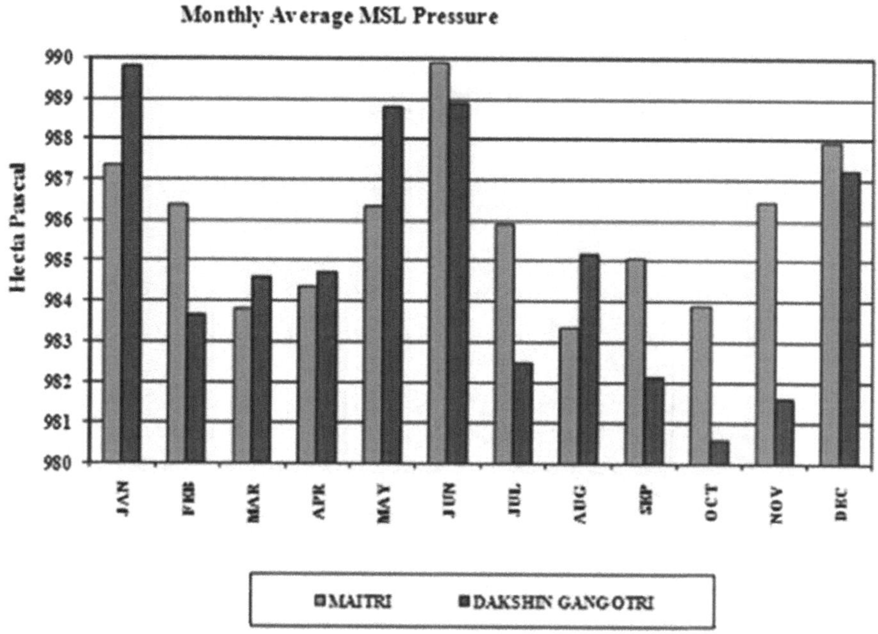

FIGURE 3.8
Monthly variation of average mean sea level pressure at Dakshin Gangotri and Maitri, Antarctica.

Snowfall

The first time a snow gauge was installed at Maitri was in the year 1998 for measuring snowfall during snowfall and blizzard days (Figure 3.9). With this instrument, we can measure the collected snow equivalent to water. The measuring of snow at Antarctica is very important because of the alarm of global warming.

Snowfall was observed almost in all months. The number of days with snowfall at Dakshin Gangotri and Maitri is shown in Figure 3.10. There are on average 88 days with snowfall at Dakshin Gangotri and 60 days at Maitri. Figure 3.11 shows the monthly average snowfall recorded at Maitri.

The maximum snowfall is recorded in the month of August, followed by June and September. Snowfall occurs on average on some 7 days at Dakshin Gangotri and 5 days per month at Maitri through the year. As with other Antarctic stations in strong wind areas, precipitation has not been successfully measured quantitatively because of the difficulty of distinguishing between falling and drifting snow as collected in a gauge and because of the low density and moisture content of polar snow. A few

FIGURE 3.9
Snow gauge installed at Maitri.

very rare cases of rainfall have occurred at Maitri, resulting in a usually short-lived glaze or rime covering exposed rocks. On January 17, 1991, Maitri experienced light rain and drizzle around 1930 to 2030 HRS local time. Similarly, during 1996, on February 2, Maitri experienced light rain equivalent to about 2 mm. The intensity of rain was so much that the roof of Maitri station started leaking.

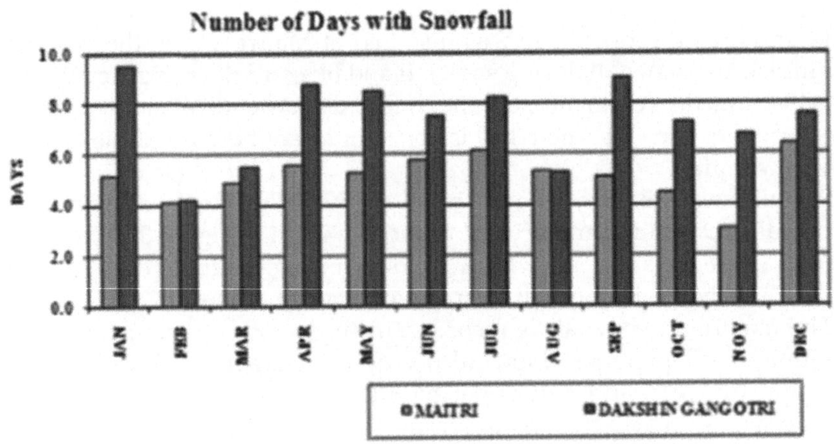

FIGURE 3.10
Monthly variation of number of days with snowfall at Dakshin Gangotri and Maitri, Antarctica.

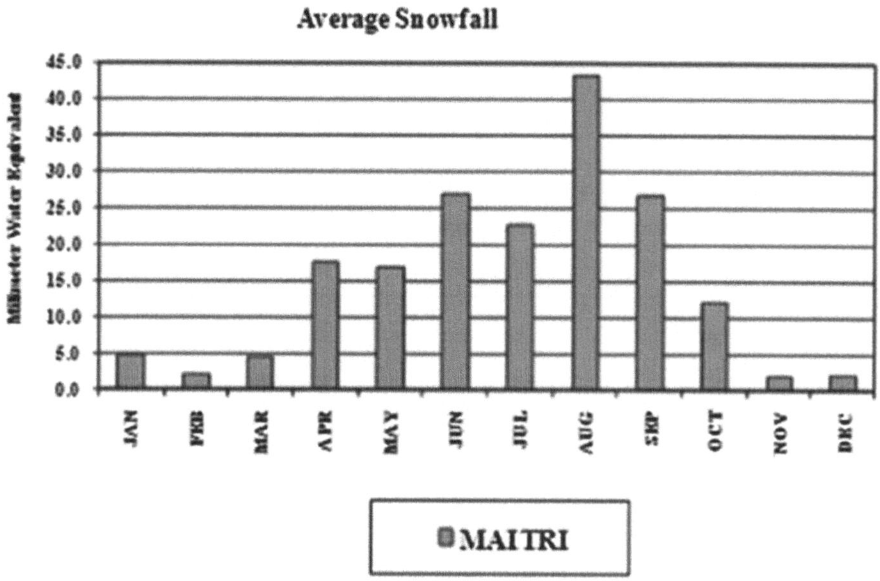

FIGURE 3.11
Average monthly snowfalls in millimeter of water equivalent for Maitri for the period 1998–2005.

Blizzards and Drifting Snow

Weather in Antarctica is subject to frequent and sudden changes. Strong winds and blizzards dominate Antarctic weather. When wind speed exceeds 23 knots with drifting/blowing snow reducing the surface visibility to less than 1 km, the weather phenomenon is defined as a blizzard for our study purposes. On average during the year about 52 blizzards affect Dakshin Gangotri for 115 days, and about 23 blizzards affect the station for 47 days during the year at Maitri. The monthly variation of number of days with blizzards is shown in Figure 3.12.

At Maitri station, the maximum number of blizzards occur in the month of August, with about 7 blizzard days. The average wind speed recorded during blizzards is about 52 knots, but it exceeded 100 knots on several occasions. The duration may vary from hours to days, with an average of 25 hours. The longest duration, 168 hours, was recorded in June 1997. During this period, the pressure dropped by 53 hPa and the temperature rose by 17°C, with the wind reaching 87 knots during the blizzard. There are about 12 such occasions when blizzards lasted more than 72 hours.

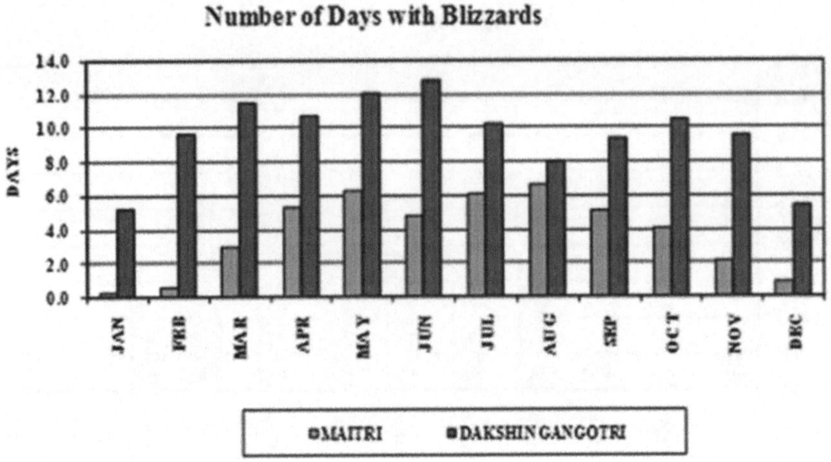

FIGURE 3.12
Monthly variation of number of days with blizzards at Dakshin Gangotri and Maitri, Antarctica.

Clouds

The mean cloudiness throughout the year averages 4.5 and 4.4 octas for Dakshin Gangotri and Maitri, respectively. It is highest in summer. The frequency of clear days is somewhat higher in winter than in midsummer when the retreat of sea ice brings the station under more maritime conditions. Most cloud forms are a variation of stratus, although stratocumulus is often seen in summer time. Cumulus cloud is very rare, but cirrus cloud is often widespread and forms at much lower levels than in temperate regions. Reliable cloud observations are very difficult during blizzards since the sky invariably remains obscured.

Clouds are almost always stratiform at different levels, ranging from high cirrus to more common stratus. The cloud types most commonly observed are stratus (stratus fractus and stratus fractus of bad weather), altostratus (altostratus translucidus and altostratus opacus), altocumulus (altocumulus translucidus and altostratus duplicatus), and cirrostratus (cirrostratus fibratus and cirrostratus nebulosus).

Radiation Balance Studies: Total Global Solar Radiation Measurement

Total global solar radiation was continuously recorded at Dakshin Gangotri and Maitri stations using a thermo-electric pyranometer installed on

FIGURE 3.13
Pyranometer installed at Maitri.

the roof of the station (Figure 3.13). Radiation measurements were taken throughout the year. Global solar radiation at Dakshin Gangotri has been studied earlier (Kulkarni et al., 1988; Koppar and Kedare, 1993). Average monthly total global solar radiation values for Dakshin Gangotri and Maitri are shown in Figure 3.14.

At Dakshin Gangotri, the mean daily global solar radiation around 33 MJ/m^2 is maximum in December when the sun never sets over the station. At Maitri,

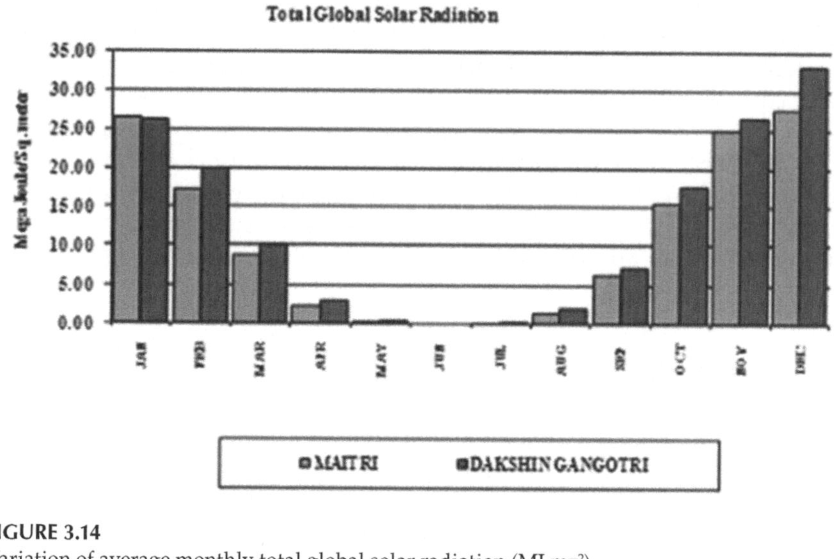

FIGURE 3.14
Variation of average monthly total global solar radiation (MJ m^{-2}).

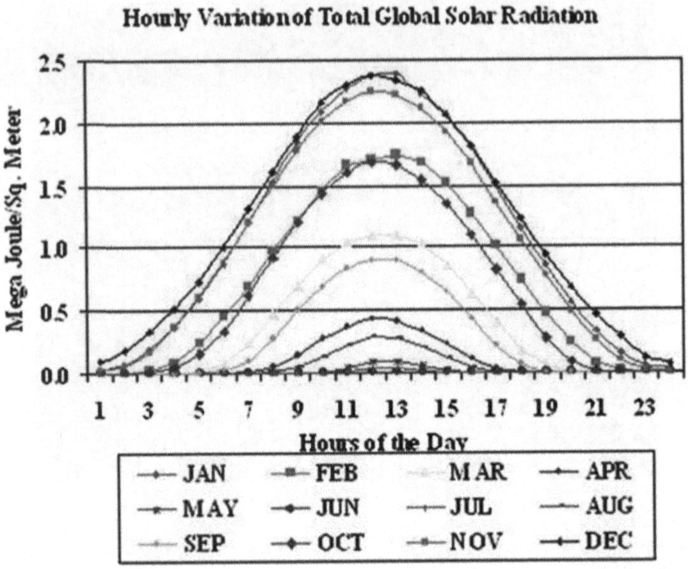

FIGURE 3.15
Monthly variation of average hourly total global solar radiation (MJ m^{-2}).

the total global solar radiation was found maximum in the months of January, November, and December during the year due to long periods of cloud-free sky, insolation, and high solar elevation. As the solar elevation and duration of sunshine start reducing, total global solar radiation also reduces to zero by the end of the month of May. Solar radiation again starts from the last week of July and reaches the highest value by the beginning of the polar day.

The distribution of mean hourly global solar radiation with hours in local apparent time (LAT) is shown in Figure 3.15. The irradiance received is in direct correspondence to the duration of sunshine and solar elevation. The highest value of 2.39 MJ m^{-2} occurs around noon in January and December, obviously, because the solar elevation reaches a maximum angle of 45° in the second fortnight of December. Noontime lowest values of 0.04 MJ m^{-2} occur in the month of July when radiation intensity is weakest, the solar elevation being less than 0.5°, since annual sunrise occurs in the third week of this month.

Atmospheric Ozone and Ozone Hole Studies

Surface Ozone

Surface is the main sink for ozone and since the mean concentration of ozone at different levels of atmosphere remains substantially the same, the rate of production of ozone at higher levels must be equal to the rate of destruction near the surface. Therefore, continuous monitoring of ozone was done, its hourly values are computed manually from the autograph charts, and the results are under process. The main objective of the measurements was to find out the nature of variations of surface ozone during the rapid depletion and fast recovery of total ozone in late winter and early spring over the Antarctic.

Measurement of surface ozone started at Dakshin Gangotri in 1989 and has continued at Maitri since 1990. The electrochemical Brewer bubbler ozone sensor used in ozonesonde for measurement of ozone was modified and used for continuous recording of ozone near the surface. The equipment has an accuracy of 10%. At Antarctica, the sensor is kept inside the laboratory, warm enough to prevent freezing of the electrolyte, with the short polytetrafluoroethylene (PTFE) intake tube outside the building for taking in free air.

From the record of daily surface ozone concentration, half-hourly values were computed for a study of the hourly and day-to-day changes and annual cycle. The results of observations at Dakshin Gangotri have been discussed by Shreedharan et al. (1993). The surface ozone values show a summer minimum on the order of 20 ppbv and a winter maximum on the order of 42 ppbv. The transition from the low summer value to the high winter value occurs during May–June. From October to November, there is a sharp fall in the surface ozone values. This fall takes place at a time when the total ozone values are sharply rising from the lowest annual value. Other notable features of the surface ozone are the near absence of diurnal variation, even during the summer months, and the absence of significant changes during blizzards. The results agree fairly well with the measurements reported from Syowa in Antarctica, a station located at nearly the same latitude.

Vertical Distribution of Atmospheric Ozone

Regular measurement of ozone at the Indian Antarctic Station, Dakshin Gangotri, started in 1987 and continued at Maitri since 1990 with the help of an Indian electrochemical ozonesonde. Weekly ozone soundings are taken throughout the year, and during the month of September–October, more frequent soundings are taken to keep a close watch on fast depletion of atmospheric ozone. Ozone density in the vertical column of the atmosphere is calculated with the help of weekly ozonesonde ascents (Figure 3.16). The concentration of ozone at standard and significant levels is computed from raw values. After the preliminary computation, each sounding is corrected against total ozone value. The result of these ascents taken at Dakshin Gangotri and Maitri has been discussed by various workers (Shreedharan et al., 1986; Faruqui et al., 1988; Shreedharan et al., 1989; Koppar and Nagrath, 1991; Shreedharan et al., 1993; Peshin et al., 1996; Peshin et al., 1997; Tiwari, 1999). The average pattern of the ozone profile for the months of July (normal profile) and October (depletion period profile) is shown in Figure 3.17.

FIGURE 3.16
Ozonesonde launching in process at Maitri.

Following results can be concluded from these ascents:

i. Tropospheric ozone below 250 hPa levels has remained practically unchanged. The lowest tropospheric value of 10 nb occurs between 300 and 500 hPa. Drastic changes in ozone amount occur between 100 and 30 hPa year after year. This is the region where maximum ozone (~150–170 nb) is usually recorded from January to June every

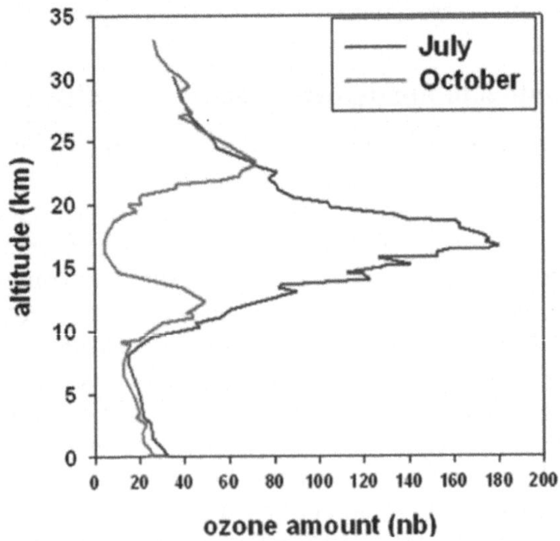

FIGURE 3.17
Average vertical profile of ozone for the month of July (normal) and October (depletion).

year. However, a region of low ozone (10–50 nb) concentration is also reported at the same level during the month of September–October every year. This is popularly called the "ozone hole." The maximum ozone is destroyed in this layer.

ii. In January to June, the change in the ozone partial pressure profile is not significant.

iii. Although the depletion of atmospheric ozone starts gradually with the onset of polar winter, sharp depletion took place in the month of September. Thereafter, recovery of atmospheric ozone started quite early from the middle of the month of October, and the normal value of ozone partial pressure was recovered by the last week of the month. The depletion period was shorter than normal.

iv. During the depletion period, atmospheric ozone was found considerably depleted at some levels between levels 110 hPa and those above. The minimum ozone value in this layer, which was on the order of 50 nb during 1987–1988, has further decreased to 5–10 nb during recent years, and the vertical depth of the ozone hole and period of occurrence have been increasing from year to year. Ozone observations from other stations over Antarctica also support the above view that ozone hole has been increasing in horizontal and vertical area and that total column ozone is becoming thinner.

v. Two maxima were observed during the depletion period separated by the depletion zone. The first maxima during the depletion were found around 140–170 hPa and the second maxima around 15–20 hPa. Ozone recovery started from the higher levels and descended gradually to lower levels.

Brewer Spectrophotometer Observations

Brewer Spectrophotometer System

The Brewer spectrophotometer MKIV version is a scientific instrument that measures atmospheric ozone and sulfur dioxide by examining the absorption of specific ultraviolet wavelengths from the solar spectrum. The instrument can be switched to visible wavelengths and measure nitrogen dioxide.

The Brewer is an optical instrument designed to measure ground-level intensities of the attenuated incident solar ultraviolet (UV) radiation at five specific wavelengths in the spectra of ozone and sulfur dioxide. The MKIV version allows switching to nitrogen dioxide operation at 430–450 nm (second order). The Brewer contains a modified Ebert f/6 spectrometer with a 1200 line/mm holographic diffraction grating operated in the third order for $O3/SO_2$ and second order for NO_2.

Sunlight enters the fore optics system through the quartz inclined window adjacent to the zenith-adjustment knob. The incoming light is directed through the fore optics by a director prism, which may be rotated to select light from the zenith sky, the direct sun, or one of the two calibration lamps. A mercury lamp provides a line source for wavelength calibration of the spectrometer; a halogen lamp provides a well-regulated light source so that the relative spectral response of the spectrometer may be monitored. The elements in the fore optics provide adjustment for field-of-view, neutral-density attenuation, ground quartz diffusion, and selection of a film polarizer for zenith sky measurements.

A modified Ebert grating spectrometer disperses ultraviolet light onto a focal plane. Six slits are positioned along the focal plane at the operating wavelengths: 303.2 nm (302.1 nm for mercury-wavelength calibration), 306.3, 310.1, 313.5, 316.8 nm in ozone mode with 0.6 nm resolution and 426.4, 431.4, 437.3, 442.8, 448.1, 453.2 nm in NO_2 mode with 0.85 nm resolution. Wavelength is adjusted by rotating the grating with a stepper motor, which drives micrometer acting on a lever arm.

The exit-slit plane is shielded by a cylindrical mask, which exposes only one wavelength slit at a time. The mask is positioned by a stepper motor, which cycles through all five operating wavelengths approximately once per second. The light passing through the exit slits is collected on the cathode of a low-noise EMI9789QB05 photo-multiplier. The photon pulses are amplified, discriminated, and divided before being transmitted to a counter. The resulting photon count is registered in one of six wavelength channels.

The ultraviolet-B monitor is an optical assembly, which enables the Brewer to measure UV-B irradiance using a thin disc of Teflon (cosine response) as a transmitting diffuser. The disc is mounted on the top of the instrument under a 5-cm diameter quartz dome and is thus exposed to the horizontal UV irradiance. Beneath the disc is a fixed reflecting prism, which is located so that the disc is in the spectrometer field-of-view when the zenith prism is set for a zenith angle of −90. The UV measurement software routine supplied scans from 290 to 325 nm on slit 1 in 0.5-nm increments and then scans back to 290 nm. The irradiance at each wavelength is integrated to produce a damaging UV value (DUV) weighted to the DIFFEY action spectrum.

Operation and Maintenance

From 1999 to 2005, a Brewer spectrophotometer was in operation at Maitri. The Brewer data for the above period is being processed and results regarding total ozone, DUV, and SO_2, and NO_2 measurements for individual years have been published by Lal and Krishnamurthy (2002), Lal and Machnurkar (2007), and Kulandaivelu and Peshin (2003), and year-to-year

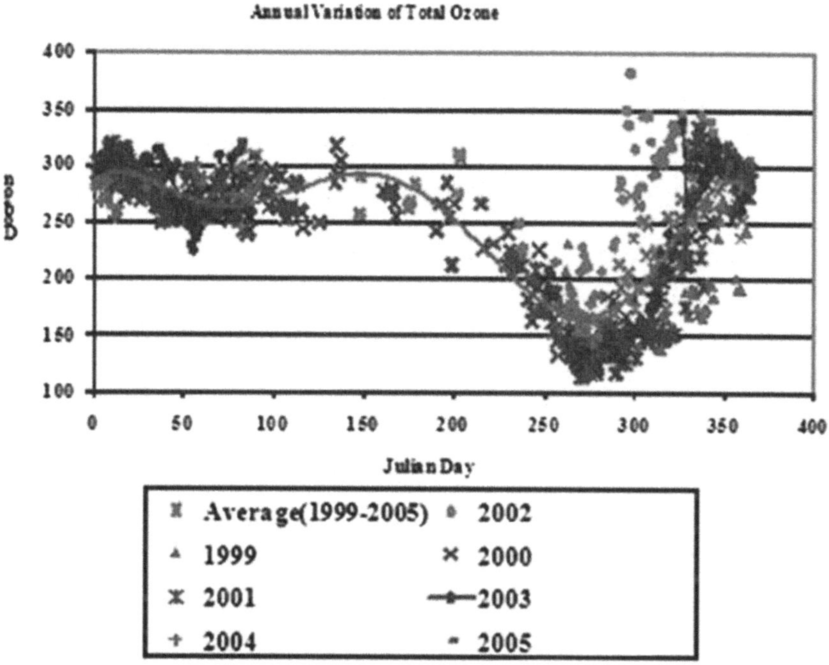

FIGURE 3.18
Day to day variation of total columnar ozone measured by Brewer spectrophotometer at Maitri.

studies have been published in DOD technical publications. Day-to-day variation of total ozone, along with the average for the years 1999–2005, is shown in Figure 3.18.

Direct Sun Total Ozone Measurements

The total ozone measurement in direct sun measurement mode is taken on the maximum number of days except on days with blizzard, snowfall, gale-force wind, and thick overcast sky. Measurements are taken for all possible durations of sunlight to find diurnal variation in total ozone amount. The total ozone values depicted a diurnal variation with maxima during morning and evening and minima near local noontime. There was occasional shooting up of the total ozone value due to intense solar activity and the arrival of ozone-rich subtropical air with the passage of extra tropical storms. A sharp decrease on some days was also observed when polar air was inducted in the area. The days with abrupt rises in total ozone value were also associated with auroral activity.

Focused Sun Observations

Total ozone observations in focused sun observation mode were taken for the period when the solar zenith angle remains above 73 until the mu value reached equal or above 8. Since these observations were made near the ozone maxima period, the values were found higher than the mean daily total ozone value measured in DS mode. The difference was about 10 DU during the normal total ozone value and about 20 DU during the depletion period.

Focused Moon Observations

In the months of May, June, and August, focused moon mode of operation for the measurement of total ozone was attempted when lunar mu was less than 3 and more than 2/3 of the lunar disc became visible. The total ozone measured in this mode was found to be very close to the daily mean total ozone measured in direct sun mode, but the values were towards the higher side by 5 DU. These (focused moon) observations are useful for total ozone measurements during the polar nights and the period of low solar elevation.

Zenith Sky Observations

The Brewer is operated in Zenith Sky mode of operation for measurement of total ozone, SO_2, and NO_2. The value obtained in this mode was not compatible with the values observed in DS mode at lower elevations of the sun. The main reason for this error is the nonavailability of Zenith Sky Map for this location. However, the values measured during high solar elevation were found to be in agreement with the DS mode of measurements.

Damaging Ultraviolet Radiation Measurements

DUV measurement was done on the maximum number of days from dawn to dusk during the period March to October. When solar days became longer, on every 10th day, DUV measurements were taken around the clock. On other days, the observation was restricted for mu values less than 8.

As discussed earlier, in UV measurement mode, diffused sunlight is scanned from 290 to 325 nm in 0.5 nm increments and then scanned back to

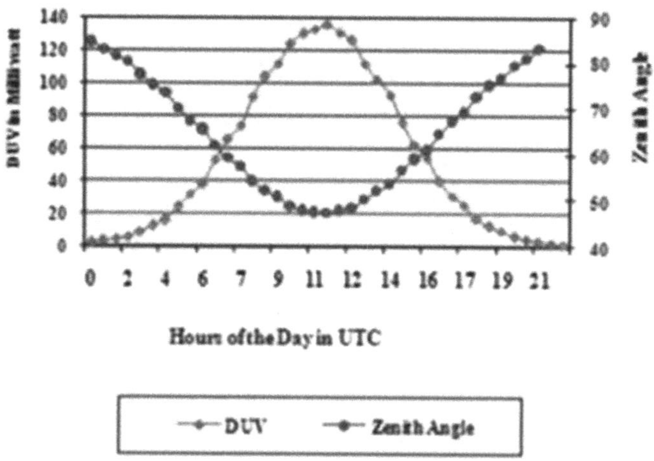

FIGURE 3.19
Hourly variation of DUV on a day as recorded by Brewer spectrophotometer at Maitri.

290 nm. From the analysis it is evident that the major contribution of irradiance comes from the wavelengths 310 to 325 nm, with maxima around 317 nm. DUV values measured in mW/m^2 show diurnal and seasonal variation. Daily DUV values show a maximum near the local noon, and its value changes according to solar zenith angle. A sample observation of a day's variation is shown in Figure 3.19. It was found to reach zero after 1 hour of sunset and starts rising from the zero value before 1 hour of sunrise.

SO_2 and NO_2 Observation

SO_2 measurement is done in direct sun measurement mode as a subroutine of ozone mode, whereas NO_2 measurements are done by a separate independent measurement mode. The system also measures and calculates the stratospheric and tropospheric amount of NO_2 in m-atm-cm or Dobson units for each morning and evening twilight periods if series of ZS measurements are available between angles of 84.5 and 96.5. Figure 3.20 shows day-to-day variation of SO_2 and NO_2 at Maitri for the year 2002.

Conclusions

A continues record of surface meteorological observations at Dakshin Gangotri and Maitri indicates that the weather over Antarctica experiences

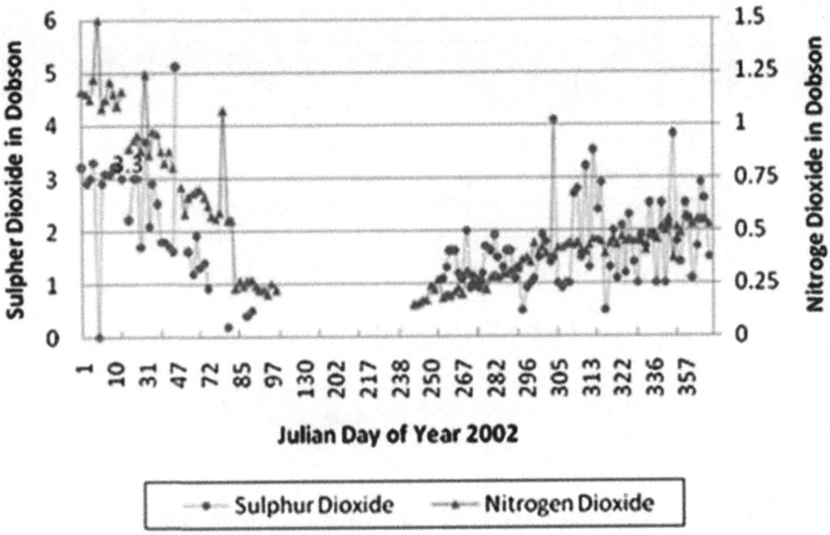

FIGURE 3.20
Daily mean variation of sulfur dioxide and nitrogen dioxide at Maitri, Antarctica.

large seasonal variations. The climatology of the station Maitri gives several interesting indications concerning the climate of Schirmacher Oasis. The study of temperature has shown a cooling trend of 0.26°C per decade. A similar significant negative trend has been observed in other meteorological parameters such as MSL pressure, wind, number of days with precipitation, and number of blizzard days. In view of these facts, it is apparent that continuing polar records of the unchanging environment of Maitri will be extremely valuable in monitoring of the global climate in foreseeable future. Thus, the long-term data being archived by IMD has tremendous potential and significance for understanding the meteorology and climate of Antarctica.

Acknowledgments

The author is thankful to the Director General of Meteorology, India Meteorological Department, for his encouragement and kind permission for pursuing this study and use of Antarctic data from the archives of IMD. The efforts put up by all IMD Antarctic Expedition team members for taking meteorological observations under adverse hostile environmental conditions are also acknowledged. Without their support and contribution, this study was not feasible.

References

Bishnoi B. N. and Singh M. 2000. *Meteorological and climatological studies at Maitri, Antarctica*. DOD Technical Publication No. 15, pp. 51–62.

Datta R. K. 1989. Temperature anomaly over Antarctica and monsoon activity. *Mausam*. v. 40, pp. 95–100.

Faruqui A. A., Koppar A. L. and Nagrath S. C. 1988. Antarctic ozone-hole investigation programme—Preliminary result of ozonesonde ascents during January–February 1987. *Mausam*. v. 39, pp. 313–316.

Gulhane P. M. and Katariya S. S. 1995. *Meteorological studies carried out at Maitri during winter period of Tenth Indian Scientific Expedition to Antarctica*. DOD, Technical Publication No. 8, pp. 71–90.

Hosalikar K. S. and Machnurkar P. N. 1998. *Report on meteorological and climatological studies at Maitri, Antarctica*. DOD Technical Publication No. 12, pp. 25–46.

Koppar A. L. and Nagrath S. C. 1991. Seasonal variation in the vertical distribution of ozone over Dakshin Gangotri, Antarctica. *Mausam*. v. 42, pp. 275–278.

Koppar A. L. and Kedare P. L. 1993. Short term climatology of global solar radiation at Dakshin Gangotri, Antarctica. *Mausam*. v. 44, pp. 111–114.

Koppar A. L. 1995. *Report on the meteorological programme carried out during Eleventh Indian Expedition to Antarctica*. DOD, Technical Publication No. 9, pp. 53–78.

Kulkarni N. V., Desai C. S. and Iyer N. V. 1988. Global solar radiation at Dakshin Gangotri in Antarctica. *Mausam*. v. 39, pp. 113–115.

Kulandaivelu E. and Peshin S. K. 2003. Measurements of total ozone, D-UV radiation, sulphur dioxide and nitrogen dioxide with Brewer spectrophotometer at Maitri, Antarctica during 2000. *Mausam*. v. 54, pp. 505–510.

Kulandaivelu E. and Venkteshwar Rao S. 2005. *Meteorological observations and ozone hole investigations during 2000 at Indian Antarctic Station, Maitri*. DOD Technical Publication No. 17, pp. 115–146.

Lal B. 1987. *Meteorological studies during the fourth Indian Scientific Expedition to Antarctica: to March 1986*. DOD Technical Publication No. 4, pp. 107–136.

Lal R. P. and Krishnamurthy L. 2002. *Study of Ozone hole phenomena and meteorological parameters at Schirmacher Oasis Antarctica during 18th Indian Antarctic Expedition 1998–2000*. DOD Technical Publication No. 16, pp. 17–50.

Lal R. P. and Machnurkar P. 2007. *Study of meteorological parameters and ozone hole phenomena at Schirmacher Oasis Antarctica during 21st Indian Antarctic Expedition 2002–2003*. Twenty First Indian Expedition to Antarctica, Scientific Report, Ministry of Earth Sciences, Technical Publication No. 19, pp. 73–93.

Peshin S. K., Rao G. S. and Rao P. R. 1996. Ozone soundings over Maitri (Antarctica)–1994. *Vayu Mandal*. v. 26, pp. 21–23.

Peshin S. K., Rao P. R. and Srivastav S. K. 1997. Antarctic Ozone depletion measured by balloon sondes at Maitri–1992. *Mausam*. v. 48, pp. 443–446.

Ram S. and De M. K. 2006. *A report on meteorological observation made during Eighteenth Wintering (2001–2002) at Indian Antarctic base Maitri, Central Dronning Maud Land, East Antarctica*. Twentieth Indian Expedition to Antarctica, Scientific Report, Ministry of Earth Sciences, Technical Publication No. 18, pp. 75–90.

Rao G. S. and Gaonkar S. B. 1997. Meteorological studies at Maitri during Thirteenth Indian Antarctic Expedition.

Rao G. S. 1997. *Meteorological studies at Dakshin Gangotri during 1988*. DOD, Technical Publication No. 11, pp. 165–180.

Rao T. V. P. B. 1988. *Meteorological studies at Antarctica during the period March 1985 to February 1987*. DOD Technical Publication No. 5, pp. 259–281.

Rasal A. S. and Mohar D. R. 1999. *Meteorological studies carried out during the 15th Indian Scientific Expedition to Antarctica 1996*. DOD Technical Publication No. 13, pp. 39–62.

Rizvi S. R. H. 1986. *Meteorological studies at Dakshin Gangotri (Antarctica) March 1984 to February 1985*. DOD Technical Publication No. 3, pp. 49–73.

Singh S. and Kulandaivlu E. 1994. *Establishment of a new meteorological observatory at Maitri and study of meteorological parameters, ozone depletion in Antarctic spring and solar radiation during 9th Antarctic Expedition*. DOD, Technical Publication No. 6, pp. 35–56.

Shreedharan C. R., Chopra A. N. and Sharma A. K. 1986. Vertical distribution of ozone over Dakshin Gangotri, Antarctica. *Indian Journal of Radio and Space Physics*. v. 15, pp. 159–162.

Shreedharan C. R., Gulhane P. M. and Kataria S. S. 1993. Ozone soundings over Antarctica. *Current Science*. v. 64, pp. 634–636.

Shreedharan C. R., Rao Sudhakar G. and Gulhane P. M. 1989. Ozone measurements from Dakshin Gangotri, Antarctica, during 1988–1989. *Indian Journal of Radio and Space Physics*. v. 18, pp. 188–193.

Shreedharan C. R., Russel D. G. and Basarkar S. S. 1993. Surface ozone measurements from Dakshin Gangotri, Antarctica. *Current Science*. v. 64, pp. 632–634.

Tiwari V. S. 1999. Measurement of ozone at Maitri, Antarctica. *Mausam*. v. 50, pp. 230–210.

Venkateswarlu S. and Lal R. P. 1996. *Meteorological studies carried out by IMD team during XII Antarctic Scientific Expedition at Maitri*. DOD, Technical Publication No. 10, pp. 23–40.

Venkateswarlu S. and Singh U. P. 2000. *A report on meteorological studies carried out during XVI Indian Antarctic expedition winter*. DOD Technical Publication No. 14, pp. 41–74.

4

Mathematical Modeling of Atmosphere and Ocean Processes around Antarctica

A. C. Pandey, S. Rai, A. P. Mishra, V. K. Pandey, S. Dwivedi, A. Mitra, and K. C. Tripathi

CONTENTS

Introduction

The Ministry of Earth Sciences (MoES), Government of India, pursues programs in line with the Ocean Policy Statement of 1982. Among the important objectives of the MoES are:

- Capacity Building Programme toward self-reliance through basic scientific research in the ocean sector, human resource development, and establishing centers of excellence in academic institutions.
- Multi-institutional and multidisciplinary polar science programs and Antarctic Expeditions with scientific and geopolitical significance.

Under the above mandate, MoES agreed to establish a center of excellence in oceanic studies specializing in mathematical modeling at Allahabad. The center is dedicated to the memory of Prof. Kedareshewar Banerjee, former Head of the Physics Department of the University of Allahabad during 1952–59. The academic programs (research only) of the center are run in collaboration with the University Departments of Physics, Mathematics, Statistics, and Computer Science.

Field of Investigations

Climate and ocean circulation modeling studies of Antarctica and the Southern Ocean.

Scientific Importance

The K. Banerjee Centre of Atmospheric and Ocean Studies (KBCAOS) is involved in mathematical modeling of various atmospheric and ocean processes around Antarctica. The problem taken is unique in the sense that KBCAOS is the only academic unit involved in modeling of atmospheric and oceanic processes of polar regions where as many national organizations participate in Antarctic expedition and related research work. Various studies undertaken so far are related to the study of processes and parameters vital for the understanding of oceanic system memory and its role in weather and climate such as anomalous sea surface height, heat flux and sea surface temperature, sea ice thickness, and so on, the details of which are given in the following sections. Although several ocean general circulation models (OGCMs) have been developed and used for investigating various oceanic phenomenon, no model accurately simulates and predicts oceanic processes of polar regions and their teleconnection with a variety of real phenomenon, albeit the resolution in time and space is insufficient to understand the finer details of phenomena such as the effects of small turbulence at high latitudes or the physico-chemical modifications. The knowledge of such phenomena provides insight into better parameterization of large-scale ocean models. These scientific questions are being addressed in the program of KBCAOS.

Various Research Projects

The following three research projects were sanctioned by MoES for implementation in collaboration with National Centre for Polar & Ocean Research (NCPOR),formerly known as the National Centre for Antarctic & Ocean Research (NCAOR) under the broad heading of:

- Realistic simulation of Antarctic circumpolar current and other circulation features of the southern Indian Ocean
- Intercomparison of Semtner's model (1976) with Winton's three sea–ice layer model (2000) and configuring the GCM for Antarctica
- Effect of throughflow in the Indonesian archipelago region of connectivity between the Pacific and Indian Oceans on the southern Indian Ocean

Scientific Aims and Objectives

The vision of KBCAOS is to undertake academic programs for

- Understanding the various geophysical processes responsible for climate variability both on the local as well as global scale.
- Investigating the role of the Antarctic zone of the Southern Ocean in climate variability.
- Developing and standardizing a mathematical model for numerical weather prediction including the coupled atmosphere–ocean–land system.
- Establishing a state-of-the-art computational facility for atmospheric and ocean modeling.

KBCAOS has a multidisciplinary and multi-institutional character, which makes it a national center working in close collaboration with other national/ international institutions involved in ocean-atmosphere-land-interaction research.

Region of Investigations

The Princeton ocean model (POM) is configured for the region 70°S to 25°S and 0°E to 136°E. Study for the region 65°S to 30°N and 30°E to 120°E is done from the modular ocean model (MOM3.0) for sea surface height (SSH) and sea surface temperature (SST) and their covariance. The region covered so far is around Antarctica in the Indian Ocean. Time series analysis of the Antarctic oscillation (AO) index, which is characterized by pressure anomalies of one sign centered in the Antarctic and anomalies of the opposite sign centered about 40°S–50°S, has been performed. The AO index is the dominant pattern of nonseasonal tropospheric circulation variations south of 30°S. Semtner's three-layer sea ice model and Winton's reformulated three-layer sea ice model have been implemented for the Arctic region. Winton's model has also been configured for the Antarctic region (77°S 162°52°E).

Detailed Review of the Work

Simulation study of the Southern Indian Ocean (SIO) and Antarctic circumpolar current (ACC) by the Princeton ocean model.

Introduction and Model Settings

The model used here is the free surface, bottom following sigma coordinate Princeton ocean model (Blumberg and Mellor, 1987), incorporating turbulence closure to provide a realistic parameterization of the vertical mixing processes (Mellor and Yamada, 1982). Two simplifying approximations are used; first,

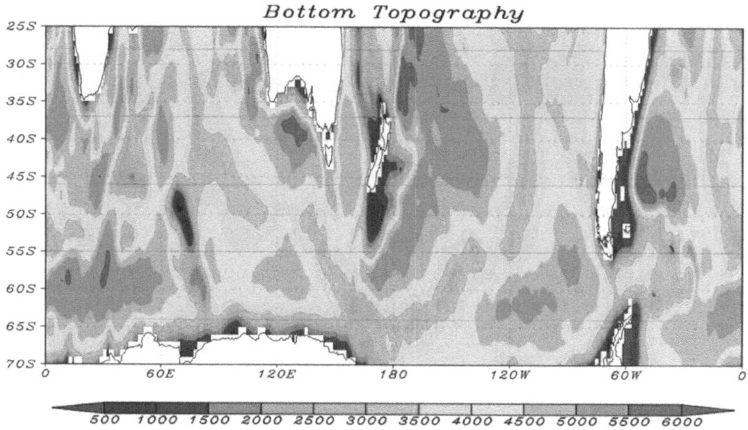

FIGURE 4.1
Model domain and bottom topography (in meters).

it is assumed that the weight of the fluid identically balances the pressure (hydrostatics approximation), and second, density differences are neglected unless the differences are multiplied by gravity (Boussinesq approximation).

Model bottom topography is shown in Figure 4.1. Bottom topography of 5′ resolution of the terrain base is used and interpolated to a model grid. The sigma coordinate has 14 vertical layers ($\sigma = 0.0, -0.002, -0.004, -0.009, -0.018, -0.036, -0.071, -0.143, -0.286, -0.429, -0.571, -0.714, -0.857, -1.0$); the resolution is fine for the upper eight layers and coarse for the lower five layers. The model has two-dimensional external time step of 25 s and a three-dimensional internal mode time step of 750 s. The model grid extends from the deep ocean to 10 m depth on the coastal region to the shallow region in the model simulation. The maximum grid between two adjacent grid points is $\Delta H/H < 0.2$.

The model was initialized with the yearly climatology from the Levitus and Boyer (1994) atlas for temperature and salinity and forced with monthly climatology wind stress of $1° \times 1°$ of the Comprehensive Ocean-Atmosphere Data Set (COADS) analyzed by Da Silva et al. (1994). The model domain extends from the Antarctic continent to the northern boundary that includes major subtropical gyres of the Southern Ocean, an area extending from 0°E to 360°E and 70°S to 25°S. The model is run for 15 years for a proper spin-up, but only the circulation pattern of the SIO is discussed.

Simulation of Southern Indian Ocean Currents with σ and z Coordinate Ocean Models

The simulation was done with two different OGCMs, MOM3.0 and POM. The coordinate system, mixing schemes, model resolution, and surface forcing are different. MOM3.0 is globally configured, but POM is configured for the Southern Ocean region. Forcing in POM is wind stress only, and it alone captures all the major current systems in SIO, confirming that the wind is the main driving force for the ACC region. Although some buoyancy forces also drive these currents, these effects are small. Another experiment is also performed with MOM3.0 by applying heat flux forcing since MOM3.0 includes heat flux and salt flux forcing.

In recent years, much attention has been drawn to the effect of topography and stratification on the transport of the ACC, but their influence remains unclear. The need to perform sensitivity experiments by changing wind stress in this part of the ocean was assessed to answer the underlying issues such as how the ACC responds to changes in the southern hemisphere winds and how the momentum input by the surface wind stress can be transferred down to the ocean floor. To that end, POM was used to simulate the SIO region, and it is shown that wind is the main driving force (Rai et al., 2006).

Simulation study of SST, SSH, and heat flux in SIO and ACC region from MOM3.0

Introduction and Model Description

The effect of temperature fluctuations as produced by model output in the southern part of the Indian Ocean shows moderate climatic variability with respect to the observation, and it remains to be fully understood. In the past, some studies were made to compare the SST from model output to that of observations, along with some other fields such as salinity. However, the lack of observational data over the data-sparse region has limited the scope of understanding the tropical ocean atmosphere interaction processes and their simulation by means of numerical models.

Any model employed to simulate any individual output fields should be validated against its reproducibility of these features. Therefore, the opposite comparison between model output and observation under different cycles gives us a proper understanding of the efficiency of the model and thus assumes importance. Consequently, studies of the seasonal cycle of temperatures and heat flux are being studied and the performance of model output as compared to observations is assessed. As a result, interannual variability of SST, SSH, and heat flux anomaly is studied as well.

The ocean general circulation model used in this study is a version of the Geophysical Fluid Dynamics Laboratory (GFDL) modular ocean model (Pacanowski et al., 1993; Huang and Schneider, 1995; Schneider et al., 1999). Its domain is that of the world oceans between 74.25°S and 65°N. The zonal resolution is 1.5°. The meridional resolution is 0.5° between 10°N and 10°S, gradually increasing to 1.5° at 20°N and 20°S.

There are 25 levels in the vertical. The model is forced by the monthly averaged surface wind stress from the National Centers for Atmospheric

Prediction (NCEP)-National Center for Atmospheric Research (NCAR) reanalysis for the period 1958–1998. The original surface reanalysis is on an irregular grid with a zonal resolution of 1.875° and Gaussian latitudes of grid spacing less than 2°, which is linearly interpolated to the OGCM grids. The model surface salinity has been taken from Levitus's (1982) monthly climatology. Here sea-ice interaction is not considered.

We have computed mean and seasonal states for temperature at the surface and subsurface from MOM3.0 output for the region 74°S to 15°N and 30°E to 120°E and compared the values estimated thus with the observations in the World Ocean Atlas (Levitus monthly and seasonal climatology) for the same fields, covering mostly the southern part of Indian Ocean for the time period of 24 years (1975–1998). The seasonal mean in the present study is defined as 3-month mean for January-March (JFM), April-June (AMJ), July-September (JAS), and October-December (OND).

The surface heat flux is one of the components of the ocean heat balance and has been studied here. Heat flux is neither easily measured in situ nor by any space-based sensor, yet efforts have been made to measure it since the temperature of the ocean varies spatially and temporally and such variation indicates heat transfer by currents (advection) and consequently heat loss and gain at the surface. Mean and seasonal states for heat flux are also computed for the region 74.25°S–0.25°N and 15.75°E–120.75°E and compared with Southampton Oceanography Centre (SOC) climatological flux (Josey et al., 1997) and CLIO ocean model (Netherlands Center for Climate Research: Deleersnijder and Campin, 1995). The resolution of CLIO is 3° in latitude and longitude, and there are 20 unevenly spaced vertical levels in the ocean.

Results and Discussion

Comparison of Model Output with Observation

The computed fields agree with the observed field, and the ocean model captures most of the main oceanic features and their mean cycle. SST monthly mean climatology computed from the model for January and July matches with the observation, but the temperature in the region 55°S to 70°S is 1°C higher than that of observation (figures not shown).

Model-produced subsurface intermediate water mean at the level of 100-meter depth for January and July have also been compared with the observation. The water temperature in the belt 10°N–10°S drops to 7°C compared to the surface temperature. In fact, the temperature distribution in the subsurface water in the Indian Ocean, especially in the SIO, is more complicated, and the isotherm situation is related to individual closed formations, but the cold intermediate water tends northward. However, our model output at the subsurface obviously shows the same pattern with the observation at the same level (figures not shown). Therefore, model efficiency is good about the observed values. Seasonal mean SST from 24-year (1975–1998) model output of different seasons, January-March, April-June, July-September, and October-December, matches nearly with Levitus's seasonal climatology for the same seasons, indicating the response of the model with respect to observation for the respective field. It is found that the model output maximum SST for January–March lies between 10°N and 20°S and again decreases as the latitudes move southward, whereas the same pattern is observed with Levitus's observation for January-March, while the maximum SST in April-June from model output lies between 10°N and 10°S, but in observation it lies between 10°N and 15°S, approximately the same trend. It is noticeable that in January-March, the temperature simulated by the model is about 1°C higher than that observed by Levitus's climatology between the region 55°S and 70°S. This error may be due to the approximation owing to neglect of sea-ice interaction, which may play an important role in this region.

Anomalous Variation of Sea Surface Height in the Southern Indian Ocean and Its Connection with Sea Surface Temperature

Changes in SST are governed through the heat balance in the surface mixed layer of the ocean. This is influenced by turbulent fluxes of momentum and buoyancy (heat and freshwater) at the sea surface, entrainment processes at its lower boundary, and advective and diffusive processes within the mixed layer itself (Leeuwenburgh and Stammer, 2001).

Some effects of interior variability on SST have been observed and already studied in the early 1980s. They were assumed to be small in magnitude and were mostly treated as small-scale noise. A striking large-scale correlation has been reported between SST and sea surface height fields (Leuliette and Wahr 1999; Nerem et al., 1999). This leads to serious questions regarding our understanding of processes responsible for observed SST and SSH variation, their relation to surface heat fluxes, and changes in the oceanic heat content and ocean dynamics. Further analysis thus leads to added insight into the role the ocean plays in setting the SST (observed) pattern and for better understanding of processes governing SST and SSH changes and the air-sea coupling mechanism.

Here, empirical orthogonal function (EOF) analysis of SST anomaly (SSTA) and SSH anomaly (SSHA) is used to identify dominant modes of interannual variability. The anomaly of each variable is calculated by subtracting respective monthly mean climatology from data, derived from 1975 to 1998 model run data. The region of our study is 65.25°S to 30.75°N and 29.25°E to 120.75°E.

The spatial patterns in Figure (4.2a, b) are re-dimensionalized by multiplying constants back. Its principal components (PCs) are shown in Figure (4.2c), which are normalized by its maximum. The PC for the first mode demonstrates interannual oscillation for a small span of time, 1983–1987; prior to 1983, oscillatory behavior is found, but the values fluctuate in the negative

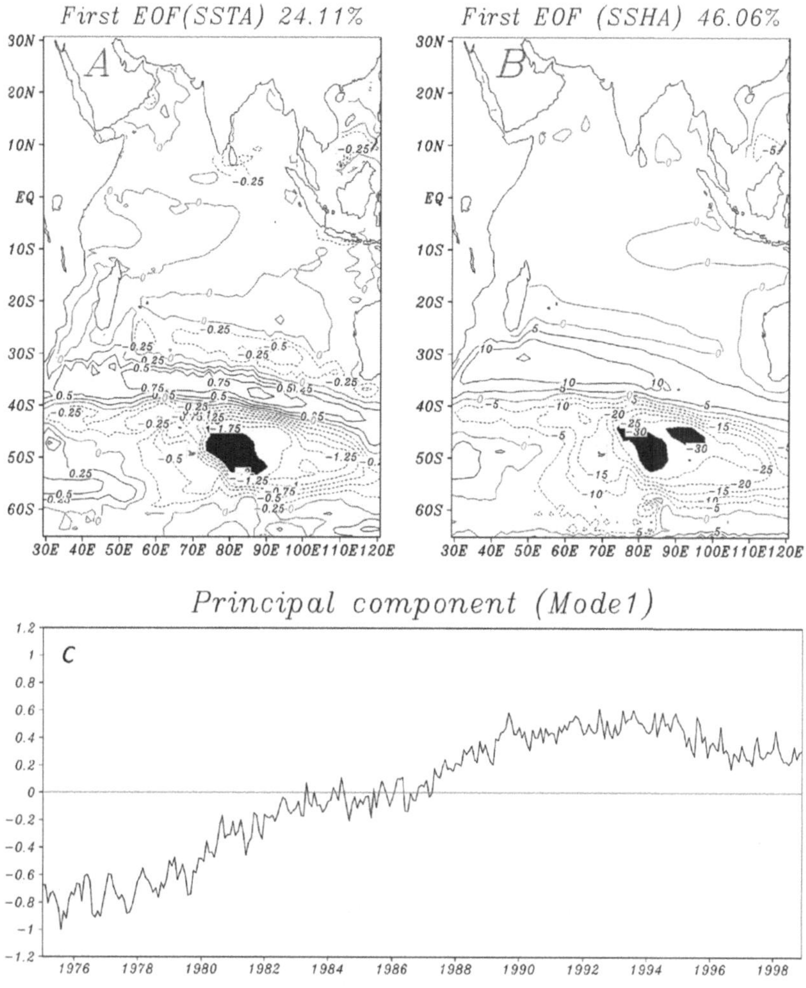

FIGURE 4.2
The spatial structure of first EOF mode for monthly mean anomaly (a) SST and (b) SSH from 24-year (1975–1998) model output (contour interval is 0.25°C and 5 cm, respectively). Regions with SSTA lower than −1.75°C and SSHA lower than −30 cm are darkly shaded. (c) Represents principal component (non-dimensional).

direction only; a similar case is obtained after 1987. SSTA is symmetrical about 40°S; however, it is perceptible that SSTA to the south of 40°S is elongated up to 60°S and between 60°E to 115°E and its intensity is more than twice that of its northern counterpart. The region between the Antarctic polar front and sub-Antarctic front shows a maximum decrease in sea level of up to 30 cm, which is remarkable. The vigorous inter-basin exchange accomplished by the ACC admits the possibility of oceanic teleconnections, where an anomaly formed in one basin may be carried around the globe to influence climate at remote locations (White and Peterson, 1996). The cold SSTA is elongated and occupies a larger domain of the basin with weak warm SSTA in the smaller region between 30°S and 40°S, which may be attributed to the strong Agulhas current system at the southern coast of Africa. The corresponding pattern of SSHA (Figure 4.2b) is consistent with that of SSTA, suggesting a connection between them; however, there are some slight differences. Cold SSTA represents a decrease in sea level, and warm SSTA represents sea level rise, but the region 35°S to 25°S and 80°E to 100°E shows cooling; nevertheless, there is no sign of deepening of sea level in this region.

Figure 4.3 represents the second mode EOF of SSTA and SSHA, which explains 8% and 9% of the total variance for SSTA and SSHA, respectively. The PC of the

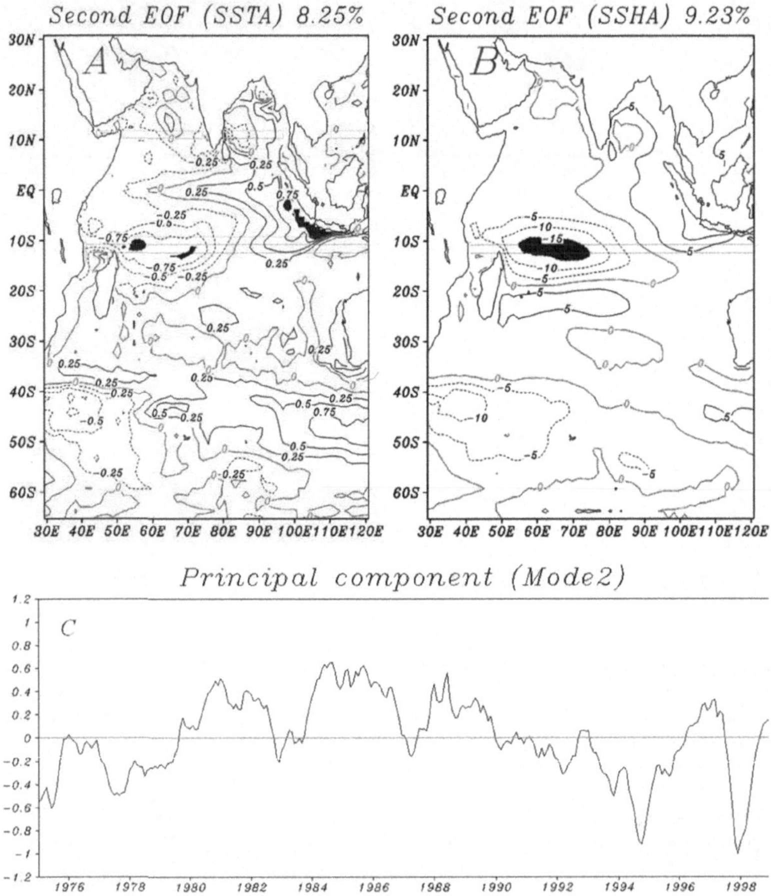

FIGURE 4.3
The spatial structure of second EOF mode for monthly mean anomaly (a) SST and (b) SSH from 24-year (1975–1998) model output (contour interval is 0.25°C and 2.5 cm, respectively). Regions with SSTA lower than –1.75°C and SSHA lower than –20 cm are darkly shaded. (c) Represents principal component (non-dimensional).

second mode demonstrates an oscillation with a time scale of 2–4 years. Cold SSTA is seen in two regions, one in the tropical Indian Ocean (10°S–20°S and 40°E–75°E) and other in the region 40°S–60°S to 30°E–50°E; corresponding to this cooling, it is found that sea level also deepens, as shown in Figure 4.2b. The equatorial and southern trade wind anomalies together form an anomalous cyclonic atmospheric circulation over the Southern Ocean, which decreases the SST in the southwest part (Huang and Kinter, 2002). Warm SSTA is evidenced mainly in two regions, one in the region 10°N to 10°S and 80°E to 100°E and the other in the region 40°S to 55°S with the same zonal direction; a rise in sea level is also found in these regions, but it is not as strong as the deepening.

The surface geostrophic streamlines ψ_g of the model can be calculated by the equation given below:

$$\psi_g = \frac{gh}{f} \tag{4.1}$$

where g is the gravitational acceleration, f is the coriolis parameter, and h is the SSHA. Zero contours are coincident with the axis of ACC; positive streamlines lay poleward and negative equatorward. The contours that bound circumpolar flow correspond to streamline heights of $\psi_g = -1.0 \times 10^4$ and $\psi_g = 3.0 \times 10^4$ m/sec, respectively (here $g = 10$ m sec^{-2} and $f = 10^{-4}$). These patterns almost match with the observation from TOPEX/Poseidon altimetry data over the period October 1992–February 1995 (Karsten and Marshall, 2002).

Similar spatial pattern matching between SSTA and SSHA was obtained for mode 3, which explains 6% and 5% of total variance, respectively (figure not shown). PC of the third mode gives oscillation ranging from 1 to 3 years. It is clear from all three modes that cooling of 0.1°C represents a deepening of 20–30 cm, whereas warming of 0.1°C corresponds to a sea level rise of 10 cm. These SSHA are much higher compared to other parts of the ocean, suggesting that processes other than mere changes in the upper-ocean heat content sway observed variability in sea level there.

Figure 4.4 represents PCs for the three modes of SSTA and SSHA. Modes 1 and 2 show a higher correlation coefficient of 0.94 and 0.87, respectively, whereas mode 3 shows a negative correlation of −0.6; the negative correlation of the third mode may be due to some unknown error. These correlations pass the 95% significance level test.

Estimation of Heat Flux and Its Variability in the Southern Indian Ocean from the Modular Ocean Model

It is found that between the equator and 15°S, heat flux from the model is within 0–50 W/m^2, while heat flux from the SOC climatology lies within 60–80 and 80–100 W/m^2 (figure not shown). The observed heat gain is from 100–150–250 W/m^2 for the region 50°E–70°E, 45°S–50°S, which agrees with the SOC flux as compared to 140–160–200 W/m^2. In the Southern Ocean, south

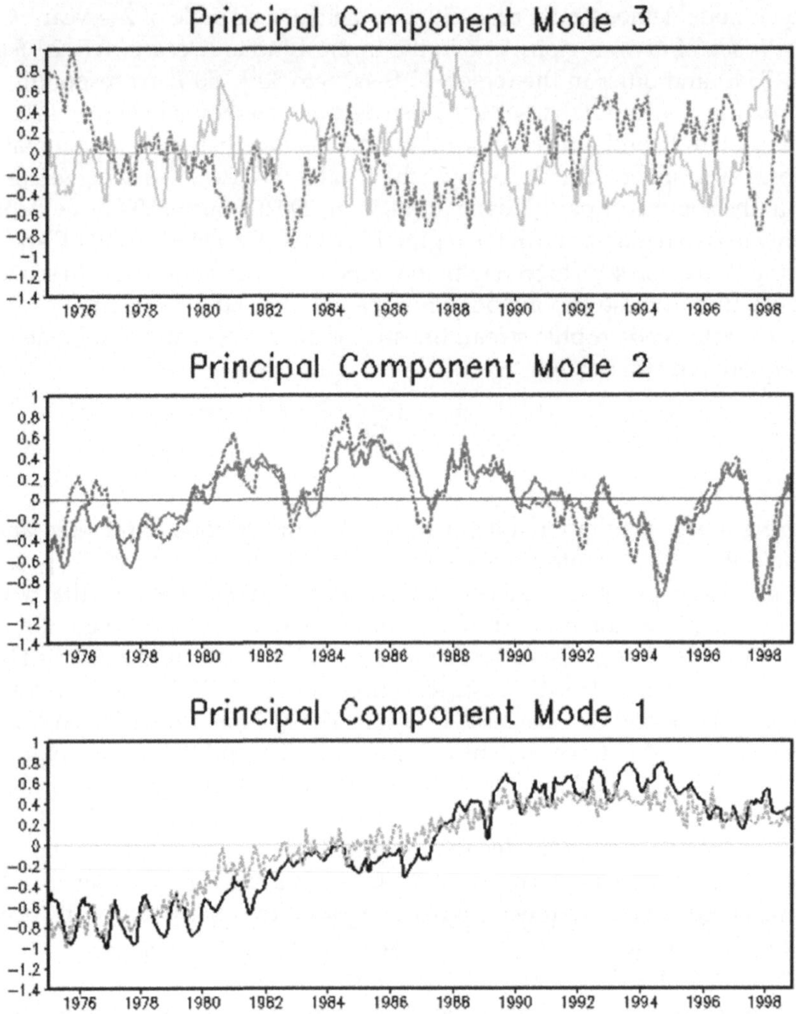

FIGURE 4.4
The principal component of SSTA (dark line) and SSHA (dashed line) computed separately for (a) mode 1, (b) mode 2, and (c) mode 3, respectively (non-dimensional).

of 60°S, there is always heat loss from the ocean to the atmosphere (−50 to −100 W/m^2), as simulated by the model, while in the SOC climatology, there is heat gain by the ocean to the atmosphere (100–140 W/m^2). There is stronger heat loss observed in the western boundary near the African coast region in the model output that is not reflected in SOC climatology. Further, net heat flux fields were found to be underestimated in the model output as against the SOC climatology, which is more than the corresponding model values at various locations. It may be pointed out that these biases arose primarily because of underestimation of gain in shortwave and overestimates of latent heat loss (figures not shown).

To delineate the major modes of variability in the analysis of 24 years (1975–1998), we have performed an EOF analysis and obtained the first three leading EOFs, which are able to account for a total variance of 52%. Since these patterns of first three EOFs are enough to explain the spatial anomalies of whole data, it suffices to focus only on the first three EOFs.

Seasonal mean SST from 24 years (1975–1998) model output of different seasons, JFM, AMJ, JAS, and OND, almost go with the CLIO model. However, there are a few spatial inconsistencies in the model outputs (figures not shown).

The first EOF mode explains 34% of the total variance and PC is normalized by its maximum. Heat flux anomaly (HFA) is mostly distributed in SH in between 25°S and 65°S. Positive HFA is elongated between 60°E and 120°E and between 40°S and 60°S. Negative HFA is seen in a very small portion, 32°S–38°S, 30°E–120°E, and the western Southern Ocean (Figure 4.5).

In Figure 4.5a, it is obvious that large areas of positive loadings (Shabbar et al., 2003) in the eastern SO (40°S–60°S, 55°E–120°E) signify a net decrease in the energy transfer from the atmosphere to the ocean. Figure 4.5a also identifies a negative loading pattern elongated in the zonal direction (longitudinally covering whole domain) in the south of 30°S to 40°S and near 50°S–60°S, 25°E–55°E, which shows a net increase in the net transfer of energy from atmosphere to the ocean. The internal circulation in the SO and related ocean may be responsible for such a role in this region.

Further, if EOF 1 mode is showing negative for certain years, then PC > 0 corresponds to a negative heat flux anomaly and PC < 0 corresponds to a positive heat flux anomaly (Lorenz, 1956). So, as evidenced from time series from 1975 to 1987 (Jan to Dec), PC is always negative and reaches a minimum value of −1.0, which implies the positive HFA over this period with a heat flux gain component to south of 40°S, confined in the eastern Southern Ocean after 1987. PC is always positive from Dec 1987 to Dec 1998, which implies negative HFA over this period with a net heat flux loss component between 30°S and 40°S and between 30°E and120°E. Figure 4.6a,b represents the second EOF, which explains 11% of total variance and its principal component. A second eigenvector (EOF2) indicates few contour lines close to the equator and from the south of 35°S, which may be described as negative HFA residing in the eastern Indian Ocean between 35°S and 50°S and getting elongated from 40°E–120°E, whereas positive HFA is seen at two different places in the larger domain. First, positive HFA is seen near the western part of the southern Indian Ocean, equator to 20°S, and second, positive HFA is seen near 40°S–60°S (western part of the Southern Ocean), and the corresponding PC seems to be quasiperiodic with a period of a few years. Thus, we identify EOF2 associated with some oscillation patterns.

The above facts indicate the appearance of two regions having components with either net heat flux gain or net heat flux loss in the domain. The spontaneous decadal variability (~11 yrs) with a positive heat flux gain component in one oscillation and negative heat flux loss component in another is found to exist in the present case and is of importance. We thus identify that fluctuations of this pattern (EOF1) have decadal time scale variability.

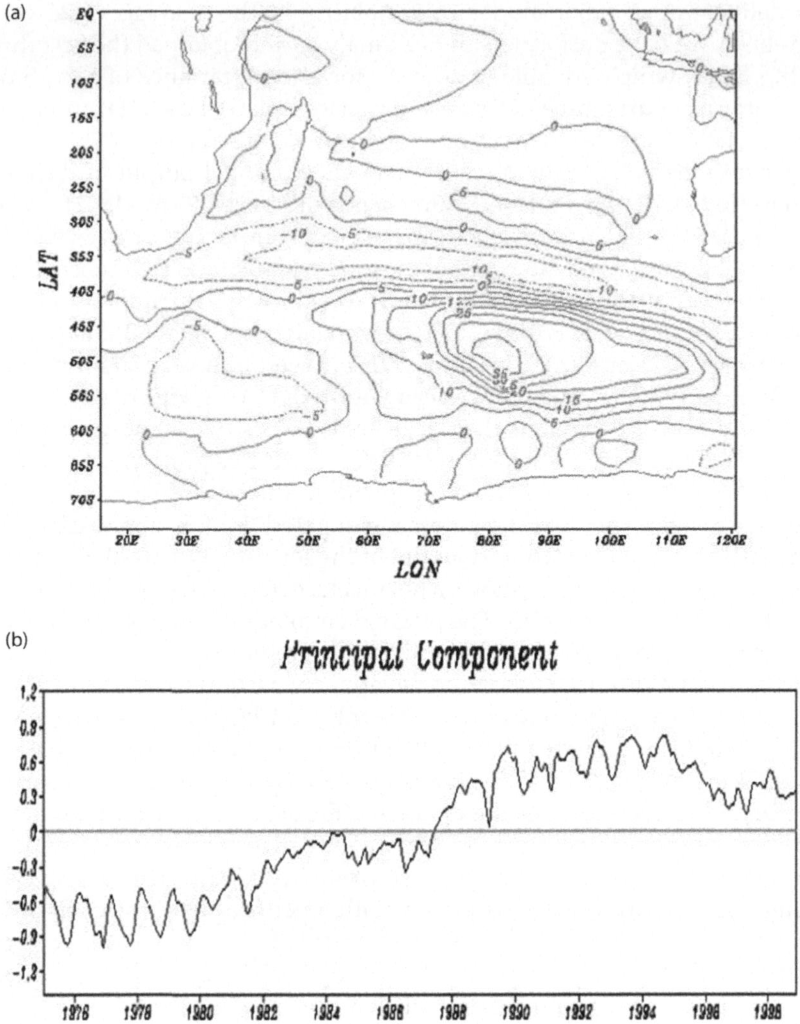

FIGURE 4.5
(a) The spatial structure of first independent EOF mode for monthly mean anomaly heat flux from 24-year (1975–1998) model integration. The contour interval is 5 Wm^{-2}. (b) Principal component (non-dimensional).

Conclusions

Two oceanic parameters, temperature and heat flux from the modular ocean model, output under different cycles have been analyzed and compared with the observations. It has been found that the areas of observed SST and subsurface temperature match broadly with those estimated by the ocean model, but it is noticeable that in Jan-Feb-Mar (summer), the temperature

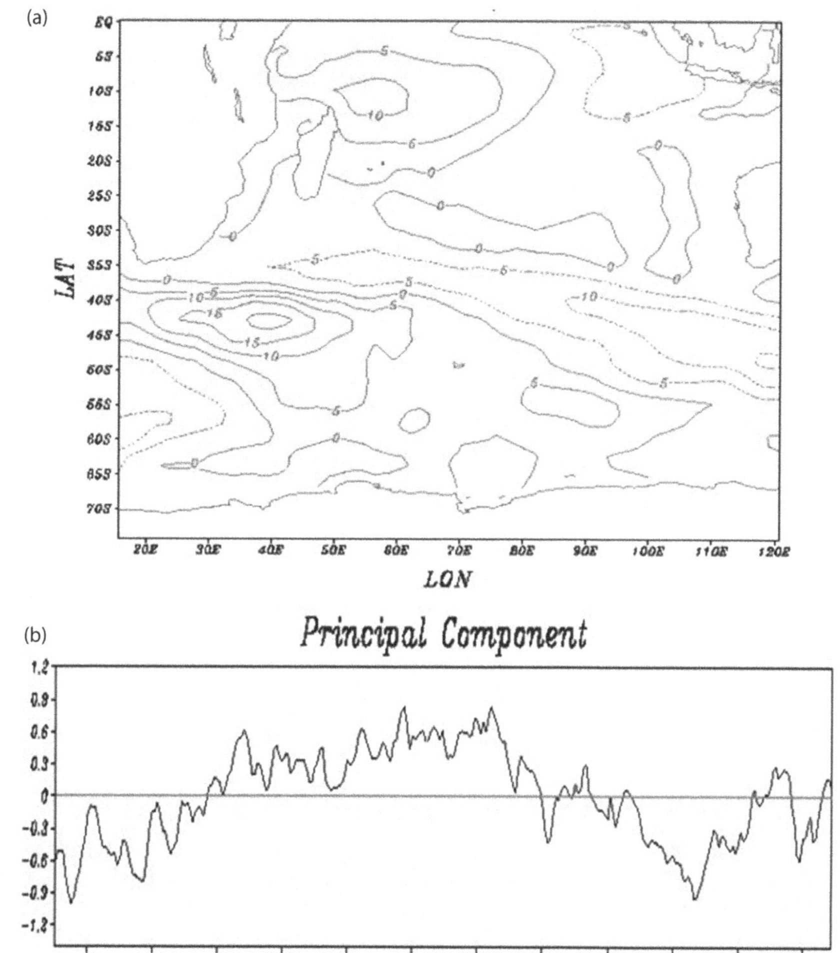

FIGURE 4.6
(a) The spatial structure of second independent EOF mode for monthly mean anomaly heat flux from 24 years (1975–1998) model integration. The contour interval is 5 Wm^{-2}. (b) Principal component (non-dimensional).

simulated by model is about 1°C higher than that observed by Levitus's climatology between the region 55°S and 70°S. This discrepancy may be due to the approximation that neglects sea-ice interaction, which may play an important role in this region, and the study of sea-ice interaction at higher latitudes is a matter of further investigation.

Comparisons of the net heat flux are also done from our model (MOM3) with the SOC climatology at the ocean surface and seasonal climatology of the surface heat flux with the CLIO ocean model. The net surface heat flux from the model agrees with SOC climatology (within +50 W/m^2) with spatial uncertainty involved in the two cases at various locations. The SOC climatology heat gain is found to overestimate the model heat flux in the present domain. The mechanism responsible for these biases may be due to the contribution of the systematic uncertainties that tend to be larger than random uncertainties at most latitudes. Further, the seasonal climatology from the model agrees mostly in seasons SON, MAM, JJA and moderately in DJF with CLIO model output. The uncertainty involved in the two model outputs can be explained, as model biases may be primarily due to the combination of underestimated shortwave gain and overestimated latent heat loss. However, it is reasonable to mention that quality of field has a strong spatial dependence, which decreases in the southern hemisphere, particularly south of 40°S, and the errors have a propensity to be large; therefore, model results and climatology have differences at some locations in the region (Mishra et al., 2007).

The change in SSH has been observed from the SST anomaly (T) of 1°C calculated by the equation given below:

$$\eta' = \int_{-h}^{0} \alpha(z)T'(z)dz \tag{4.2}$$

where a is the thermal expansion coefficient, and z is the ocean depth.

If we take h = 200 m, that is, only the upper 200 m depth of the ocean, then SSH rise is on the order of 2 cm, but we are getting a much higher sea level rise, demonstrating that there are processes other than upper-ocean heat content.

A sea level rise of about 5–10 cm is obtained along the east coast of Africa, spreading over southwest Australia through Madagascar, and it is due to the Agulhas current system near Africa, flowing along the east coast of Africa. High correlation between SST and SSH anomalies shows a connection between them that is verified by many altimetry observational studies.

The spread in percent variance explained for heat flux appears to be well separated. As far as trends in the model are concerned, the first EOF seems to be a strong trend of the surface heat flux field in the region of the southern polar ocean and ACC and dominant patterns represent a decadal time scale of variability over the period of the integration. So, a heat flux gain component to the south of 40°S, confined in the east of Southern Ocean, and a net heat flux loss component in between 30°S–40°S, 30°E–120°E have been identified. The spatial pattern of EOF 2 looks complicated and quasi-periodic. The motivation of the present study is that more attention should be given to the role of the *oceanic systemic memory*, its time-dependent response, and possible feedback to the atmosphere with reference to SST and SSH, and surface heat flux data in the Southern Ocean suggests that the feedback might be both positive and negative, depending on location, but certainly not inconsequential.

Study of Volume Transport and Heat Transport through Indonesian Throughflow Region

Introduction

The effect of Indonesian throughflow (ITF) contributes to maintaining the circulation system around eastern Australia and the Southern Indian Ocean. The Indonesian archipelago region is composed of three major passages (Figure 4.7) called the Lombok strait to the south of Java, Savu Strait to north of Timor Island, and Timor Strait between Timor Island and Australia. The ITF carries warm and fresh Pacific waters through the Indonesian archipelago region into the Indian Ocean (IO). This interocean exchange of water masses occurs only through these channels in the tropical region. There are few observational studies in this region and most determinations of net transport and regional dynamics have been made from indirect observations. Gordon (1986) suggested that the ITF could be important to global ocean thermohaline circulation, as it might return the North Atlantic deep water that up-welled and warmed in the Pacific back to the Atlantic. The potential role of the ITF to the global climate system continues to be an important area of investigation.

The ITF strongly influences the heat and freshwater budgets of these two oceans and therefore may be considered a key component in the El Nino and southern oscillation (ENSO) and monsoon climate phenomena (Webster et al., 1998). The meridional circulation stratification, sea surface temperature, and sea level of the oceans would be altered if the ITF were zero (Verschell et al., 1995). Large-scale observation-based studies reveal significant Pacific export of freshwater and heat into the IO. Increased oceanic heat and freshwater flux into the IO at the expense of the Pacific affect atmosphere-ocean coupling,

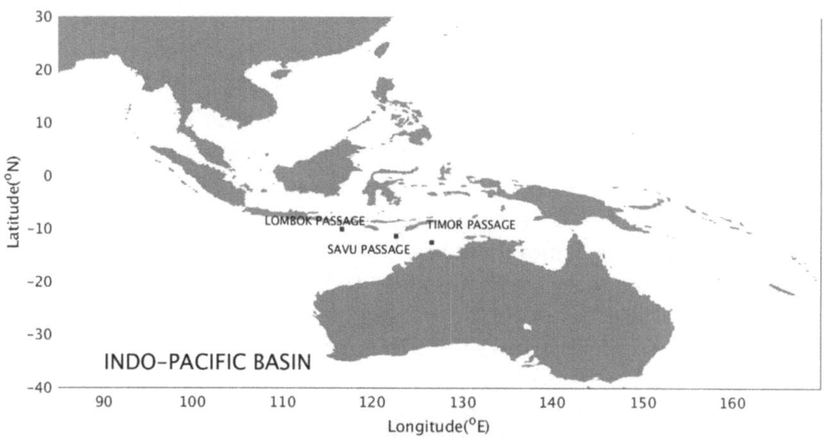

FIGURE 4.7
Indonesian throughflow region (Lombok, Savu, and Timor Straits).

with potential impacts on the ENSO and monsoon phenomena (Vranes et al., 2002). Studies have shown that heat transport varies with ENSO, as it is lower during El Nino and higher during La Nina (Meyers, 1996). The ITF heat transport remains a part of the ocean dynamics to give an overview of the ITF effect on the SIO.

As evidenced by the literature, the ITF affects the Pacific and Indian Ocean, and many issues still need further investigation, especially those related to effects of the ITF on the climate and dynamics of the SIO. In the present study, the main objective is:

a. To find the effect of the ITF on the mean circulation and dynamics of the SIO.
b. To evaluate the ITF's effect on seasonal variability.

For this purpose, the problem is investigated from a purely oceanic context and a model (POM) is integrated over period of 1 year. Model solutions with open and closed geometries of Indonesian passages are compared to highlight the effects of the ITF.

The transport through the Pacific and Indian Oceans in the Indonesian Archipelago (PACIO) region and its influence on the dynamics of the upper tropical Pacific and Indian Oceans is examined. This is done using two different model geometries: the Indonesian domain 100°E–150°E, 40°S–0°N, with an open PACIO region, and the same domain with a closed PACIO region, that is, very moderate Pacific throughflow is permitted. The heat transport of the ITF and its effect on the SIO is of paramount importance and therefore it is also discussed in subsequent sections.

Model Configuration

The basic model configuration and geometry of the Princeton ocean model are the same as given previously. Model topography is generated from the ETOPO5 terrain base data set of 1/12° resolution and interpolated to a model grid. In the topography, the westernmost Lombok Strait is found near (115° 50′ E, 8°30′S) and Savu Strait near (122° E and 9° 30′S), and the easternmost Timor Passage is nearby (128° E and 11° S). There are very small gaps near these passages that are very shallow, and small throughflow occurs.

In the first experiment, the model has been integrated with the open channel of the ITF for the domain 100°E–150°E and 40°S–0°S for 1 year when a complete flow field is generated. In the second experiment, the model has been run for the same domain with a closed channel of the Indonesian archipelago region for the same period. The model is initialized by Levitus's 1994 annual mean field for temperature and salinity. Surface forcing includes monthly climatological wind stress of the Comprehensive Ocean–Atmospheric Data Set analyzed by Da Silva et al. (1994). The above experiment is performed on the fact that the basic idea of masking scheme is to use topographic height, that is, depth = 0 at a grid point and zero velocity, that is, velocity components (u and v), are zero.

The result of two experiments will be referred to as the open experiment and experiment with forcing. The experiments differ in the sense that they have different geometries, that is, open and closed cases of these three near-equatorial passages between the Pacific and Indian Oceans.

Results and Discussion

Transport of Indonesian Throughflow and Its Effect on the Southern Indian Ocean

One-year time series of transports through Lombok, Savu, and Timor Straits, along with the combined transport through all three passages, were calculated from the model run with open PACIO and closed PACIO geometry, as shown in Figure 4.8. The maximum westward transport was 27.0 Sv during a January 1994 episode, which may be neglected because the model was not in equilibrium at this time; the average transport over the 12-month experiment was 10.5 Sv from the Pacific to Indian Ocean, with most of the transport values being between 5.0 and 17.0 Sv. Transport through these three passages is highly correlated as expected. The largest mean transport under open PACIO is through Savu

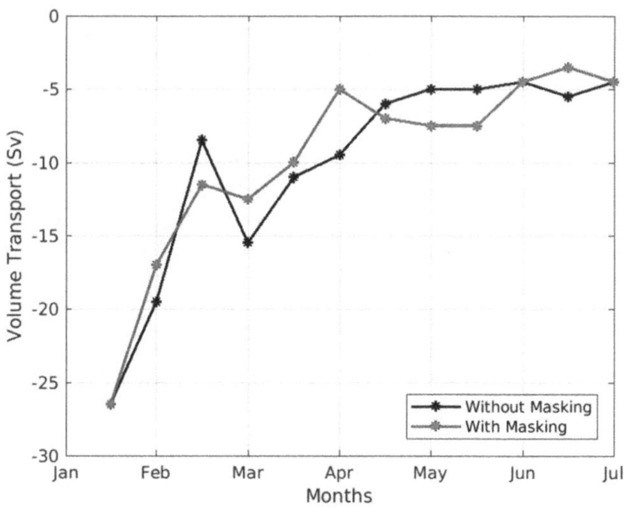

FIGURE 4.8
Monthly variation of volume transport (in Sv) without masking and with masking.

Strait (11.9 Sv), closely followed by Timor Strait (10.7 Sv). Also, we observed that the mean transport of Lombok strait is 9.3 Sv. Average transports for both Savu Strait and Timor Strait have been found significantly higher than that of Lombok Strait. One signal is that transport is mostly westward throughout the model run. It was also observed that there are limited regions where eastward transport occurs during the year. Transport values are smaller during the months of October to December, with an average of 6.0 Sv.

The focus of this study is to investigate what effect the PACIO region has on the tropical Pacific and Indian Oceans by closing off the passages between the two oceans. The average transport in the Indian and Pacific Ocean basin between 100°E to 150°E and 40°S to the Equator were calculated for both cases. The tropical Pacific Ocean average transport differs slightly, while SIO transport differs significantly in the closed experiment case as compared to the open case.

In the first month, the sea surface elevation varies from –1.0 to 4.0 meters, with maximum elevation around 25°S in the fourth month. Two set of numerical experiments are performed which are listed below: (a) Model run with open channel and surface forcing. (b) Model run with closed channel and surface wind forcing.

The sea surface elevation ranges from −0.6 to 1.4 meters, with maximum elevation near 20°S, again in the seventh month (July), and the sea surface elevation lies between −0.5 and 1.0 meter, with maximum variation again near 17°S but covering a smaller region, and in the tenth month, the sea surface elevation values are in the range of −0.6 to 0.6 meter, with maximum elevation around 15°S covering a somewhat larger area. The higher value for the first few months shows that the model is not in equilibrium, but after 6 months of the model, elevation is established to the equilibrium position (figures not shown), while in the closed PACIO experiment, these surface elevation values slightly differ from the open experiment, but there may be more cooling in the SIO than in the open PACIO experiment.

The general surface circulation scheme in the month of January illustrates that South equatorial current (SEC) is noticed clearly and cyclonic circulation is observed in SIO from 10°S to 30°S. Further, the circulation in April still shows SEC and slightly noteworthy Leeuwin current and less prominent cyclonic circulation, and the same features are observed for July and October as in April. Thus, SEC at the surface is most dominant current in all months, while the Leeuwin current is less intense (figures not shown), while in the closed PACIO experiment, it is observed that blockage of the ITF weakens the Indian Ocean's south equatorial current and southward-flowing Leeuwin current. Therefore, the ITF is found to strengthen the SEC as well as the Leeuwin current in the SIO.

Indonesian Throughflow Heat Transport via Lombok, Savu, and Timor Straits

One-year time series of heat transports through Lombok, Savu, and Timor Straits, along with the combined heat transport through all three passages, were calculated from the model run with open PACIO geometry. The maximum southward heat transport is 2.828 PW during the January 1994 episode. Most of the transport values were between 0.424 and 1.414 PW. Transport through these three passages is highly correlated as expected. The largest mean heat transport under open PACIO is through Savu Strait, which is 1.104 PW, closely followed by Timor Strait at 1.086 PW. Also, mean heat transport through Lombok Strait is observed to be 1.072 PW. The average heat transport is similar for all the straits. It is suggestive of the signal that heat transport is mostly southward throughout the model run.

Heat transport values are smaller during the months of October to December, with an average of 0.597 PW, and largest during January to June 1994, with an average of 1.514 PW. Maximum heat transport occurs from mid-January to mid-March, and these higher values may be neglected since the model is not in equilibrium up to 3–4 months, while the minimum heat transport is between October and November. The average heat transport in the Indian and Pacific Ocean basin between 100°E–150°E and 40°S–equator were calculated for open PACIO cases with and without wind forcing as 1.087 and 1.134 PW, respectively.

From Figure 4.9, it follows that with the wind forcing the heat transport decreases from the first to second month in all three channels and then increases slightly in the fourth month. The heat transport of ITF evaluated as previously also shows abnormal transport for the first month followed by some fluctuations in the last month. The seasonal variation in heat transport is an indication of change in heat flux in the region that has a direct impact on the climate of the Indian and Pacific Oceans.

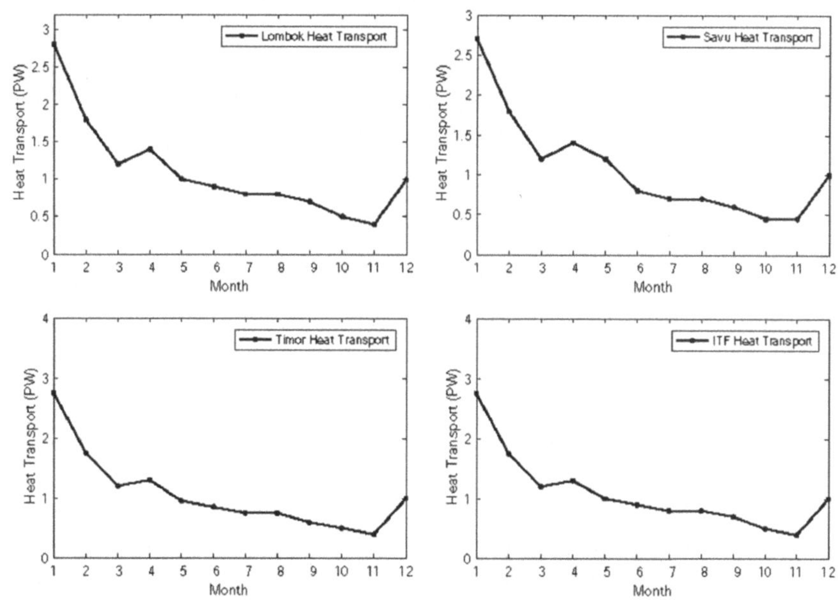

FIGURE 4.9
Heat transport with wind forcing.

Summary and Conclusions

The ITF is found to affect the circulation around the western tropical Pacific to the SIO. The blockage of the ITF weakens the SEC and southward-flowing Leeuwin current and strengthens the East Australian current. The blockage of the ITF cuts off the heat transport from the Pacific to Indian Oceans. This causes an overall warming in the Pacific and cooling in the SIO, resulting in the shoaling of the thermocline in these two oceans.

The model has been used to investigate the regional effect of variation in the magnitude of the ITF on the IO upper layer heat content. Consequent changes in SST associated with changes in the ITF transport shift the position of the deep atmosphere convection region of the western tropical Pacific and, due to changes of SST in the Indian Ocean, net evaporation within the Indian Ocean is affected, which in turn governs the onset of monsoon. Although investigations of the model output reveal the insignificant contrast between open and closed ITF, they do not explicitly consider the effects or causes of variations in the ITF transport and temperature profiles, yet changes in their relationship, even without change in net temperature, may be expected to alter the SST and heat budget of the Pacific and IO and associated climate phenomenon.

Is There a Strange Attractor for Antarctic Oscillations?

Introduction

Chaotic time series analysis of time series data of North Atlantic oscillation (NAO), Southern oscillation (SO), Pacific Decadal oscillation (PDO), and Antarctic oscillation (AO) has been performed to find low-dimensional strange attractors. The existence of low-dimensional strange attractors suggests that a deterministic subsystem is responsible for generating the time series. This should enable better short-term predictions than those based on stochastic models. Whereas we could identify strange attractors for SO and PDO/NPO, the same has not been found in NAO and AO. This could be due to insufficiency of data for the Antarctic oscillation index (AOI). There is a need for development of improved techniques for estimating important geophysical parameters from satellite data on one hand and in-situ data observations at a greater number of data stations in the Antarctic region on the other.

Sea-level pressure anomalies periodically reach the western Ross Sea from the north and propagate eastward through the Amundsen and Bellingshausen Seas up to the Weddell Sea. Sea-ice concentration and ice drift anomalies also propagate eastward in tandem (and phase-shifted) with atmospheric circulation anomalies. This is an expression of coupled air-ice-sea interactions. Such oscillations are significantly correlated with the El Nino-southern oscillation phenomenon. Pressure anomalies originating in the western tropical Pacific on ENSO timescales are suggested to reach the subpolar region through atmospheric teleconnections (Venegas, 2001). Thus, the natural climate variability in both the Arctic and the Antarctic regions is shown to involve regional coupled interactions between the ocean, the sea-ice, and the atmosphere, which are strongly linked to global-scale circulation patterns such as the NAO in the Arctic and ENSO in Antarctica. Recent studies of the variability of surface air temperature and sea-ice climatologies in the western Antarctic peninsula region show long-term persistence in surface-air temperature and sea-ice anomalies, where two to four low-temperature/high-ice years are followed by one to three high-temperature/low-ice years, a pattern coherent with the Southern oscillation index (SOI), indicating that there may be an ENSO teleconnection between the western Antarctic Peninsula region and lower latitudes (Smith et al., 1996). To look for the possible teleconnection between different oscillations found in the southern and northern hemisphere, we have done a chaotic time series analysis of the different oscillation indexes. The objective of time series analysis is to build a model for the unknown dynamical system that generated the time series informative of a chaotic system because often the only information we have about these types of systems is in the form of single-variable time series. Several nonlinear time series analysis tools have been developed in mathematical physics in the recent past, which will allow us under certain circumstances to decide whether the chaotic behavior of a time series results from the nonlinearities inherent in the system under consideration or is due to some external noise source/systematic error.

Results and Discussion

The time series we took for our analysis are

i. NAO index, which is the normalized pressure difference between a station on the Azores and one in Iceland. This time series is monthly anomaly for the period 1825–1999.
ii. SO index, which is the normalized pressure difference between Tahiti and Darwin. This time series is monthly anomaly for the period 1866–2002.
iii. NPO index, which is derived as the leading PC of monthly SST anomalies in the North Pacific Ocean, poleward of 200N.This time series is monthly anomaly for the period 1900–2001.

iv. AO index, which is characterized by pressure anomalies of one sign centered in the Antarctic and anomalies of the opposite sign centered about 40–50S. The AO index is the dominant pattern of nonseasonal tropospheric circulation variations south of 30S. This time series is monthly anomaly for the period 1948–2001.

To find the answer for the aperiodicity in these time series indexes, we have calculated the correlation dimension of each time series using the optimized box assisted algorithm (Grassberger, 1990). A plot between log C(r) and log (r) corresponding to different values of embedding dimensions (m) is plotted for each time series, and the slope gives the value for correlation dimension d (Grassberger and Procaccia, 1983). This is shown in Figures 4.10 through 4.13, respectively, for NAO, SO, NPO, and AO.

Generally, with increasing the value of embedding dimension, the value of the fractal dimension should also increase until it reaches a saturation value, after which an increase in the value of m should not increase the fractal dimension (d). We find that the saturation correlation dimension (fractal dimension) exists only for the Southern oscillation index and Pacific decadal oscillation index time series. This is shown in Figures 4.14 through 4.17 for

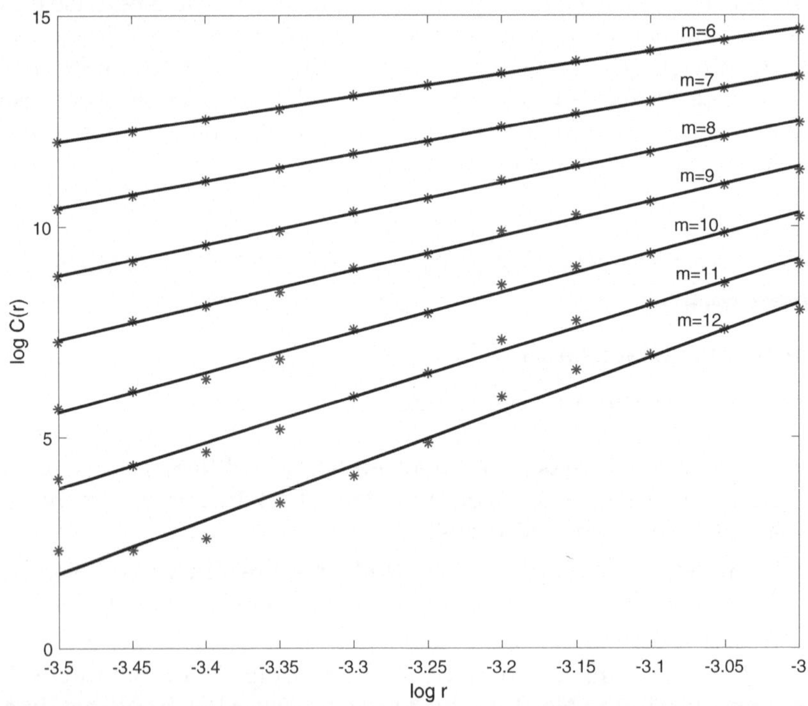

FIGURE 4.10
Plot for the calculation of correlation.

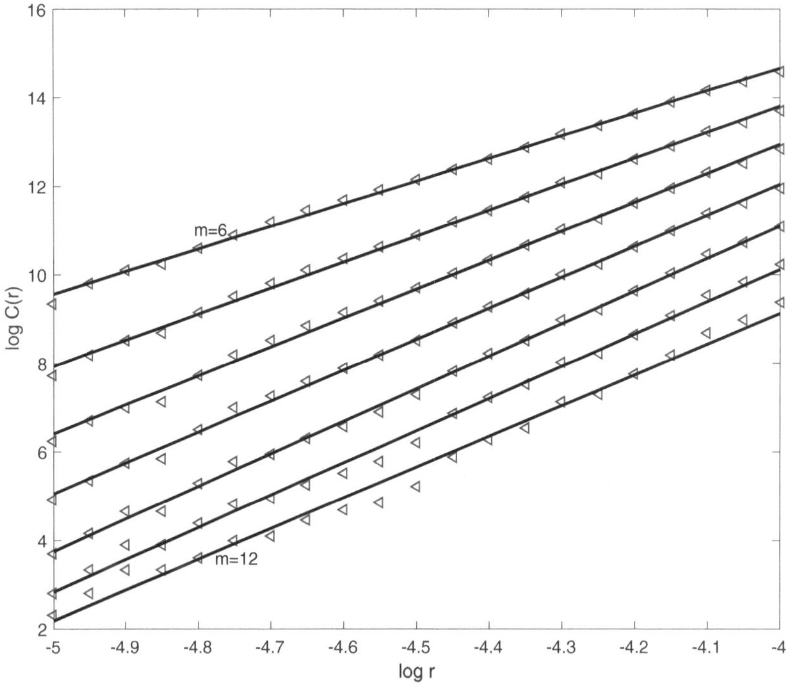

FIGURE 4.11
Plot for the calculation of correlation dimension of SOI.

NAO, SO, NPO, and AO, respectively. From the figures, it is clear there exists no strange attractor for Antarctic oscillation.

The objective of our study is to look for the possible teleconnection between the Southern oscillation index and Antarctic oscillation index. For this, we surmise that if we can get nearly the same value of the fractal dimension for SO and AO, then they might be said to be governed by the same strange attractor. Here, we could not get the fractal dimension for AO and NAO. For SOI and NPOI, we get the minimum number and maximum number of variables needed to model the dynamics of the system, which are given respectively by the integer next to saturation value of the correlation dimension and by the embedding dimension. We find the value of d_{sat} for SOI is 7.28 and m_{sat} is 11, and hence the minimum and maximum number of variables needed to describe the system are 8 and 11. Similarly, for NPOI, d_{sat} is 10.12, corresponding to $m_{sat} = 15$, and hence the minimum and maximum number of variables needed to describe the dynamics of NPOI are 11 and 15. This is not surprising, because the time series used here includes weather phenomena, the long-range processes from season to season, and the interannual and interdecadal variability. A dynamical system, which includes all these aspects, depends on many independent variables (degree of freedom) and the inherent noise. The fact that the attractor has a fractal dimensionality provides a natural explanation of the intrinsic

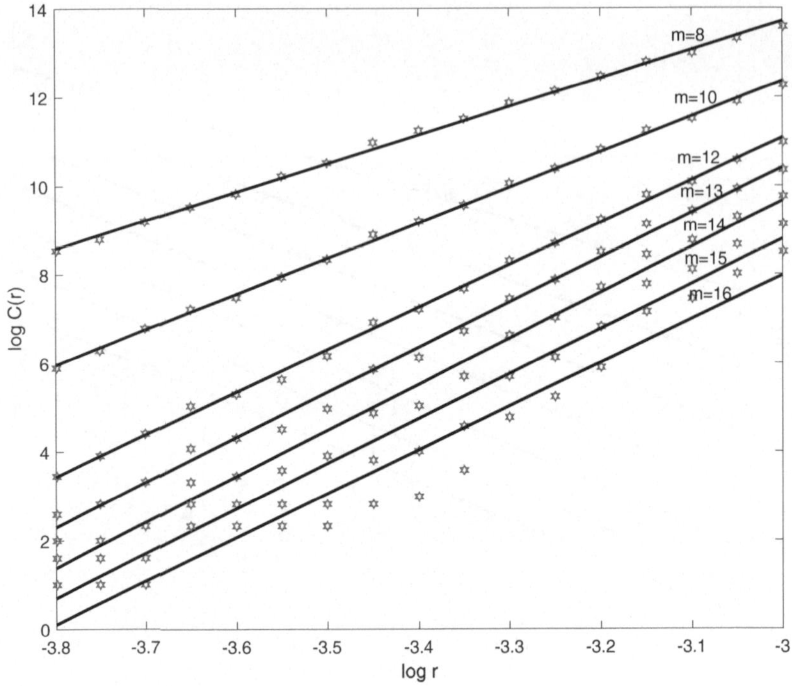

FIGURE 4.12
Plot for the calculation of correlation dimension of NPOI time series.

variability of the climate system despite its deterministic character. We got a clear indication that the attractor is sufficiently embedded and has reached its limiting dimension for the SOI and NPOI time series. The possible cause behind the non-saturation of the correlation dimension in other time series may be due to the small length of time series data; this may also be due to random noise present in the system (measurement).

Conclusions

An attractor reconstruction technique is used here for single-variable time series analysis, and we have calculated the correlation dimension and Lyapunov exponent for different time series. An optimized box assisted algorithm is used here to calculate the correlation dimension, which allows us to go for a greater number of iterations, as it takes less computation time compared to a Genetic Programming (GP) algorithm. The minimum and maximum number of variables needed to describe the dynamics of the system are obtained for SOI and NPOI time series. Although we are not able to establish the teleconnection

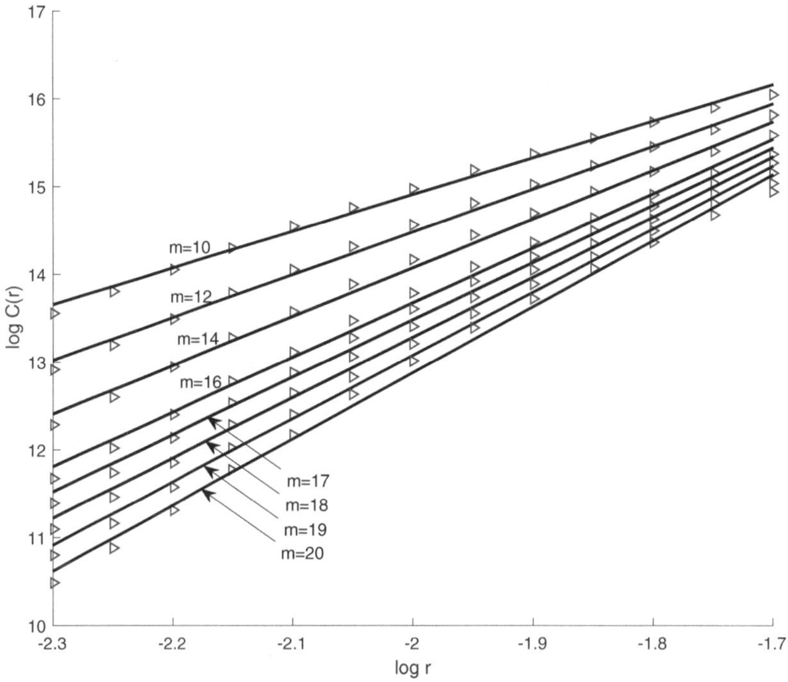

FIGURE 4.13
Plot for the calculation of correlation dimension of AOI time series.

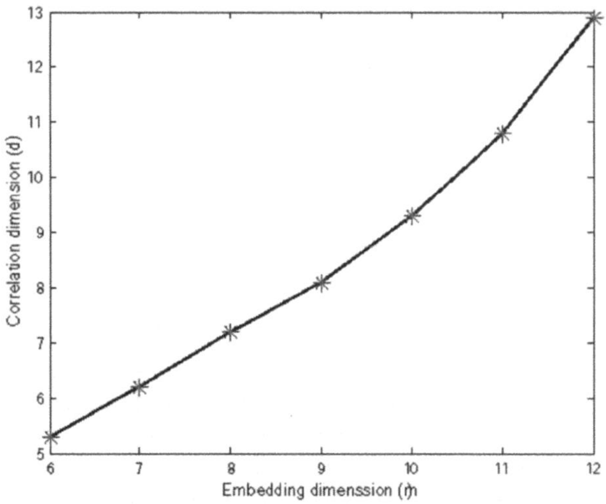

FIGURE 4.14
Plot of correlation dimension vs. embedding dimension for NAOI time series.

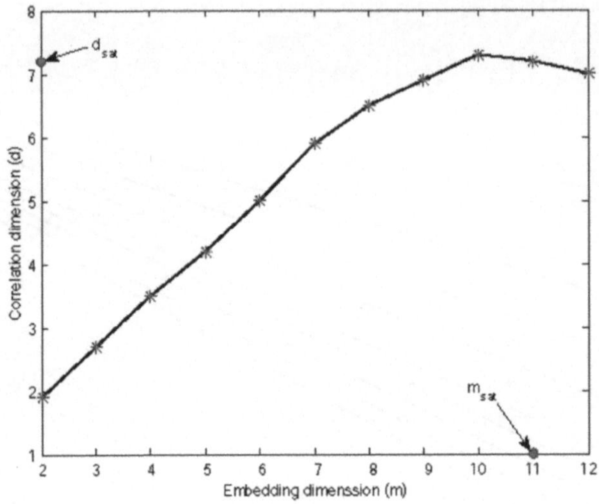

FIGURE 4.15
Plot of correlation dimension vs. embedding dimension for SOI time series.

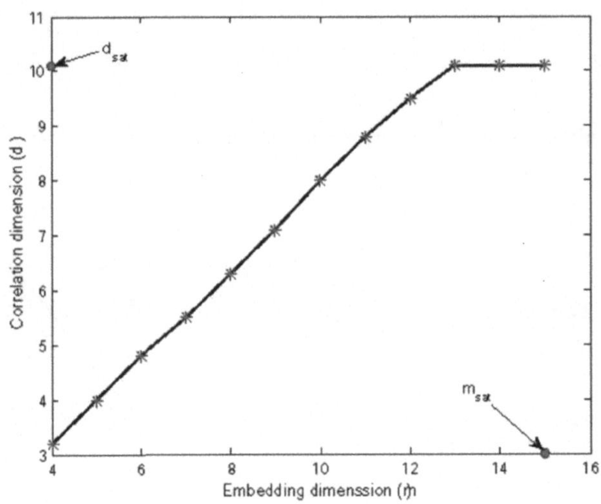

FIGURE 4.16
Plot of correlation dimension vs. embedding dimension for NPOI time series.

between SO and AO, one thing that is very clear from our observations is that AO plays the same role in the southern hemisphere as does NAO in the northern hemisphere. Similar is the case with PDO/NPO and SO because we get the fractal dimensionality in the case of NPO and SO and no such dimensionality is observed in the case of AO and NPO.

The probable cause behind the non-observance of a strange attractor for the Antarctica oscillation index may be the data scarcity in that region, and

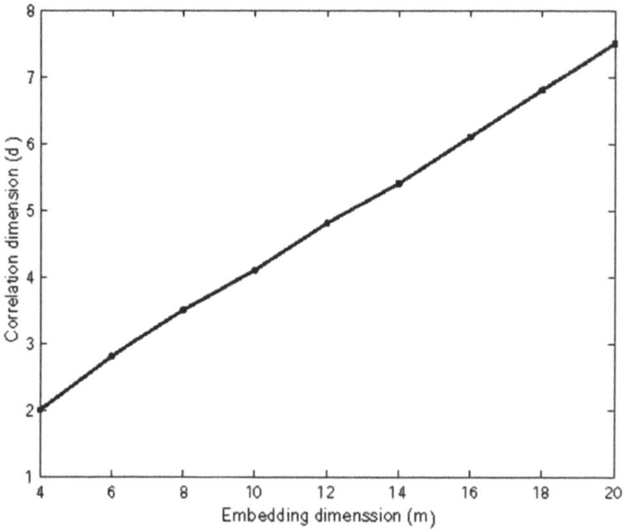

FIGURE 4.17
Plot of correlation dimension vs. embedding dimension for AOI time series.

problems also exist with the length of available records. The need, therefore, is of development of improved techniques for estimating important geophysical parameters from satellite and in-situ data observations and a greater number of data stations in the Antarctic region; then we can possibly get the desired teleconnection. We hope that the technique used here may serve as an additional tool for finding the teleconnection between ENSO and AO and thus this study can be used to understand the natural climate variability in the Antarctic region.

Sea Ice Thickness

Semtner's Model

In Semtner's model, sea ice is assumed to be a horizontal uniform slab of ice on which snow accumulates seasonally. In this model, one grid point lies in the snow and two grid points lie evenly spaced within the ice. A snow layer of thickness hS lies above two layers of ice, each of thickness (hi/2). The algorithm continually changes the resolution such that the numbers of ice points remain the same, even though the ice thickness varies. A fraction of the incoming solar radiation penetrates the ice only when the ice is snow free.

In Semtner's model, a simple formulation is used for the absorption of penetrating radiation, which is stored in a heat reservoir. Energy from this reservoir is used to keep the temperature near the top of the ice from dropping below the freezing point (unless the reservoir is exhausted). Thus, the penetrating solar energy does not cause immediate surface melting in summer but does retard internal cooling in the fall.

Formulation of the Model

The temperature within the snow cover is governed by the one-dimensional heat equation

$$(\rho c)_S \frac{\partial T}{\partial t} = k_S \frac{\partial^2 T}{\partial z^2} \tag{4.3}$$

Values for the constants are $k_S = 7.4 \times 10^{-4}$ cal cm^{-1}s^{-1} °C^{-1} and (ρc) S = 0.165 cal cm^{-3} °C^{-1}. The temperature within the ice is governed by a similar equation

$$(\rho c)_i \frac{\partial T}{\partial t} = k_i \frac{\partial^2 T}{\partial z^2} \tag{4.4}$$

here $k_i = 4.86 \times 10^{-3}$ cal cm^{-1}s^{-1} °C^{-1} and (ρc)i = 0.45 cal cm^{-3} °C^{-1}.

Once the temperature is predicted in each grid box, uniform spacing of temperature points in the ice is re-established after mass changes at the upper and lower boundaries.

Application of Semtner's Model to Arctic Region

Code was written, installed, and implemented for Semtner's three-layer model. The model was run for the monthly averaged values of the flux of latent heat (F_l), flux of sensible heat (F_t), incoming long wave radiation (F_L), and incoming short-wave radiation (F_r) taken from Fletcher (1965), and the results of the original paper of Semtner were reproduced to check the correctness of the code.

In the Arctic Ocean, the mean ice thickness is about 300–400 cm (Hartmann, 1994). Semtner's model was run for several years until equilibrium was achieved. Equilibrium was achieved after 35 years with an initial ice thickness $h_i = 300$ cm and initial snow thickness $h_S = 30$ cm. A standard forcing is prescribed with the oceanic heat flux $F_B = 1.5$ kcal cm^{-2} yr^{-1}, the fraction of penetrating radiation into ice I0 = 0.17, and the albedo of ice $a_i = 0.64$. The calculations may be simplified by taking the penetrating radiation I0 = 0 and oceanic heat flux $F_B = 0$, and the yearly snowfall was set at zero as done by

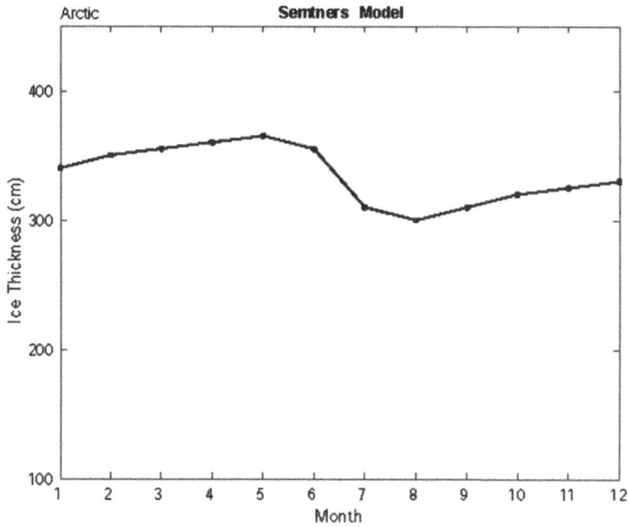

FIGURE 4.18
Monthly variation of ice thickness.

Semtner. The monthly variation of ice thickness with such simplifications is shown in Figure 4.18.

The predicted average monthly temperature T_S at the snow surface and T_i at the snow-ice interface is shown in Figure 4.19 for standard forcing with $T_S = -32°C$, $T_i = -18°C$, $T_1 = -14°C$, and $T_2 = -6°C$. Figure 4.20 shows the effect of varying oceanic heat flux on the average monthly ice thickness for the same standard forcing. Variations in the oceanic heat flux FB from 0.0 to 4.5 kcal cm^{-2} $year^{-1}$ decrease the average annual ice thickness by a factor of more than 5.5.

Winton's Model

The Winton reformulated three-layer sea ice model considers two equally thick ice layers; the upper ice layer has a variable specific heat resulting from brine pockets and the lower ice layer has a fixed heat capacity. The upper ice layer with a variable heat capacity is like the sophisticated models of Ebert and Curry (1993) and Bitz and Lipscomb (1999). A zero-heat-capacity snow layer lies above the ice. Heat fluxes at the top and bottom surfaces are used to calculate the change in ice and snow layer thickness.

Formulation of the Model

The model consists of a snow layer of thickness h_s lying above two equally thick ice layers, each of thickness (hi/2). The temperature variables used in

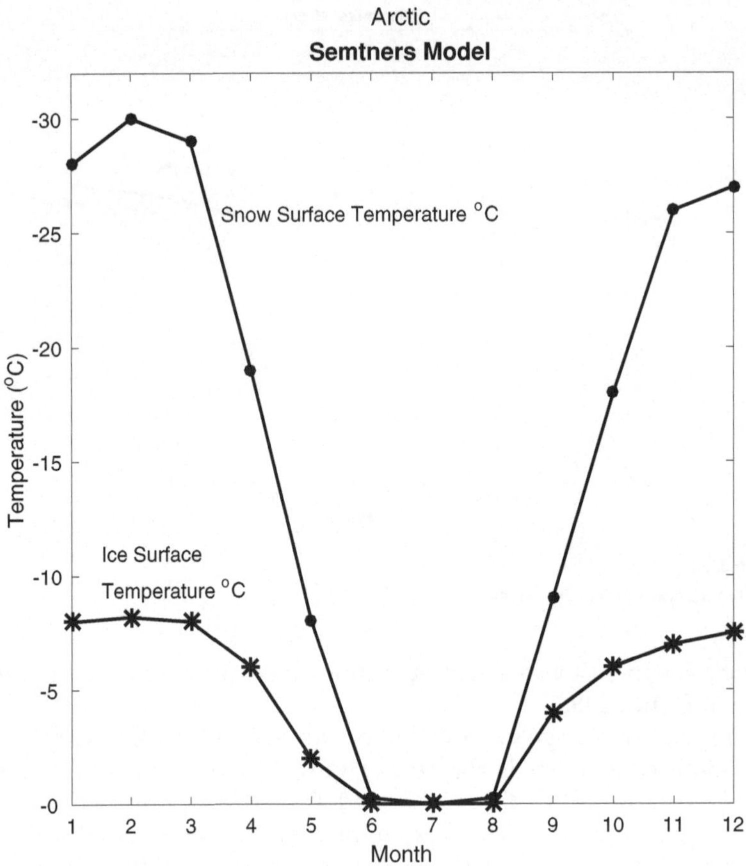

FIGURE 4.19
Variation of temperature in case of standard forcing.

the model are T_s the snow surface temperature, T1 and T2 the upper and lower ice layer temperatures at the midpoints of the two ice layers (hi/4) and (3hi/4) below the ice surface, respectively, and T_f the freezing temperature of seawater.

Application of Winton Model in Arctic Region

The Winton model was also run for the monthly averaged values of the flux of latent heat (F_l), flux of sensible heat (F_t), incoming long wave radiation (F_L), and incoming short-wave radiation (F_r) taken from Fletcher (1965), and the results of the original paper of Winton were reproduced to check the correctness of the code.

In the Arctic Ocean, the mean ice thickness is about 300–400 cm (Hartmann, 1994). Winton's model was run for several years until

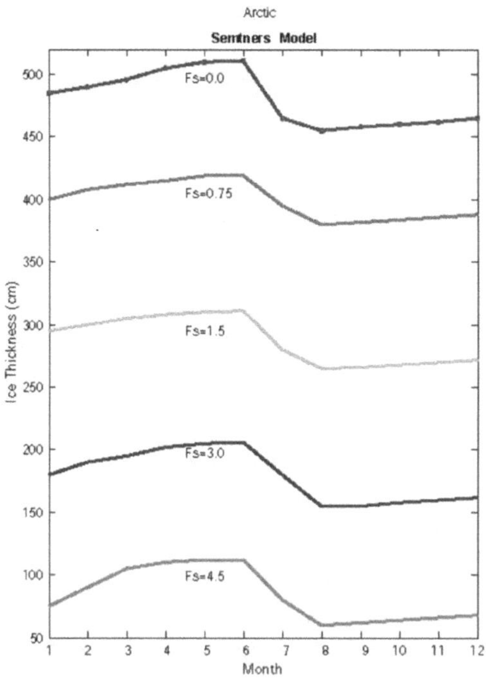

FIGURE 4.20
Effect of varying FB on monthly average ice thickness.

equilibrium was achieved. Equilibrium was achieved after 37 years, with an initial ice thickness $h_i = 300$ cm and initial snow thickness $h_S = 30$ cm. The oceanic heat flux was kept at $F_B = 1.5$ kcal cm^{-2} yr^{-1} and initial temperatures were taken as $T_S = -32°C$, $T_1 = -12°C$, $T_2 = -5°C$, and $T_f = -2°C$ with HFD = 1.3.

Figure 4.21 shows the seasonal variation of ice thickness for 30% penetrating radiation, that is, (I0 = 0.3) and a salinity of 1 per mil. During the first half of the year, ice thickness shows an increasing trend and attains a maximum value, and then in the next three months, it shows a decreasing tendency and finally starts increasing after November onward. The ice thickness is maximum during June and minimum during October. There is no significant change in ice thickness during the months of September, October, and November. Figures 4.22 and 4.23 show the variation of upper and lower ice temperatures (T1 and T2), respectively. The temperatures of the upper and lower ice layers are maximum during the months of July, August, and September and minimum during February and March.

Application of Winton Model in Antarctic Region

Winton's model was configured for the McMurdo region of Antarctica by forcing it with monthly seasonal fluxes of latent heat and sensible heat as well

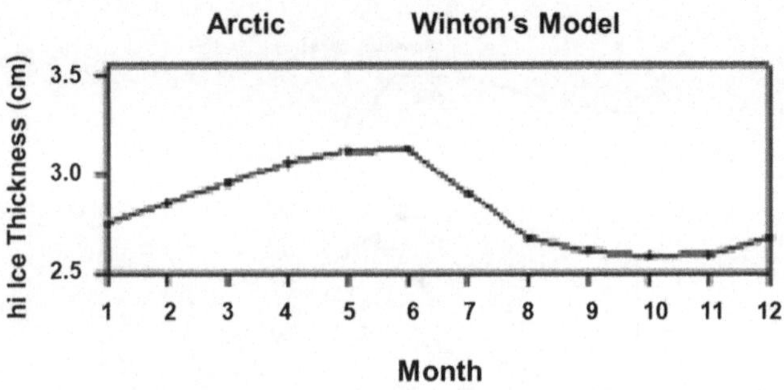

FIGURE 4.21
Monthly variation of ice thickness.

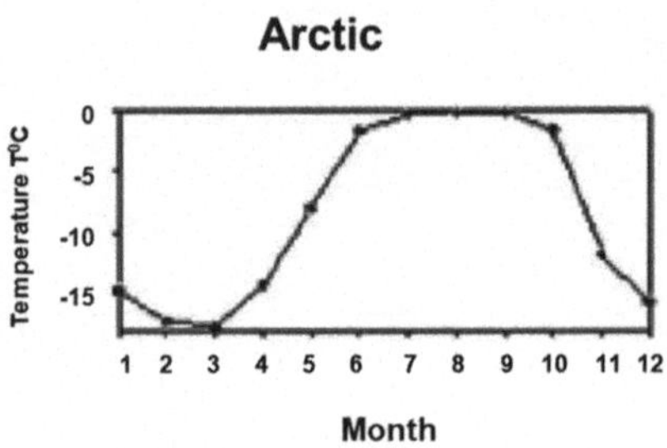

FIGURE 4.22
Monthly variation upper in ice.

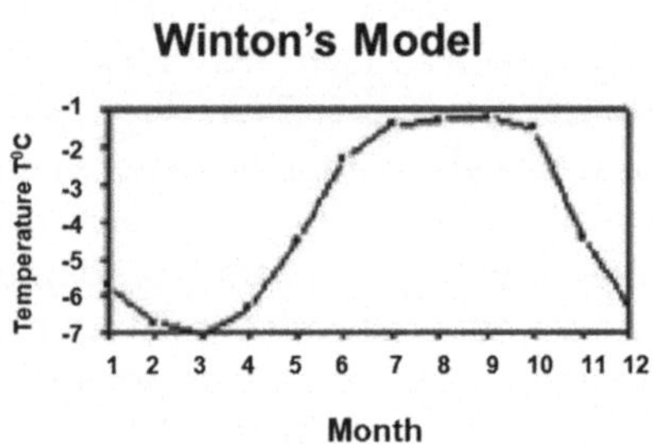

FIGURE 4.23
Monthly variation in ice temperature.

as fluxes of downward shortwave and longwave radiation. The forcing data were compiled from the data obtained from the Taylor glacier meteorological station (Dana et al., 1998) and the NCEP data taken from the Climate Diagnostics Centre website.

Winton's model was run for several years until equilibrium was achieved. Equilibrium was achieved after 5 years. The ice thickness was initially kept at

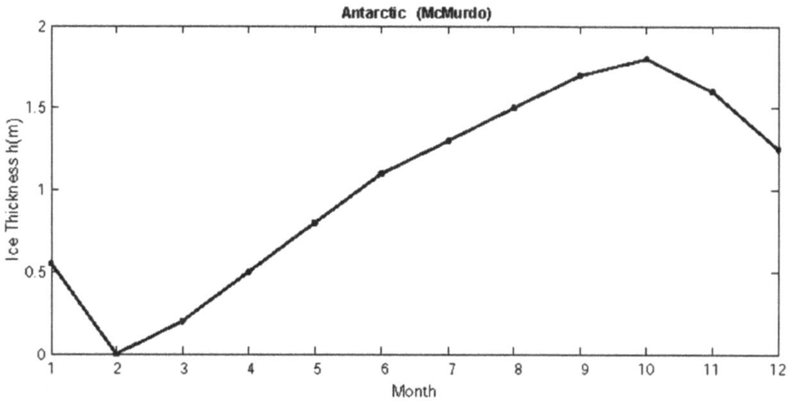

FIGURE 4.24
Monthly variation of ice thickness in Antarctica.

$h_i = 100$ cm and snow thickness at $h_S = 24$ cm. The oceanic heat flux was kept at $F_B = 18.576$ kcal cm^{-2} yr^{-1} and initial temperatures $T_1 = -3.5°C$, $T_2 = -2.5°C$, and snow surface temperature $T_S = -2.273°C$. The value of HFD was adjusted to 0.7 to keep the ice thickness in the range of 100–200 cm, as in the Antarctic Ocean the mean ice thickness is about 100–200 cm (Hartmann, 1994).

Figure 4.24 gives the monthly variation of ice thickness. Figure 4.25 shows the monthly variation of snow surface temperature. The maximum depth of ice is obtained during the month of October, while the ice is thinnest in the month of February. The lowest snow surface temperature achieved is around −65°C in the month of August. Incidentally, the minimum temperature recorded so far in the Antarctic peninsula is −89.5°C at the Russian station Vostok (Pandey and Khare, 2001). Figures 4.26 and 4.27 show the variation of upper ice temperature T_1 and lower ice temperature T_2. Both temperatures T_1 and T_2 show the same monthly trend and are minimum during the months of August and September. The minimum values of T_1 and T_2 are around −30°C and −11°C, respectively.

Comparison of Semtner and Winton Models

The difference between Semtner's model and Winton's model of sea ice lies in the fact that Semtner assumes one snow layer and two ice layers of constant heat conductivities and a simple parameterization for the brine content, while Winton treats the upper half of ice as a variable heat capacity layer, keeping the vertical layer structure of Semtner's model intact. In this way, the heat contained in the brine pockets is treated more physically and naturally in Winton's model.

Also, the number of temperature points in the Winton model is less than those used in the Semtner model; hence, the calculation part is reduced and the number of initial inputs required is also reduced.

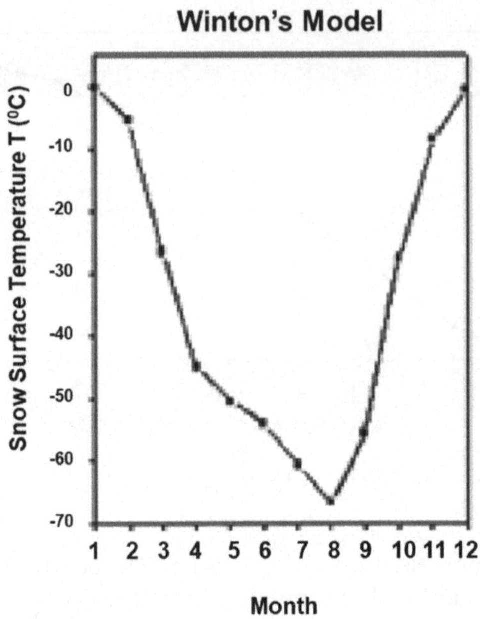

FIGURE 4.25
Monthly variation of snow surface temperature.

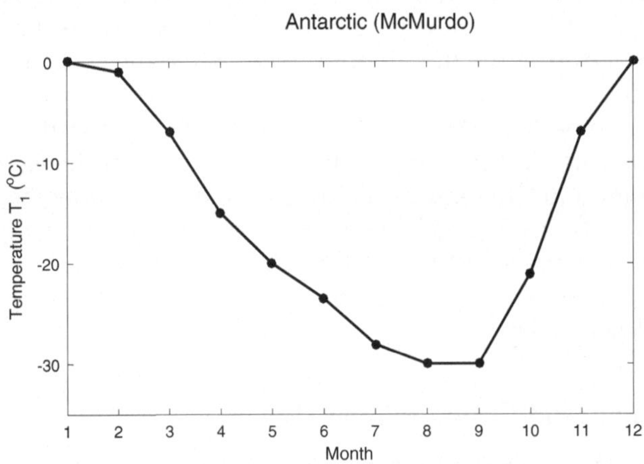

FIGURE 4.26
Monthly variation of upper ice temperature.

While implementing both these models, only the vertical thermodynamics of the sea ice was considered. Sea-ice is static and transmission of heat fluxes take place through ice and snow, taking into consideration their conductivities and heat capacities, including the latent heat of brine at each iteration of the model.

Prediction of Sea Surface Temperature in the Indian Ocean Region Using Artificial Neural Networks

Introduction

Sea surface temperature is an important parameter that contributes to the interaction of tropical atmosphere and oceans. It significantly affects the monsoon over the Indian sub-continent. Artificial neural networks (ANNs) are now widely used for the prediction of various atmospheric and oceanic phenomenon. An ANN is a massively parallel-distributed processor made up of simple processing units that has a natural propensity for storing experiential knowledge and making it available for use (Aleksander and Morton, 1990). Generally, it performs a mapping task. It first recognizes the map present in the sample input-output data (called training of the network), and then it calculates the output when an unseen input (i.e., input that is not present in the training data) is presented to it. Of the many learning algorithms available in the literature (Bishop, 1995; Haykin, 2002; Zurada, 2002), here the error backpropagation has been used. In the present analysis, a multilayer feed-forward neural network is used to predict the SST anomalies over the Indian ocean region (100°E, 30°S). The Reynolds reconstructed SST from 1 January 1950 to 1 December 2001 has been used for the analysis. To the best of our knowledge, prediction of SST anomalies has not been done using ANN.

Methodology

The method essentially involves time series prediction. We use the SST values for 52 years as mentioned above. Thus, there are 52 input-output relations. The SST values have a very low variation from the mean. Hence, instead

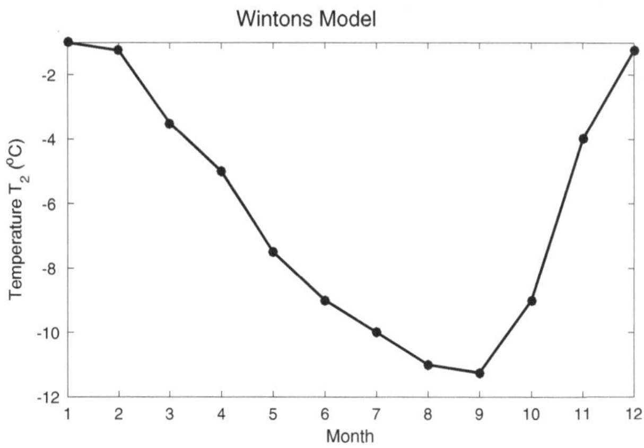

FIGURE 4.27
Monthly variation of lower ice temperature.

of working with the SST values directly, their deviation from the mean is calculated and the anomalies have been predicted.

The anomalies for two previous consecutive years form the input to predict that for the third year. To include the anomalies of all 24 months of the previous two years would require 24 neurons in the input layer. Even a small network with one hidden layer with two neurons would have 48 weights between the input and hidden layers. If we make one network for the prediction of SST of each month, then there would be two links between the hidden layer and the output layer, making the total number of weights to be adjusted to 50. Even if we use all 52 years' values for training, it would be too small a set to train 50 weights (plus the biases). Thus, as a matter of simplification, we use the anomalies of only three months (September, October, and November) of each year. For the same reason, we have restricted the number of years, which acts as input for the prediction of current value to 2 years only. Lastly, separate networks are made for the prediction of each month's anomaly, as one single network for predicting the anomalies of all months will increase the number of neurons in the output layer, leading us to similar situations.

The ANN architecture used by us consists of six neurons in the input layer and only one in the output layer. Also, it has been proved that a three-layer network can map any arbitrary function. Therefore, only one hidden layer is taken. The number of neurons in the hidden layer is fixed by trial-and-error determination. The final architecture is shown in Figure 4.28.

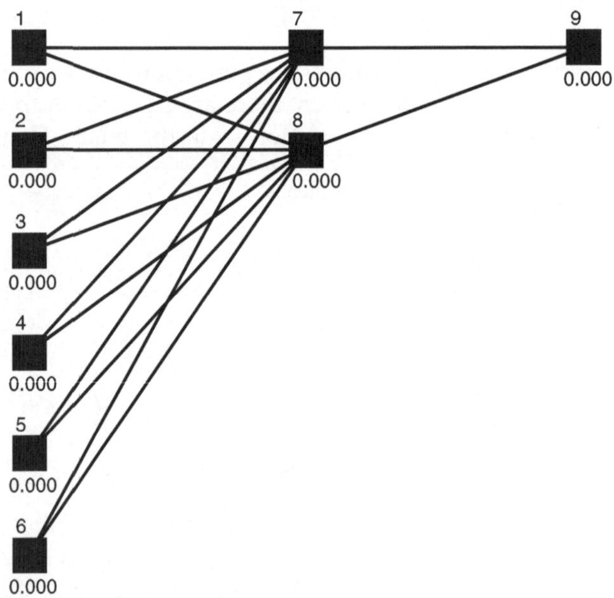

FIGURE 4.28
Architecture of the neural network used.

It is clear from the architecture that only14 weights (plus the biases) are required to be adjusted. The standard backpropagation algorithm is employed for training. Two such networks are constructed, one each for predicting the anomalies of the months of September and October of 1995–2001.

The Training Data Set

From the 52-year data set, first 45 (1950–1994)-year data is used for training. This splitting is done serially. From the 45-year set, 7 years are randomly taken for cross-validation and the remaining 38 are applied for estimation of weights.

Normalization of the Data

The anomaly data is normalized using the maximum-minimum normalization approach as given below:

$$X_n = (X - X_{min})/(X_{max} - X_{min})$$

where X is the SST anomaly, X_{max} is the maximum anomaly for a month, X_{min} is the minimum anomaly for the month, X_n is the normalized value of the anomaly. In practice, X_{min} is taken slightly less than the actual minimum and X_{max} is taken slightly greater than the actual maximum.

Criteria for Stopping the Training

The training is stopped when the cross-validation error and the estimation error become nearly equal (keeping in mind that the validation error does not increase once both errors become nearly the same). The training is done for nearly 45,000 cycles for both months. Cross-validation is performed after every 10 cycles. To avoid monotony in the training, the patterns are shuffled before the beginning of each cycle. The weights are adjusted after each pattern presentation; that is, the epoch size is kept at 1. A momentum term is not included, as the patterns are of sufficiently small size.

Results and Discussion

The SST anomalies for the test and training sets for the month of September are shown in Figures 4.29 and 4.30, respectively. Correlation coefficients of 0.72 and 0.77 between the outputs (predicted) SST and the observed SST are obtained for the test set and training set, respectively. The SST anomalies for the month of October are shown in Figures 4.31 and 4.32 for the test and training sets, respectively. A correlation coefficient of 0.67 between the output SST and the

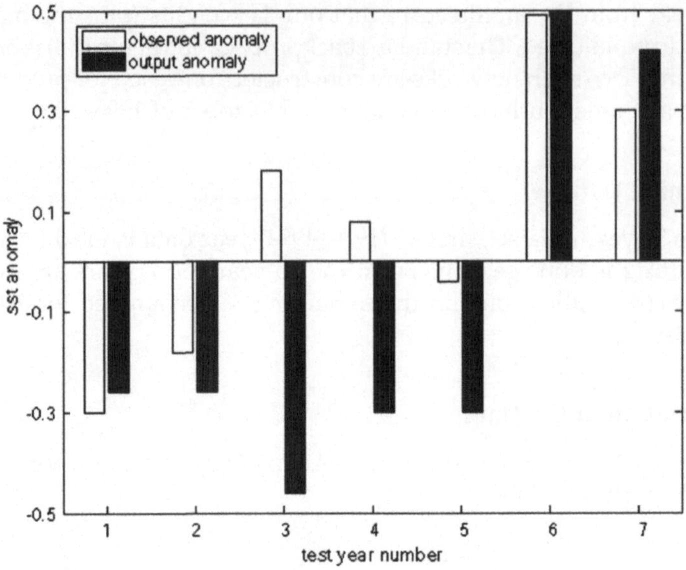

FIGURE 4.29
SST anomaly for September (test case).

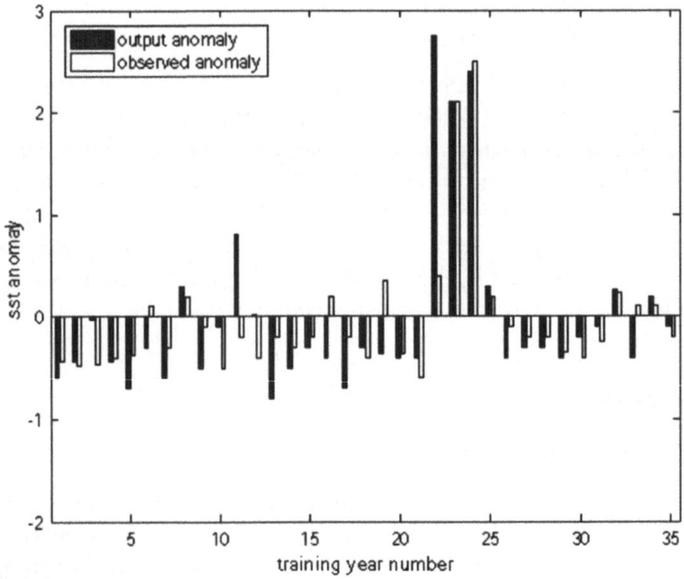

FIGURE 4.30
SST anomaly for September (training set).

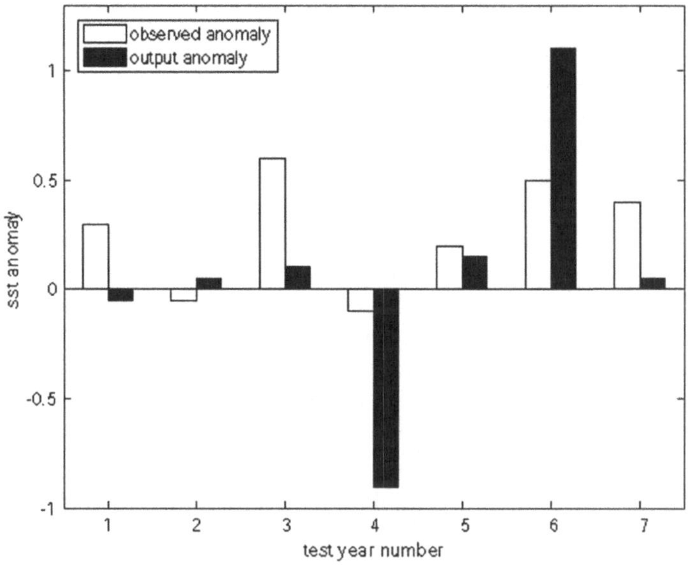

FIGURE 4.31
SST anomaly for October (test set).

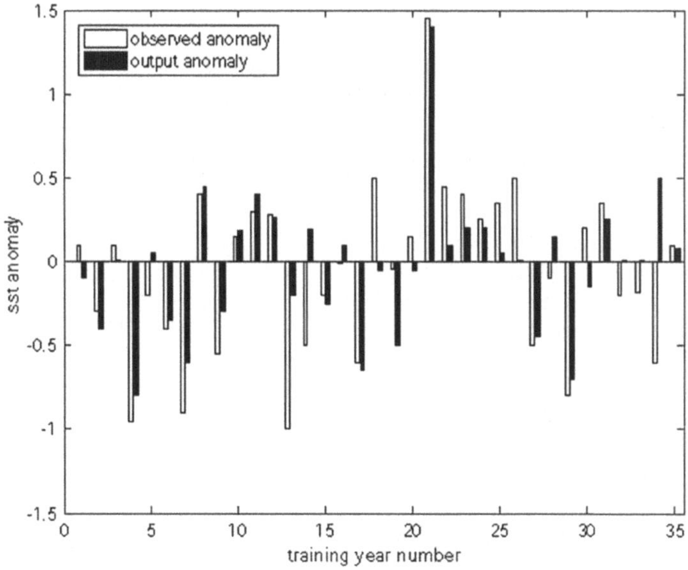

FIGURE 4.32
SST anomaly for October (training set) case.

observed SST is obtained for the test set and that of 0.803 for the training set. The results show a moderate relationship between the present and future SST anomalies. Also, it is seen that the ANN model predicts the future SST anomalies from those of the present and near past with reasonably good accuracy.

The number of months that have to be taken from the previous 2 years to predict the current monthly SST anomalies is crucial while designing the ANN.

By taking only 3 months of the previous 2 years, precisely the months of September, October, and November, an assumption is inherently made that these 3 months are the best representative of the entire year's SST anomalies. In practice, it may not be so. A better approach may have been to extract features from the 12-month SST anomalies. These extracted features are some combinations of the 12-month anomalies. This may allow us to pick the best representative features, leading to better prediction. Also, if the number of representative features comes out to be less than three (say, two), then we shall have only eight (4×2) weights between the input and the hidden layer. Then we may have six weights between the hidden and output layers, keeping the number of weights equal to 14. This allows us to predict the 3-month anomalies using a single network only.

References

Aleksander, I. and Morton, H. 1990. *An Introduction to Neural Computing*, Chapman and Hall, London.

Allan, R. J., Nicholls, N., Jones, P. D. and Butterworth, I. J. 1991. A further extension of the Tahiti-Darwin SOI, early ENSO events and Darwin pressure, *J Climate*, 4, 743–749.

Backer, S. de., Naud, A. and Scheunders, P. 2004. *Non-Linear Dimensionality Reduction Techniques for Unsupervised Feature Extraction*, Elsevier Science Inc New York, USA, 19, 711–720.

Bellerby, T., Todd, M., Kniveton, D. and Kidd, C. 2000. Rainfall estimation from a combination of TRMM precipitation radar and GOES multispectral satellite imagery through the use of an artificial neural network, *J. Appl. Meteorol.*, 39, 742–755.

Bishop, M. 1995. *Neural Networks for Pattern Recognition*, Oxford University Press.

Bitz, C. M. and Lipscomb, W. H. 1999. A new energy conserving sea ice model for climatic study, *J Geophys Res*, 104, 15669–15677.

Blumberg, A. F. and Mellor, G. L. 1987. A description of a three-dimensional coastal ocean circulation model, in three-dimensional coastal ocean models, *Coastal Estuarine Ser.*, Vol. 4 edited by N.S. Heaps, 208 pp, AGU, Washington, D.C.

Cassano, J. J., Box, J. E., Bromwich, D. H., Lin, L. and Steffen, K. 2001. Evaluation of Polar MM5 simulations of Greenland's atmospheric circulation. *J. Geophys. Res.*, 106, 33,867–33,890.

Chantasut, N., Charoenjit, C. and Tanprasert, C. 2004. Predictive mining of rainfall predictions using artificial neural networks for Chao Phraya river, *AFITA/WCCA Joint Congress on IT in Agriculture.*

Dana, G., Wharton, R. A. and Fountain, A. G. 1998. McMurdo Dry Valleys LTER: Solar radiation on glaciers in Taylor Valley Antarctica, *Antarct. J. United States – Review. (1996).* 31, 191–193.

Da Silva, A., Young, A. C. and Levitus, S. 1994. Atlas of Surface Marine Data, NOAA Atlas NESDIS 6, 83 pp.

Deleersnijder and Campin, 1995. CLIO Ocean (OGCM) model, Netherland Centre for Climate Research.

Ebert, E. E. and Curry, J. A. 1993. An intermediate one-dimensional thermodynamic sea ice model for investigating ice-atmosphere interactions, *J. Geophys. Res.*, 98, 10085–10109.

Fletcher, J. O. 1965. The heat budget of the Arctic basin and its relation to climate, The Rand Corporation, Santa Monica, Calif., R-444-Pr.

Gairola, R. M. 2002. Rainfall Retrieval from TRMM-TMI and Precipitation Radar Data Using Neural Network Approach, Megha-Tropiques, MOG/SAC.

Gang, L., Verma, B. and Kulkarni, S. 2002. Experimental analysis of neural network based feature extractors for cursive handwriting recognition, *IEEE trans., The Technical Program for the 2002 IEEE World Congress on Computational Intelligence.*

Gong, D. Y. and Wang, S. W. 1998. Antarctic Oscillation: Concept and applications, *Chinese Sci Bull*, 43, 734–738.

Gordon, A. L. 1986. Interocean exchange of thermocline water, *J. Geophys. Res.*, 91, 5037–5046.

Grassberger, P. 1990. An optimized box-assisted algorithm for fractal dimensions, *Physics Letters A*, 148, 63–68.

Grassberger, P. and Procaccia, I. 1983. Measuring the Strangeness of Strange Attractors, *Physica D*, 9, 189–208.

Guo, Z., Bromwich, D. H. and Cassano, J. J. 2003. Evaluation of Polar MM5 simulations of Antarctic atmospheric circulation, *Mon. Wea. Rev.*, 131, 384–411.

Hartmann, D. L. 1994. *Global Physical Climatology*, Academic Press, pp. 269.

Haykin, S. 2002. *Pearson Education Asia*, second Edition.

Hossain, F., Agarwal, N., Dinku, T. and Anagnostou, E. N. 2002–2003. Assessment of various neural network schemes for classification of cloud data, *J. Env. Sys.*, 29(2), 151–173.

Hsieh, W. H. 2000. Nonlinear canonical correlation analysis of the tropical pacific climate variability using a neural network approach, *J. Climate*, 14, 2528–2539.

Huang, B. and Kinter, J. L. III, 2002. The interannual variability in the tropical Indian Ocean, *J. Geophys. Res.*, 107, 3199, doi:10:1029/2001JC001278.

Huang, B. and Schneider, E. K. 1995. The response of an ocean general circulation model to surface wind stress produced by an atmospheric general circulation model, *Mon. Weather Rev.*, 123, 3059–3085.

Josey, S. A., Kent, E. C. and Taylor, P. K. September. 1997. Southampton Oceanography Center (SOC), U.K., surface flux climatology (version 1.1). http://www.coaps.fsu.edu/RVSMDC/SAC/SOC.

Karsten, R. H. and Marshall, J. 2002. Constructing the residual circulation of the ACC from observations, *J. Phys. Oceanogra.*, 32, 3315–3327.

Krasnopolsky, V. M., Chalikov, D. V. and Tolman, H. L. 2002. A neural network technique to improve computational efficiency of numerical oceanic models, *Ocean Model.*, 4, 363–383.

Leeuwenburgh, O. and Stammer, D. 2001. The effect of ocean currents on the sea surface temperature anomalies, *J. Phys. Oceanogr.*, 31, 2340–2358.

Lerner, B., Guterman, H., Aladjem, M. and Dinstein, I. 1996. Feature extraction by neural network nonlinear mapping for pattern classification, *13th International Conference on Pattern Recognition*, ICPR13, Vienna, 4, pp. 320–324.

Leuliette, E. W. and Wahr, J. M. 1999. Coupled pattern analysis of sea surface temperature and TOPEX/Poseidon sea surface height, *J. Phys. Oceanogr.*, 29, 599–611.

Levitus, S. E. 1982. *Climatological Atlas of the World Ocean*, NOAA Professional Paper 13, US Government Printing Office, Washington DC.

Levitus, S. and Boyer, T. P. 1994. World Ocean Atlas 1994, Volume 4: Temperature, NOAA Atlas NESDIS 4, 11 pp.

Levitus, S., Burgett, R. and Boyer, T. P. 1994. World Ocean Atlas 1994, Volume 3: Salinity. NOAA Atlas NESDIS 3, 99 pp.

Lorenz, E. N. 1956. *Empirical orthogonal functions and statistical weather prediction*, Scientific Report No. 1, Statistical Forecasting Project, MIT, Department of Meteorology.

Masters, T. 1993. *Practical Neural Network Recipes in C++*, Academic Press.

Mellor, G. L. and Yamada, T. 1982. Development of a turbulence closure model for geophysical fluid problems. *Rev. Geophys.*, 20, 851–875.

Meyers, G. 1996. Variation of Indonesian throughflow and the El Nino-southern oscillations, *J. Geophys. Res.*, 101, 12255–12263.

Mishra, A. P., Rai, S. and Pandey, A. C. 2007. An estimation of heat flux and its variability in the Southern Indian Ocean (SIO) using ocean general circulation model on model (OGCM), *J. Ind. Geophys. Union*, 11(4), 217–228.

Nerem, R. S., Chambers, D. P., Leuliette, E. W., Mitchum, G. T. and Giese, B. S. 1999. Variation in global mea sea level associated with the 1997-1998 ENSO event: Implications for measuring long term sea level change, *Geophys. Res. Lett.*, 26, 3005–3008.

Pacanowski, R. C., Dixon, K. and Rosati, A. 1993. *The GFDL modular ocean model user guide*, version 1.0, Tech Rep. 2, p. 16, Geophys. Fluid Dyn. Lab. , Natl Oceanic and Atmospheric Admin., Princeton, N.J.

Pandey, P. C. and Khare, N. 2001. Antarctica: An emerging vision, *Proc. Nat. Acad. Sci. India*, 71(A), Spl. Issue, 51–62.

Rai, S., Mishra, A. P. and Pandey, A. C. 2006. Simulation of Southern Indian Ocean currents with σ and z coordinate ocean models, *Western Indian Ocean J. Mar. Sci.*, 5(1), 89–94.

Ropelewski, C. F. and Jones, P. D. 1987. An extension of the Tahiti-Darwin Southern Oscillation Index, *Mon Wea Rev*, 115, 2161–2165.

Sahai, A. K., Satyan, V. and Kulkarni, J. R. 2003. *General Circulation Model Systemic Error Correction and Seasonal Prediction Using Artificial Neural Network*, Annual Report 2002-03, IITM-Pune, pp. 46.

Sahai, A. K., Soman, M. K. and Satyan, V. 2000. All India summer monsoon rainfall prediction using an artificial neural network, *Clim Dyn.*, 16, 291–302.

Sammon, J. W. 1969. A nonlinear mapping for data structure analysis, *IEEE Trans. Comput.*, 18, 401–409.

Schneider, E. K., Huang, B., Zhu, Z., Dewitt, D. G., Kinter III, J. L. Kirtman, B. P. and Shukla, J. 1999. Ocean data assimilation, initialization, and prediction of ENSO with a coupled GCM, *Mon. Weather Rev.*, 127, 1187–1207.

Semtner, A. J. 1976. A model for the thermodynamic growth of sea ice in numerical investigations of climate, *J. Phys. Oceanogr.*, 6, 379–389.

Shabbar, A., Higuchi, K. and Yuen, C. W. 2003. *Recent NAO Changes and their association with Net Surface Heat Flux;* Meteorological Service of Canada, Environment Canada, pp. 4.

Shin, D. B. and Kummerow, C. 2003. Parametric rainfall retrieval algorithms for passive microwave radiometers, *J Appl. Meteorol.*, 42, 1480–1496.

Singh, R., Vasudevan, B. G., Pal, P. and Joshi P. C. 2004. Artificial neural network approach for estimation of surface specific humidity and air temperature using multifrequency microwave radiometer, *Proc. Indian Acad. Sci. (Earth Planet. Sci.),* 113, 89–101.

Smagorinsky, J., Manabe, S. and Holloway, J. L. 1965. Numerical results from a nine-level general circulation model of the atmosphere, *Mon. Weather Rev.*, 93, 727–768.

Smith, R. C. Stammerjohn, S. and Baker, K. S. 1996. Surface air temperature variations in the western Antarctic peninsula region. In R. M. Ross, L. B. Quetin, and E. E. Hofmann (eds.), *Foundations for Ecological Research West of the Antarctic Peninsula,* 70, AGU Antarctic Research Series.

Stephenson, D. B. 2000. *What is NAO? Keynote Opening Speech at AGU NAO Conference,* Ourense, 28 Nov–1 Dec.

Trenberth, K. E. and Hurrell, J. W. 1994. Decadal atmosphere-ocean variations in the Pacific, *Clim. Dyn.*, 9, 303–319.

Tsintikidis, D., Haferman, J. L., Anagnostou, E. N., Krajewski, W. F. and Smith, T. F. 1997. A neural network approach to estimating rainfall from spaceborne microwave data, *IEEE Trans. Geosci. Remote Sens.*, 35, 1079–1093.

Venegas, S. A. 2001. Danish Center for Earth System Science, *Ph.D. Thesis,* Niels Bohr Institute, University of Copenhagen, Copenhagen, Denmark.

Verschell, M. A., Kindle, J. C. and O'brienl, J. J. 1995. Effects of Indo-Pacific throughflow on the upper tropical Pacific and Indian Oceans, *J. Geophys. Res.*, 100, 18409–18420.

Vranes, K., Gordon, A. L. and Ffield, A. 2002. The heat transport of the Indonesian throughflow and implications for Indian Ocean heat budget, *Deep-Sea Research II*, 49, 1391–1410.

Wang, S. W. 1992. Studies on the atmospheric centers of action, Report in State University of New York at Stony brooks, 32 pp.

Webster, P., Magana, V. O., Palmer, T. N., Shukla, J., Tomas, R. A., Yanai, M. and Yasunari, T. 1998. Monsoons: Processes, predictability and the prospects for prediction, *J Geophys. Res.*, 103, 14451–14510.

White, W. B. and Peterson, R. 1996. An Antarctic circumpolar wave in surface pressure, wind, temperature and sea ice extent, *Nature*, 380, 699–702.

Winton, M. 2000. A reformulated three-layer sea ice model, *J. Atmos. Ocean Tech.*, 17, 525–531.

Zurada, J. M. 2002. *Introduction to Artificial Neural Systems,* Jaico Publishing House, fifth edition.

5

Very Low Frequency Emissions Observed at High Latitude Indian Station Maitri, Antarctica: A Review

A. K. Gwal, S. Choudhary, and S. K. Singh

CONTENTS

Introduction

The importance of very low frequency (VLF) (0.3–30 kHz) radio waves propagating in the magnetosphere has long been recognized in understanding plasma physics aspects of wave generation, amplification, and damping through linear and nonlinear interaction with particles. In addition, the role of waves in the exchange of energy between different regions of geospace is also important. The propagation of VLF waves through various levels is strongly controlled by the plasma density gradient and wave normal direction relative to the magnetic field (Smith and Jenkins, 1998). Typically, these waves travel in whistler mode. The popularity of this phenomenon as a subject of investigation in geospace research stems largely from the fact that they can propagate along the geomagnetic field aligned density irregularities (ducts), penetrate the ionosphere, and be received by a simple audio frequency receiver on the ground.

An extensive survey of whistler and VLF emissions was carried out immediately following the International Geophysical Year (IGY) 1957–1958, with some more than 50 monitoring stations extending from equator to poles. It soon became clear that Antarctica is an ideal place for VLF research. Further, favorable factors in Antarctica include low levels of both human-made and natural electromagnetic noise. In 1992, well-calibrated long-term VLF monitoring equipment called VELOX (Smith, 1995) was set up at Halley, Antarctica, and some of the results from this experiment have been reported by Smith and Jenkins (1998). Periodic and quasi-periodic emissions, chorus, and various other transient discrete emissions such as VLF risers, fallers, and hooks have been reported at Antarctica (Sazhin et al., 1993; Morrison et al., 1994; Smith, and Nunn, 1998). Smith and Jenkins (1998) have reported periodic and quasi-periodic VLF emissions observed at Halley and South Pole stations, which are approximately in the same meridian and geomagnetic latitude of 61° and 74°, respectively.

The wave and wave particle interactions occurring in the magnetosphere generate a wide variety of emissions in the VLF range. VLF emissions are basically mid- and high-latitude phenomena. These emissions have become a very important diagnostic tool for probing the plasmasphere and beyond. These emissions, although less understood than whistlers, are believed to have their origin in the ionosphere-magnetosphere coupled system and may be due to plasma instabilities or in-situ electromagnetic radiation from high-energy particles. VLF emissions are characterized by their trigger source. The emissions are observed to be triggered from the top of hiss bands, by power line harmonic radiation (PLHR), by lightning whistlers (Helliwell, 1965; Sazhin et al., 1993; Nunn and Smith, 1996; Smith and Nunn, 1998), and by wave particle interactions in the magnetosphere. Helliwell (1965) has classified these emissions as hiss, chorus, hook, periodic, quasi-periodic and triggered emissions, and so on. The group of VLF emissions is further divided into two subgroups: (i) continuous emissions in both the time and frequency, which tend to maintain a steady state such as hiss, resonance bands, and noise band near the ion-cyclotron frequency, and (ii) discrete emissions, often with periodic and quasi-periodic natures. VLF emissions, like whistlers, are localized geographically and are observed most commonly at mid and high latitudes. At locations that are geomagnetically conjugate to one another, emissions are often closely related in form as well as in time of occurrence. Emission activity generally increases with magnetic activity at low latitudes and decreases at high latitudes like Antarctica. A close association exists between auroral phenomena and certain types of VLF emissions.

Discrete emissions are often seen simultaneously at stations less than a few hundred kilometers apart (Allcock and Martin, 1956), but when the separation exceeds 1000 km, correlation is considerably reduced. Occasionally the same emission may be observed at widely separated stations. Although discrete emissions appear to be restricted in geographical coverage, it has been observed that the intensity of emissions integrated over a minute or more correlates well at stations spaced up to several thousand kilometers apart (Ellis, 1959; Pope, 1959; Dowden, 1961). In view of the limited geographical extent of emissions, it is not surprising that long-term averages of emission occurrence show marked temporal and geographical variations. Most statistical studies of these variations have been limited to chorus. In one study, it was found that all types of emissions recorded at Macquarie Island (geomagnet. lat., 61° 10′) showed about the same diurnal variation, with a peak in seasonal distributions of different types of noise.

Periodic VLF emissions are observed only in regions of closed magnetic field lines (Sazhin and Hayakawa, 1994). Their period was typically in the range 1–7 sec., but most frequently in the range 2–6 sec. The changes in period during observed events often did not exceed 1% (Helliwell, 1968). Recently, for the first time, we have recorded daytime periodic VLF emissions at our newly installed VLF station at Maitri (70° 46′ S, 11° 44′ E, L ~4.6), Antarctica. The new experimental findings of periodic VLF emissions observed at Maitri during the daytime are presented in this chapter with their possible generation mechanism. These studies were carried out during January–February 1998, 1999, and 2002 as a part of the XVII, XVIII, XIX, XX, and XXII Indian Scientific Expeditions to the Antarctic. These expeditions are organized and funded by the National Centre for Antarctic and Ocean Research (NCAOR), Ministry of Earth Sciences (MoES), Government of India, New Delhi.

Scientific Importance of Antarctica

The region above the ionosphere is not accessible by conventional methods like rockets and balloons, but it can be explored by the whistler technique. Augmentation of whistler observations taken at Maitri station in Antarctica will give important clues about electron density, ionization flux and electric field variations with altitude, and also about the control of equatorial anomalies on the propagation mode of low-latitude whistlers. Experimental significance of such study will reveal the effect of the magnetosphere and space weather.

Various Research Projects So Far Undertaken

1. Coordinated Study of Very Low Frequency (VLF) Phenomena at Antarctica (1997–2002), funded by Department of Ocean Development, Govt. of India, New Delhi.
2. Space Weather Programme at Antarctica (2006–2009), funded by NCAOR, Ministry of Earth Sciences, Goa.

Scientific Aims of Objectives

a. To study the diurnal variations of whistlers and VLF phenomena at Antarctic and other latitudes.
b. To study the propagation path of whistler at various latitudes from the observed data.
c. To study the propagation mechanism of low- and high-latitude whistler and VLF phenomena from the observed data.
d. To study the generation and origin mechanisms of whistlers from the observed data.
e. To study the mapping of electric fields from low-latitude Bhopal to high-latitude Antarctica.

Very Low Frequency Recording during January 1998 in XVII Indian Antarctic Expedition

A typical frequency vs. time spectrogram of riser emission is shown in Figure 5.1. This rising emission was observed on January 22, 1998, in the morning at 0931 universal time (UT). The riser signal had a maximum

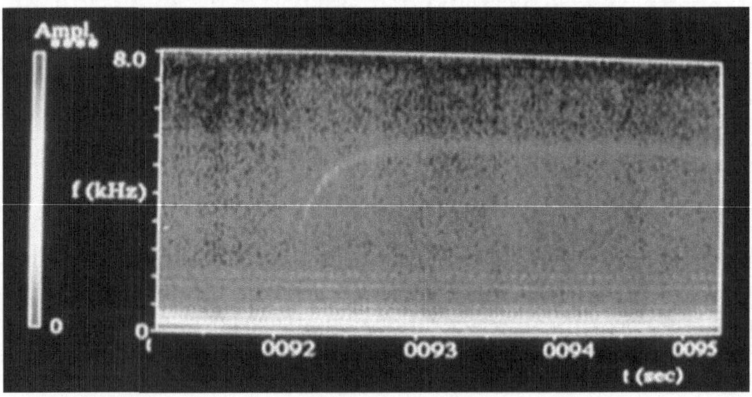

FIGURE 5.1
Spectrogram of risers observed at Maitri, Antarctica on 22 January 1998 at 0931 UT during XVII Indian Antarctica Expedition.

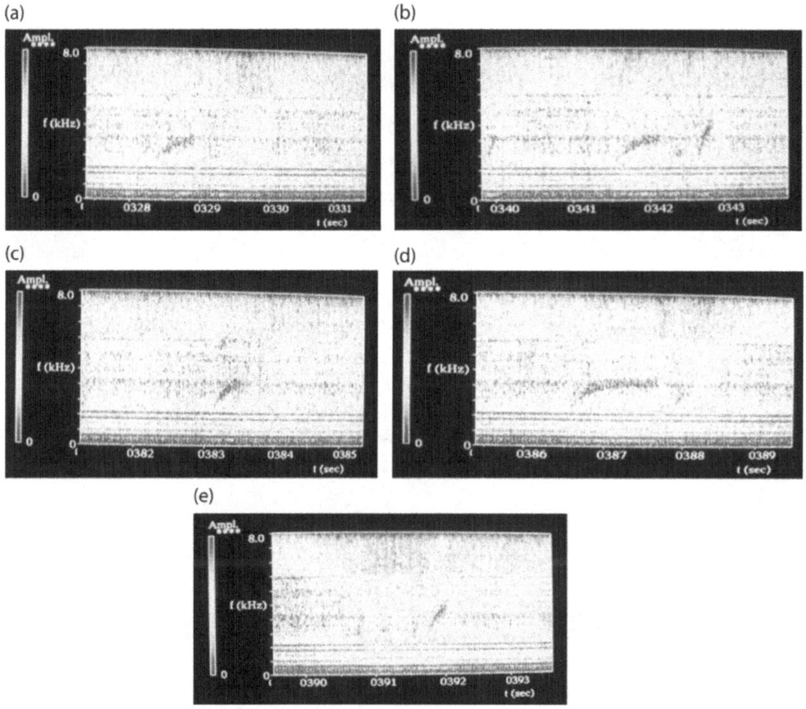

FIGURE 5.2
Spectrograms of chorus risers observed at Maitri, Antarctica on 22 January 1998 at 1035 UT during XVII Indian Antarctic Expedition.

frequency of 5.6 kHz. Figures 5.2a–e show the frequency-time spectrogram of chorus riser emissions observed on 22 January 1998 at 1035 UT. The total duration of these signals is about 70 seconds. The mean upper boundary frequency for these signals is 3.5 kHz. The events shown in Figures 5.1 and 5.2 correspond to a moderate day, with a daily sum of Kp and Ap indices of 17+ and 6, respectively, and 3-hourly values are 0+ for this duration. Morphological features of ground and satellite-based observations of discrete chorus emissions have been investigated by several workers (Hayakawa, 1975; Nunn, 1979; Nunn and Sazhin, 1991; Sazhin and Hayakawa, 1992; Smith and Jenkins, 1998). The studies have revealed that the chorus emissions are observed mainly at L-values between plasmapause and magnetopause at all local times. The Maitri provides a good situation for the observation of chorus emissions.

Figure 5.3 shows discrete rising emissions recorded on January 22, 1998, at 10:50 UT, near a constant frequency of about 2.4 kHz, with different cut-off frequencies. The mean upper boundary frequency for Figure 5.3 is 5.2 kHz. The rate of rise in frequency (df/dt) is also different for these emissions. It can easily be judged from this figure that df/dt is maximum for the first event and minimum for the second case. Similar types of discrete rising emissions were

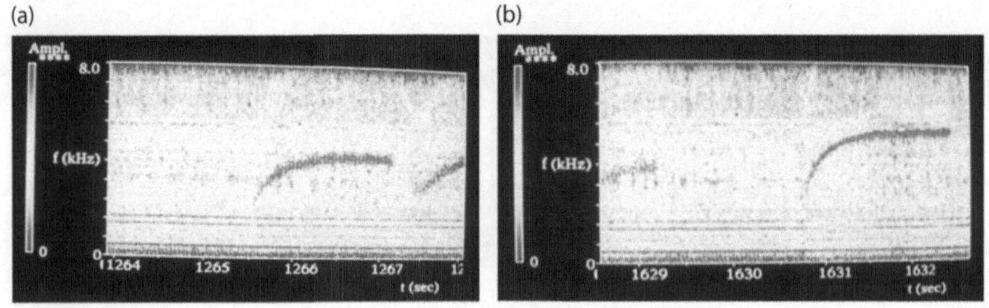

FIGURE 5.3
Spectrograms of discrete rising emissions observed at Maitri, Antarctica on 22 January 1998 at 1050 UT.

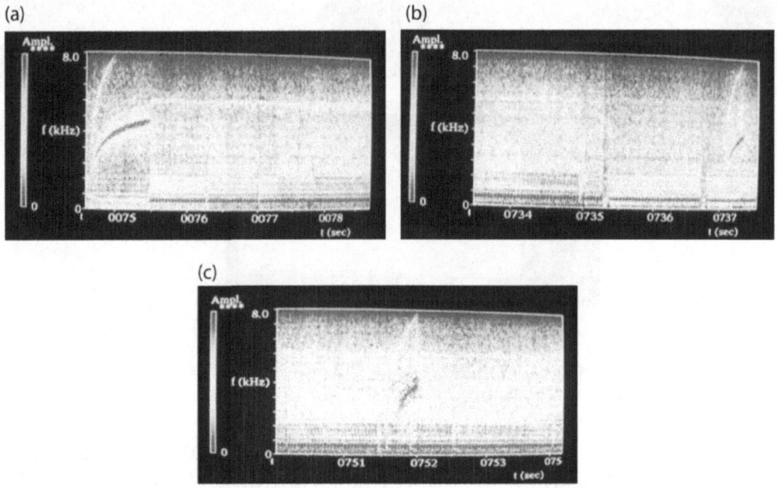

FIGURE 5.4
Spectrograms of risers observed at Maitri, Antarctica on 23 January 1998 at 0515 UT.

observed in the daytime on January 23, 1998, in the morning 0515 UT. Figures 5.4a–c display the spectrograms recorded on January 23, 1998. January 23, 1998 was a quiet day, with sum of Kp and Ap indices of –7 and 4, respectively, and more prone to observe such emissions. Also, the 3-hourly Kp value for this duration is 10. In Figure 5.5, another signal characteristic of rising VLF emission is shown. This rising emission was recorded on the same day, January 23, 1998, at 0536 UT. For this signal, the cut-off frequency was around 4.5 kHz, and the mean upper boundary frequency for this riser is 5.6 kHz.

Very Low Frequency Recording during January 1999 in XVIII Indian Antarctic Expedition

A typical spectrum of the discrete VLF emissions recorded on January 19, 1999, during daytime between 1000 UT and 1230 UT at Maitri are shown

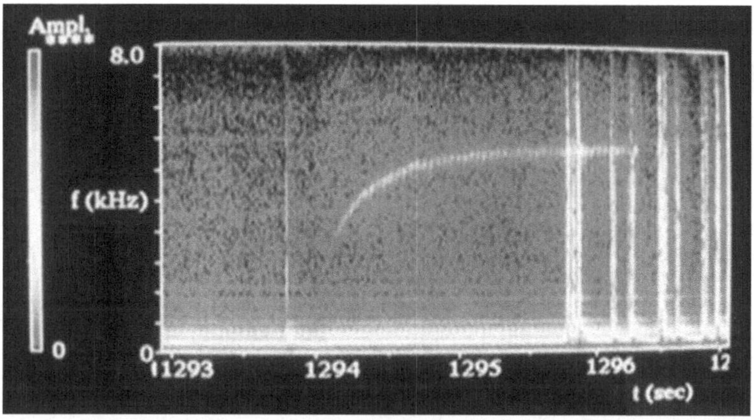

FIGURE 5.5
Spectrogram of riser observed at Maitri, Antarctica on 23 January 1998 at 0536 UT during XVII Indian Antarctica Expedition.

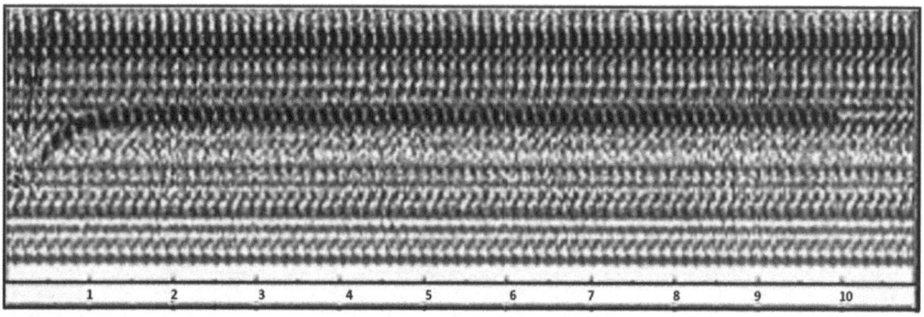

FIGURE 5.6
Spectrogram of discrete rising emissions observed at Maitri, Antarctica on 19 January 1999.

in Figures 5.6 and 5.7. Figures 5.6 and 5.7 contain two chorus risers with a background hiss band (2–4.5 kHz). The mean upper boundary frequency for this chorus is 3.5 kHz. In Figure 5.7, we find two chorus riser emissions (mean upper boundary frequency of 5 kHz). The hiss band has been observed at a frequency range 2–4.5 kHz. These figures show a simultaneous occurrence and unstructured type (hiss). It was quietest day, with 3-hourly Kp values reaching a maximum between 1+ and 0+ at the third and fifth 3-hourly average, and the sums of Kp and Ap were 8− and 4, respectively.

The event shown in Figure 5.8 is periodic emission. On January 20, 1999, the spurt in activity started around 1000 UT and lasted for about 2.5 hours, ending finally at 1200 UT. During these hours, periodic emissions of a non-depressive nature with purely discrete rising tones at about 5 kHz within a frequency band of 0.5 kHz were recorded. The measured period of these emissions is found to be very small, about 0.5 sec. The measured period of all periodic VLF emissions recorded at Maitri is found to lie typically in the range of about 0.1–0.5 sec

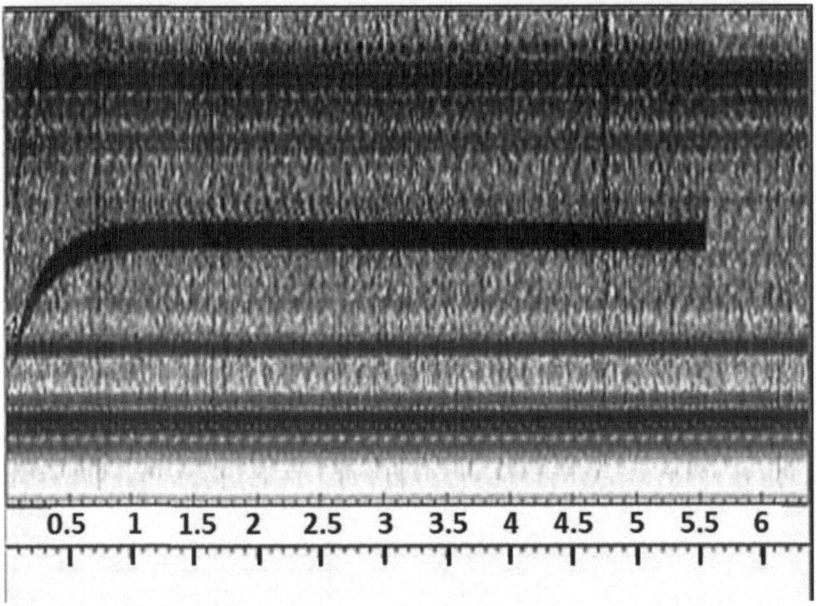

FIGURE 5.7
Spectrogram of discrete rising emissions observed at Maitri, Antarctica on 19 January 1999.

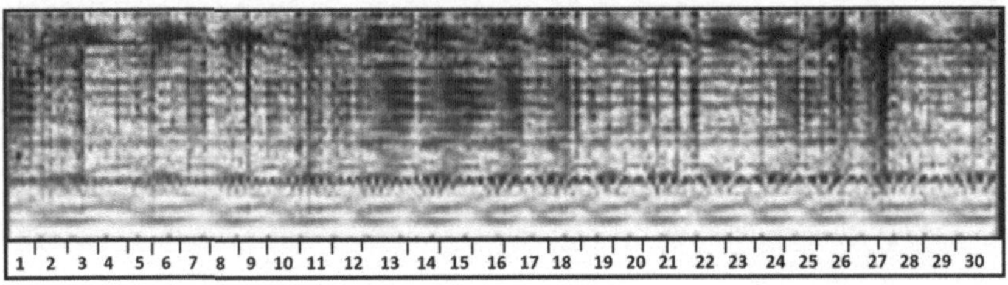

FIGURE 5.8
Spectrogram of periodic rising emissions observed at Maitri, Antarctica on 20 January 1999.

during daytime, which is much smaller in comparison to that reported from higher latitudes (Dowden and Helliwell, 1962; Sazhin and Hayakawa, 1992).

January 20, 1999, was a moderate day, with the 3-hourly Kp value reaching maximum 3o at the sixth 3-hourly average, and the sums of Kp and Ap were 13+ and 8, respectively. The noise below 1.7 kHz is due to instruments, which has been found to occur in our all records. A careful look at the spectrogram reveals the rate of rise in frequency is very high to about 20 kHz/sec (i.e., $df/dt \approx 20$ kHz/sec).

As it rises in frequency, the amplitude starts increasing from 2.7 kHz until this riser is cut off at frequency 4.2 kHz. For some other similar events not shown here, the starting and cut-off frequency and rate of risers of frequency are found to vary from event to event.

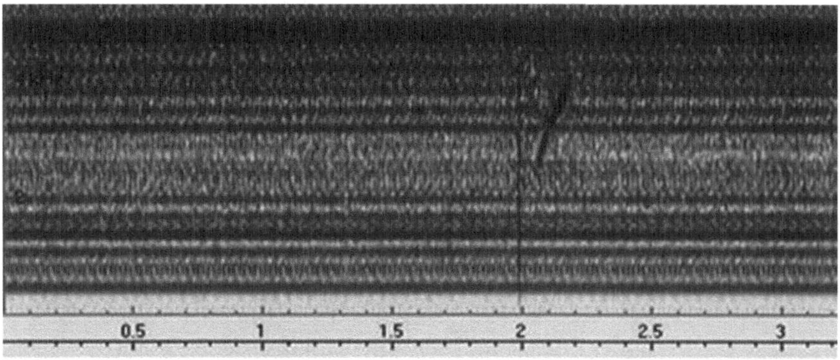

FIGURE 5.9
Spectrogram of rising emissions observed at Maitri, Antarctica on 20 January 1999.

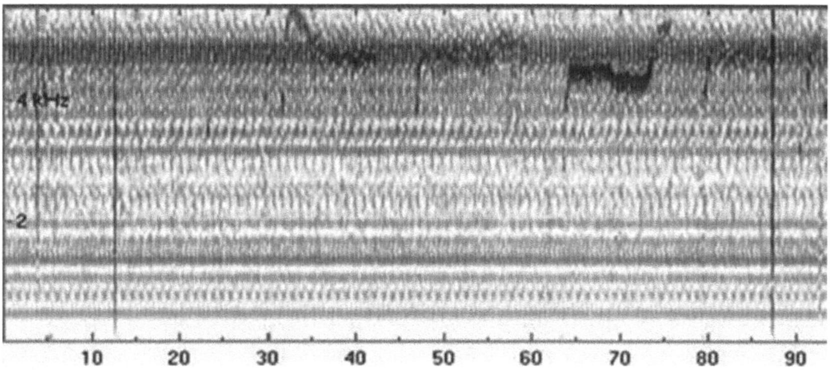

FIGURE 5.10
Spectrogram of typical hook emissions observed at Maitri, Antarctica on 20 January 1999.

The spectra in Figure 5.9 show an unusual rising emission recorded on January 20, 1999, in the frequency range 2.5–4.2 kHz. In Figure 5.10, we present an example of discrete emissions between the frequency range of 4.0–5.5 kHz observed on January 20, 1999 at 1420 UT. An interesting feature of these emissions is that they show hooks near their upper and lower cut-off frequencies. Hook emissions are known to be caused by wave particle interactions between up- and down-going energetic electrons and whistler mode waves along the field lines near the magnetic equator at higher L shells ($L > 2$); hence, they are generally observed at high latitudes. Our experimental observations shown in the figures are non-dispersive periodic VLF emissions. All periodic VLF emissions recorded at Maitri during the daytime are the non-dispersive type, having only rising spectral forms and burst-type forms. The measured period of these emissions lies typically in the range of 0.1–0.5 sec, which is much smaller than that reported from higher latitudes.

Very Low Frequency Recording during January 2000 in XIX Indian Antarctic Expedition

During the XVII and XVIII Indian Antarctic Expeditions in the years 1998 and 1999, we did VLF recording mostly during daytime, that is, from 0500 to 2000 UT. The signals observed in these periods were predominantly discrete chorus emissions, periodic and quasiperiodic emissions, rising and falling tones, hook types, hiss, and so on. We could not find any existence of whistler waves. To check the possibility of occurrence of whistler phenomena, we did VLF recording in post-midnight (2400 UT) and early morning (0500 UT) sessions during the XIX Expedition in the year 2000. In this regard, we received the same type of VLF emissions as the XVII and XVIII expeditions.

In Figure 5.11, the signal recorded at 0200 UT on January 14, 2000, is shown. This is a rising VLF emission with cut-off frequency of 5 kHz. On January 15, 2000, several discrete VLF emissions of the rising type were recorded in the subsequent time intervals. There are displayed in Figures 5.12a–f. these emissions were in the frequency range 3–5 kHz, occasionally rising above 5 kHz, as in Figure 5.12b.

A similar type of rising emission was observed at 0247 UT on January 24, 2000, as shown in Figure 5.13. January 14, 2000, was a moderate day, with 3-hourly Kp values for the duration under consideration of 30, whereas January 15, 2000, was the quiet day.

Again, January 24, 2000, was also a moderate day. A few more rising events were recorded in the early morning hours on January 27 and January 30, 2000. The relevant spectrograms are shown in Figures 5.14 through 5.16, respectively.

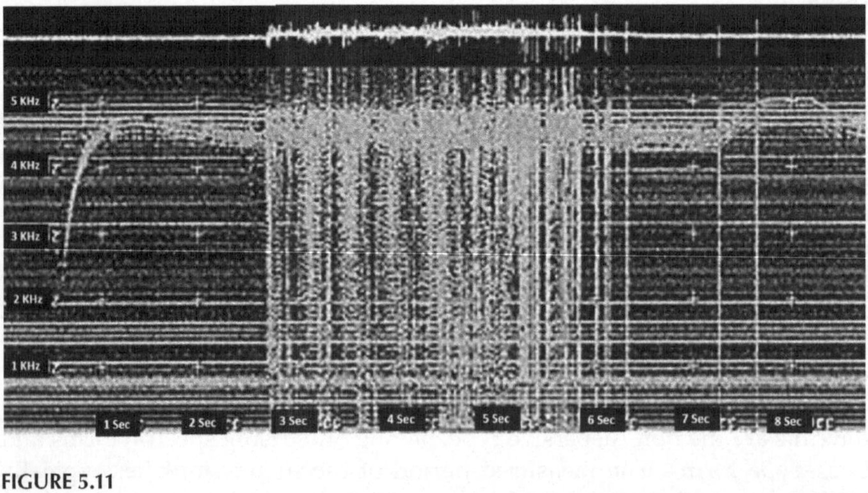

FIGURE 5.11
Spectrogram of rising emissions observed at Maitri, Antarctica on 14 January 2000 at 0200 UT.

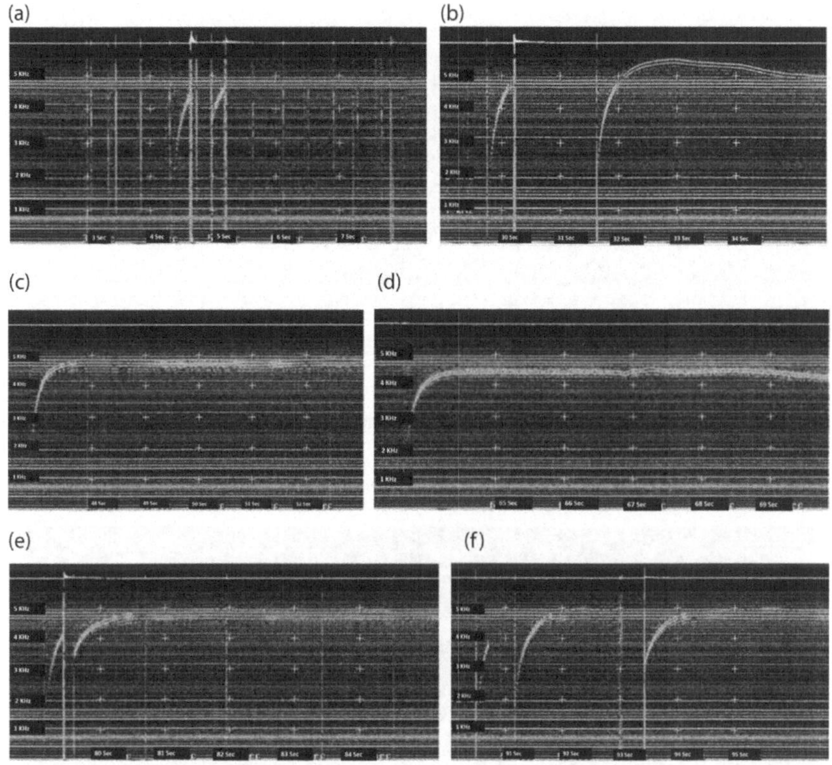

FIGURE 5.12
Spectrograms of discrete VLF emissions observed at Maitri, Antarctica on 15 January 2000 at 0200 UT.

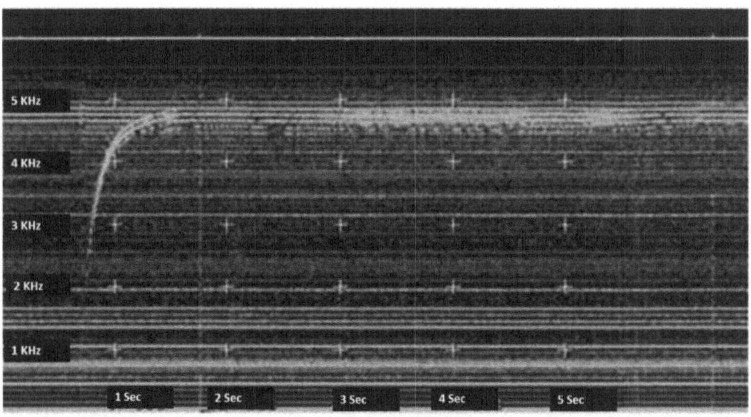

FIGURE 5.13
Spectrogram of rising emissions observed at Maitri, Antarctica on 24 January 2000 at 0247 UT.

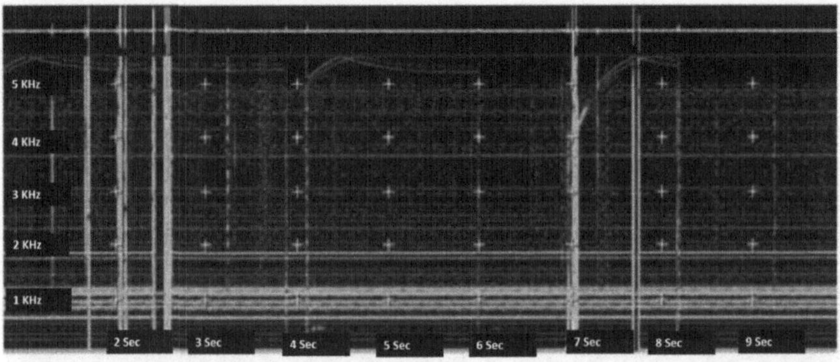

FIGURE 5.14
Spectrogram of riser emissions observed at Maitri, Antarctica on 27 January 2000 at 0400 UT.

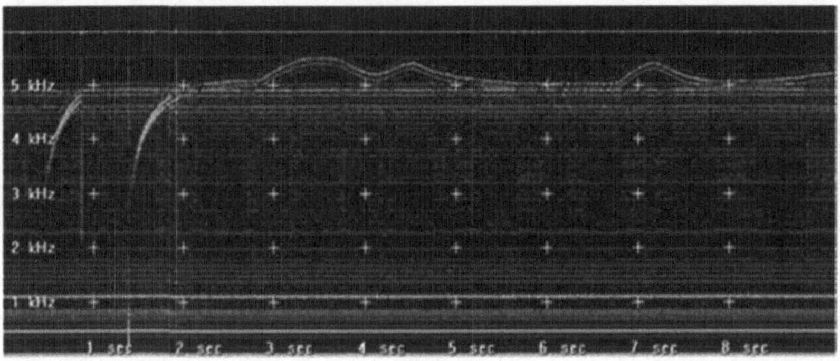

FIGURE 5.15
Spectrogram of riser emissions observed at Maitri, Antarctica on 27 January 2000 at 0420 UT.

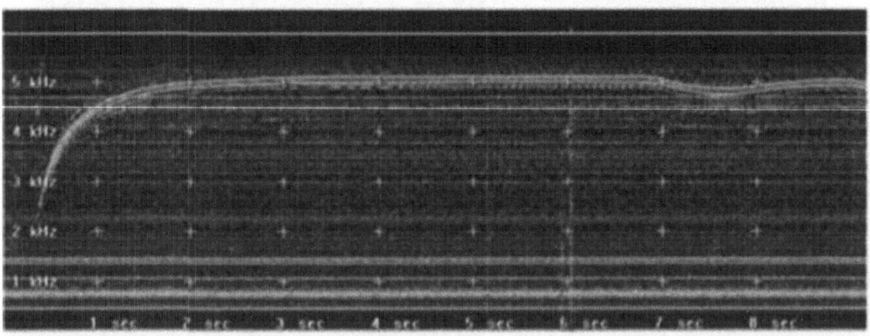

FIGURE 5.16
Spectrogram of riser emissions observed at Maitri, Antarctica on 30 January 2000 at 0400 UT.

Very Low Frequency Recording during February 2001 in XX Indian Antarctic Expedition

A VLF wave recording setup consisting of a T-type (vertical) antenna of 10 m height and 40 m horizontal length supported by two poles, a transistor amplifier, and a digital audio tape recorder is used to record VLF data at Maitri. The location of the hut containing the recording setup is geom. lat. 62°S, long. 57.2°E. The data are stored on digital audio tapes. Tapes are replayed and some of the events are analyzed. During the analysis, it was found that the noise in some cases was very high. The data with low noise levels were selected for further analysis. Figure 5.17 shows VLF hiss at relatively higher frequencies (11 kHz $<$ f $<$ 13 kHz) observed on February 17, 2001, at 11.30:35 h UT. From the figure, it can be seen that the noise level is very low. In addition to VLF hiss, three horizontal lines at 6.2, 8.0 and 9.2 kHz are seen, which were present throughout the recording period for 2 h and decrease in intensity as the increase in frequency is observed. That is, the intensity of the 6.2 kHz line is greater than the that of the other two lines. These lines are quite parallel, showing a constant frequency on the spectrogram. Similar lines on the dynamic spectra in the frequency range 2.0–5.0 kHz were reported from the ground-based stations Siple (Antarctica) and Roberval (Quebec). These lines were termed magnetospheric lines. It is reported that magnetospheric lines in the frequency range 1.5–8.5 kHz with a strong peak at 2.5–3.5 kHz. Recently, Rodger et al. (1995), analyzing ISIS-2 satellite data, have reported magnetospheric lines in the frequency range 1.8–2.3 kHz. The hiss events presented in Figure 5.17 continued for 2 h

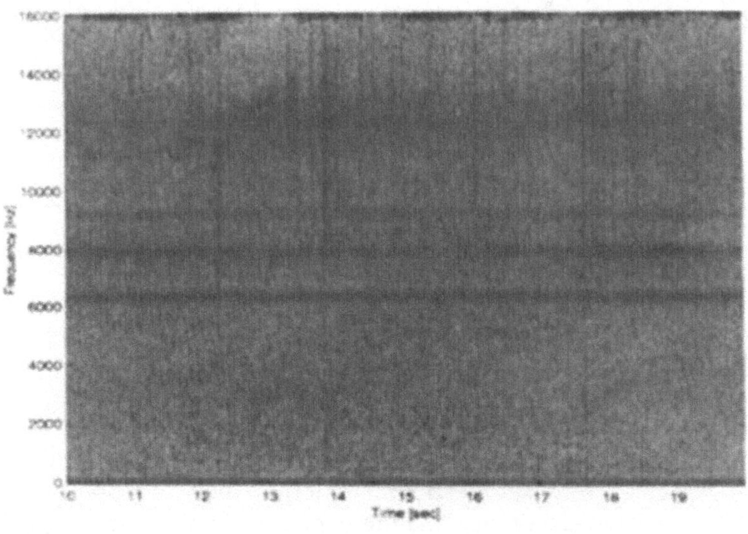

FIGURE 5.17
VLF hiss recorded at Indian Antarctic station, Maitri on 17 February 2001 at 11.30:35 UT.

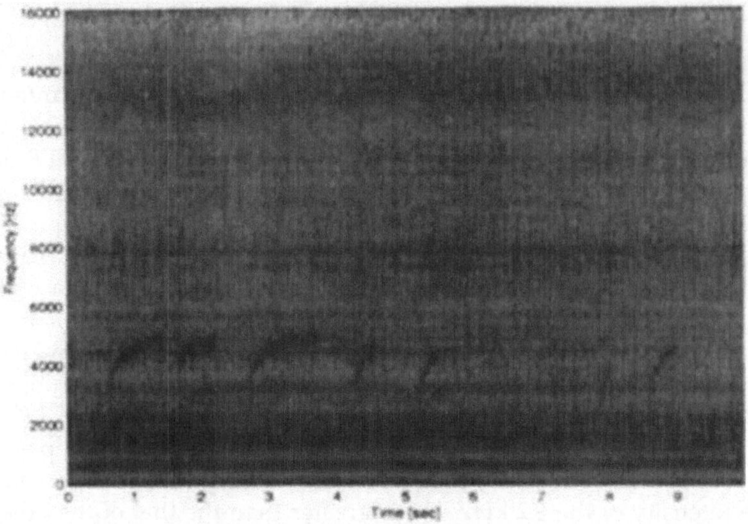

FIGURE 5.18
Example of VLF risers (3 kHz < f < 5 kHz) and VLF hiss (11 kHz < f < 14.3 kHz) recorded at Indian Antarctic station, Maitri on 3 February 2001 at 12.46:50 U T.

(11–13 h GMT), with almost the same intensity as inferred from the color of the spectrogram. Figure 5.18 shows risers in the frequency range 3–5 kHz and hiss in the frequency band 13–14.3 kHz recorded on February 3, 2001 at 12.46:50 h GMT. The possibility that these emissions were caused by the instrument has been ruled out. The hiss emission continued for more than 1 h, and risers in enough numbers were observed for about 1 h. As frequency increases, df/dt increases. It is noted that the risers originated from a narrow hiss band (3.5 kHz < f < 3.75 kHz). From the spectrograph, we also note that there is strong noise at lower frequencies (f < 2.2 kHz).

Very Low Frequency Recording during January–March 2003 in XXII Indian Antarctic Expedition

In the XXII Indian Scientific Expedition to Antarctica, scientific study have been carried out at very low frequency (VLF) to extremely low frequency (ELF) recording with triangular loop array antenna and T-type antenna to deduce the magnetospheric density from the dispersion values and to determine the emerging region of whistlers from the ionosphere. Contribution due to solar particle fluxes at high latitudes becomes dominant for VLF/ELF discrete emissions. The discrete auroral structures can be generated by wave particle interactions between up- and down-going energetic electrons and magnetospheric electric current generally observed at high latitudes. Discrete inverted-V-shaped emissions show a close association with auroral precipitation, and these emissions are

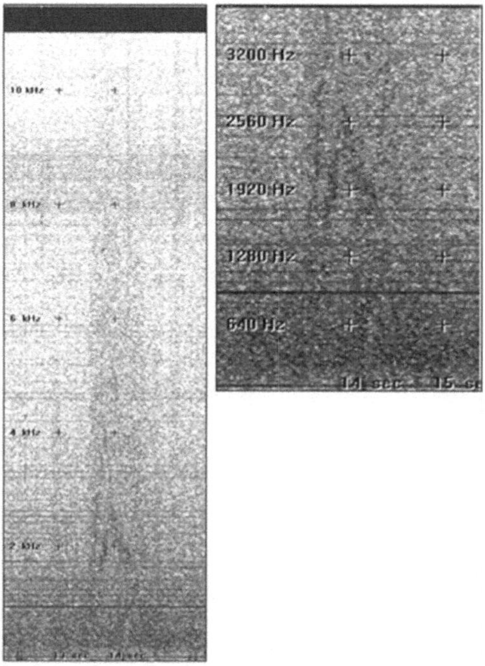

FIGURE 5.19
Inverted-V-shaped VLF/ELF emissions recorded at Maitri during XXII Indian Antarctic Expedition on 29 January, 2003, 2030 UT.

observed at the ground station. The inverted-V-shaped VLF/ELF emissions (Figure 5.15) are observed in an average frequency range of below 1000 Hz to 8 kHz. In Figure 5.19 spectrogram shown inverted-V structure appears with higher density in the ELF region and disappears in the VLF band. The electron flux, energy flux, and number density are greater inside the inverted-V structure. The complexity of their distribution suggests that the field-aligned potential drop varies temporally and/or spatially. We have observed this type of emission in moderately disturbed days and disturbed conditions, as auroral oval expansion does not take place one quiet days. The ubiquitous presence of these intense electron precipitation phenomena (such as inverted-V waves) in the auroral plasma suggests that these waves may play an important role in maintaining the auroral potential structure and in overall energy transfer processes throughout the region. The possibility of observing whistlers at Maitri during this period is rare, perhaps due to daytime ionospheric condition, but during the month of March 2003, we recorded few whistler traces associated with causative atmospherics and tweeks (Figures 5.20 and 5.21). Being daytime, it is very difficult to distinguish tweeks and whistlers, as observed dispersion of whistlers was low during this month.

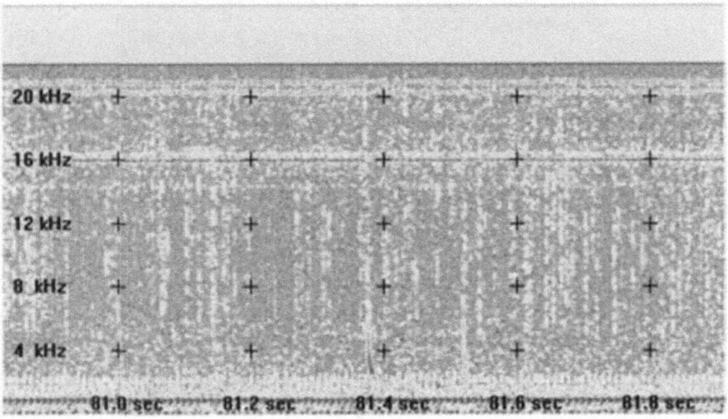

FIGURE 5.20
Tweek atmospherics recorded at Maitri during the XXII Indian Antarctic Expedition on 2 March 2003 at 2200 UT.

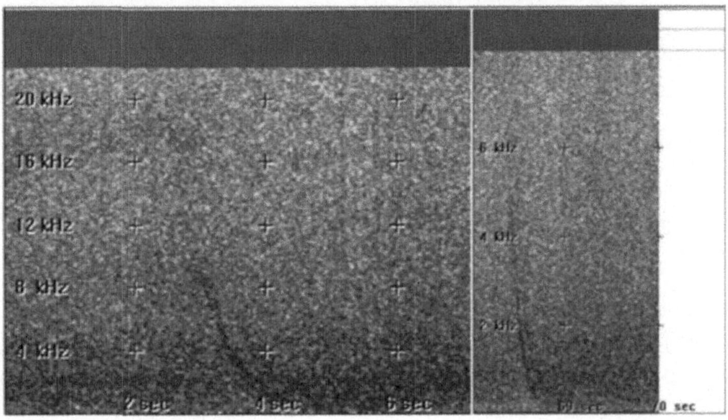

FIGURE 5.21
Whistler trace recorded at Maitri during the XXII Indian Antarctic Expedition on 2 March 2003 at 1930 and 2200 UT.

Diurnal Variation of Occurrence Number

In addition to the spectrograms displayed in this chapter, several other events were also recorded. Every year, the Indian Antarctic Expedition team used to land on Maitri Station in the first week of January. Therefore, the recording was normally started around January 10. It is seen from these daily variations that the emission activity is not the same every year.

National Status

A considerable amount of work has been done on VLF phenomena, particularly on the whistlers in the country. Under the All India Coordinated Programme of

Ionospheric and Thermospheric Studies (AICPITS) of the Department of Science & Technology, Government of India–coordinated simultaneous observations of whistlers have been carried out at Varanasi, Agra, Bhopal, and Jammu. Coordinated whistler analysis was being worked out with good results. Whistler data have been analyzed to determine electron density and radial electric fields in the plasmasphere. But no attempt has been made to determine the propagation path of whistlers using direction finding techniques. A direction finding using a goniometer technique was first used at high latitude, which determined the arrival direction of high-latitude whistlers. Unfortunately, this technique could not successfully yield the arrival direction of those whistlers whose elevation angles were high. In such cases, they verified the propagation path from noise characteristics of the whistlers.

Direction-finding techniques that were modifications of the goniometer found great application at low latitudes. Japanese workers employed this technique to determine the occurrence characteristics of low-latitude whistlers from the knowledge of their propagation path. They also determined the variation in dispersion characteristics as a function of latitude and longitude. From the real-time analysis of the data, they were able to deduce the duct structure and propagation mechanism of low-latitude whistlers.

Some ongoing experiments in the field of whistlers and related phenomena include the ground-based monitoring of signals, and also collaborative work on lightning-induced precipitation, trimpi effect, and so on has been done by South African scientists. Since India and South Africa have VLF experiments at their respective stations, Maitri and Vesleskarvet of Antarctica, this provides an ample opportunity to work on VLF phenomena from low to high latitudes. Longitudinal variations in the occurrence of VLF whistlers can also be studied.

International Status

India and South Africa operate bases in Antarctica (Maitri and Sanae) separated by about 400 km. Both institutes (the Space Physics Research Institute of the University of Natal and Space Science Laboratory, Barkatullah University) operate VLF direction finders at these bases. The 400-km separation represents an ideal baseline for direction-finding studies. The VLF field of view of each station is about 1500 km, so many events will be simultaneously recorded at both stations, and yet the angle of arrival of signals will be sufficiently different to give an accurate determination of the signal exit points from the ionosphere. The joint experiences of the two groups in this field should lead to fruitful collaboration. The study of electromagnetic pollution is already being studied at low latitude in collaboration between Hungary and South Africa, and the question arises as to whether there is significant electromagnetic pollution in the Antarctic transferred by way of the magnetosphere from the industrialized northern hemisphere. The work at Maitri and Sanae will complement the Hungarian study.

Overall Contributions and Significant Achievements

The design and constructional details of the VLF wave recorder used at our high-latitude station, Maitri, Antarctica and low-latitude station, Bhopal, are given here. The setup for VLF wave recording consists of a T-type vertical antenna, E-W and N-S direction-finding antennas (loop-type antenna), pre- and main amplifiers, and digital and magnetic tape recorders. Digital recording and analysis of whistler waves are carried out by the spectrogram version 4.1.2, Advanced Version of VLF Data Analysis System (AVDAS) and contextual long-short term memory (CSLTM) computerized system available at Space Science Laboratory, Department of Electronics, Barkatullah University, Bhopal; Radio and Atmospheric Science Laboratory, Department of Physics, Banaras Hindu University, Varanasi and Speech of Computer System Group, Central of Electronics Engineering Research Institute Centre (CEERI), New Delhi, respectively.

Antarctica is an ideal place for monitoring whistler and VLF phenomena. Earlier work in the related areas of research reveals that various transient discrete emissions such as risers, fallers, and hooks have been reported at Antarctica. At Maitri, we have observed such phenomena frequently and discuss the daily variation of occurrences. Certain types of VLF emissions can be correlated with auroral phenomena. The analysis of discrete VLF emissions observed during daytime at Maitri Station, Antarctica, reveals several facts. Discrete chorus emissions have been observed for the first time during daytime at Maitri. The computed value of df/dt is found to be 2 kHz/sec, which is within the range of observed value as reported by other workers. From detailed spectral analyses, it is found that the periodic emissions are observed in the frequency range of 2.5–6.8 kHz. The measured period of the periodic VLF emissions recorded at Maitri lies typically in the range of 0.1–0.5 sec. These periodic emissions observed during daytime show a non-dispersive character. There were typically two different types of periodic VLF signatures observed: purely "discrete rising tones" within a frequency range of 2 kHz and "burst-type" periodic emissions with the upper and lower bands of activities. It is thus observed that all the emission parameters are found to lie almost within the range of our observed parameters of discrete VLF emission at Maitri.

Most Significant Achievements

The study of very low frequency phenomena has been significantly attributed to the communication processes. The low frequency electromagnetic wave phenomena has opened up new vistas in understanding the occurrences of earthquakes. Since the Indian region is prone to earthquakes, this study will be beneficial not only for the communication and propagation of electromagnetic waves but will provide deterministic methods for the prediction of earthquakes.

Future Planning

The region above the ionosphere is not easily accessible by the conventional methods, link rockets, balloons, and so on. However, this part can be very well explored by the whistler technique. Therefore, the analysis of whistlers recorded simultaneously at multiple stations will give important clues about electron density, ionization flux, and electron field variations with altitude and about the control of equatorial anomalies on the propagation mode of low-latitude whistlers. In ongoing project, no efforts have been made so far to measure the polarization and arrival direction of whistlers. Efforts can be made to measure these parameters and to explain certain unsolved aspects of whistler propagation mechanisms such as their ducted and non-ducted propagation, duct structures.

Conclusions

The analysis of discrete VLF emissions observed during daytime at Maitri reveals the following details:

i. Discrete chorus emissions have been observed for the first time during daytime at Maitri. The computed value of df/dt is found to be 2 kHz/sec, which is within the range of observed value as reported by other workers.

ii. From detailed spectral analysis, it is found that periodic emissions are observed in the frequency range of 2.5–6.8 kHz. The measured period of the periodic VLF emissions recorded at Maitri lies typically in the range of 0.1–0.5 sec.

iii. These periodic emissions observed during daytime show a non-dispersive character. There were typically two different types of periodic VLF signatures observed: purely "discrete rising tones" within a frequency range of 2 kHz and "burst-type" periodic emissions with upper and lower bands of activities.

iv. It is thus observed that all emission parameters are found to lie almost within the range of our observed parameters of discrete VLF emission at Maitri.

v. VLF hiss, magnetospheric lines, and riser-type emissions are reported at Maitri. The waves from the source region may have propagated in ducted mode or pro-longitudinal mode along the field line.

A detailed study of periodic and discrete chorus VLF emissions observed at the high-latitude Indian ground station Maitri, Antarctica, is important from the point of view of understanding the various processes in space plasma. Based on experimental studies at Maitri, it is concluded that further detailed experimental and theoretical studies are needed to arrive at a clear phenomenological and theoretical model for the generation and propagation of daytime VLF periodic emissions.

Acknowledgments

The authors wish to thank NCAOR/Ministry of Earth Sciences, GoI, New Delhi, for funding and providing logistic support during IAEO.

References

Allcock, G. McK and Martin, L. H. 1956. *Nature*, v. 178, p. 937.
Dowden, R. L. 1961. *J. Geophys. Res.*, v. 66, p. 1587.
Dowden, R. L. and Helliwell, R. A. 1962. *Nature*, v. 195, p. 64.
Ellis, G. R. A. 1959. *Planet. Space Sci.*, v. 1, p. 253.
Hayakawa, M. 1975. *J. Atmos. Terr. Phys.*, v. 37, p. 1211.
Helliwell, R. A. 1965. *Whistlers and Related Ionospheric Phenomenon*, Stanford California, Stanford Univ. Press.
Helliwell, R. A. 1968. Physics of the magnetosphere. *Conference Proceeding* ASSL v. 10, pp. 42–49.
Morrison, K., Engebretson, M. J., Beak, J. R., Johson, J. E., Arnold, R. L., Cahili, Jr. L. J. and Carpenter, D. L. 1994. *Ann. Geophys.*, v. 12, p. 139.
Nunn, D. 1979. *Planet. Space Sci.*, v. 22, p. 349.
Nunn, D. and Sazhin, S. S. 1991. *Ann. Geophys.*, v. 9, p. 603.
Nunn, D. and Smith A. J. 1996. *J. Geophys. Res.*, v. 101, p. 5261.
Pope, J. 1959. *H. Sci.* Rep. No. 4, Univ. of Alaska College, Alaska.
Rodger, C. J. Thomson, N. R. and Dowden, R. L. 1995. *J. Geophys. Res.*, v. 100, p. 5681.
Sazhin, S. S., Bognar, P., Smith, A. J. and Tarcsai, G. 1993. *Ann. Geophys.*, v. 11, p. 619.
Sazhin, S. S. and Hayakawa, M. J. 1994. *Atmos. Terr. Phys.*, v. 56, p. 735.
Sazhin, S. S. and Hayakawa, M. 1992. *Planet. Space Sci.*, v. 40, p. 681.
Smith, A. J. 1995. *J. Atoms. Terr. Phys.*, v. 57, p. 507.
Smith, A. J. and Jenkins, P. J. 1998. *J. Atoms. Terr. Phys.*, v. 60, p. 263.
Smith, A. J. and Nunn, D. 1998. *J. Geophys. Res.*, v. 103, p. 6771.

6

Atmospheric Studies over Antarctica during Last Two Decades: An Overview

S. L. Jain, H. N. Dutta, R. Singh, B. C. Arya, S. D. Ghude, P. S. Kulkarni, S. Nath, and R. S. Tanwar

CONTENTS

Introduction

Antarctic Environment and National Physical Laboratory Atmospheric Studies

Antarctica, with its area of about 14.5 million square km, is the continent where the natural greenhouse effect is the least, as water vapor and aerosols are minimum. The low aerosol content is basically since there is no indigenous population and thus no local manmade sources of pollutants over this continent. It holds a pristine environment with an unimaginable transparency, permitting infrared energy emitted from ground to escape through the atmosphere. At the same time, over the past 200 million years, this continent has acquired a peculiar shape that is essential for its growth and survival. In the interior, ice with a thickness of 4 km from the mean sea level is at about 2000 km from the periphery. Due to radiative cooling, the icy surface becomes cooler, and the air in contact with the surface also becomes cooler than the air above it. Scientifically, we call this the formation of inversion (Philpot and Zillman, 1970; Schwerdtfeger, 1970, 1984; Parish and Waight, 1987; Parish et al., 1993; Naithani et al., 1995; Naithani and Dutta, 1995; Parish and Bromwich 1998). Thus, in the interior of the continent, the cooler air mass close to the surface, being slightly heavier than the air aloft, starts rolling down due to gravity along the icy surface of the continent, picking up a high speed to form katabatic winds. Katabatic winds are thus always directed from the interior of the continent toward the periphery (always having a southerly downwind direction) and are the most celebrated feature of the Antarctic continent. Katabatic winds can reach speeds as high as 300 km/hr and have been studied at various sites over the continent (Ball, 1957; Parish and Waight, 1987; Naithani et al., 1995; Parish and Bromwich, 1998). It is katabatic winds that transport cool airmass from the interior of the continent toward the periphery, which triggers the formation of cyclones over the ocean all around Antarctica (Van Loon, 1966, 1967, 1984; Streten, 1980;

Schwerdtfeger, 1984; Van Loon and Roger 1984; Singh et al., 1986; Naithani et al., 1995). The cyclones, in turn, transport back loss of cold air toward the interior and break pack ice over the ocean. At the same time, cyclones also mix the oxygen in the water for the survival, production, and reproduction of basic organisms (phytoplankton and the like) and throw the dirt of the ocean away from the Antarctic continent, permitting deep penetration of sunlight for the production and reproduction of phytoplankton (Parulekar et al., 1983). The cyclones move in the clockwise direction, transporting moisture from the ocean and depositing it in the form of snow to the deep interior for future use. In fact, the thickness of Antarctica is increasing by 4–5 cm/year.

At the same time, cyclones churn the ocean so that any biological life does not get a long exposure to ultra violet-B (UV-B), which may be higher in intensity due to clear skies. Also, during summer, the melted glacier water, being colder than the oceanic water, runs deeper in the ocean, throwing relatively warm oceanic water toward the surface, causing deep mixing of water and the upward movement of basic food material to attract migratory birds and mammals.

In Antarctica, contamination sources are distant, and the air is very well mixed, extremely clean, and far removed from spatial and seasonal changes. It is truly the background air of the planet compared to measurements made elsewhere around the globe. Antarctica holds the most efficiently coupled radiation-ice-air-ocean interactive system, which sustains primitive life forms over the continent. However, despite its sustenance, the Antarctic environment is influenced by global changes, which need to be evaluated to study the future of Earth's environment. Therefore, the special conditions prevailing over Antarctica make it an ideal site for the monitoring of various atmospheric parameters. It was the foresight of our honorable former prime minister the late Shrimati Indira Gandhi who conceived the idea of launching an Indian Scientific Expedition to Antarctica in 1981 for polar investigations into areas such as atmospheric, biological, medical, Earth, engineering, and communication sciences. Since then, 25 Indian Scientific Expeditions to Antarctica have been completed and various scientific experiments and investigations have been carried out by many institutions spread all over the country.

The National Physical Laboratory, New Delhi, has been participating in the Antarctic Expeditions right from their inception in 1981 and is carrying out scientific investigations related to the Antarctic atmospheric environment. With the passage of time, it has developed collaborations with other Scientific Committee on Antarctic Research (SCAR) countries, sharing the knowledge and expertise that have been gained while working at the Indian scientific stations Dakshin Ganagotri and Maitri, Antarctica. This has led the group to develop effective collaboration with the Italian, Russian, and French groups working in the Terra Nova and Schirmacher regions, Antarctica. A large variety of experiments have been undertaken during the period, which mostly center around radio and atmospheric physics. We at the National Physical Laboratory, New Delhi, are working in atmospheric sciences, specifically the planetary boundary layer (PBL), ozonosphere, ionosphere, and greenhouse gas measurements. The major experiments set up by the

National Physical Laboratory (NPL) over the years at Maitri include very low frequency (VLF); riometer; microbarograph; weather station to monitor meteorological parameters; 28-meter-high tower to install sensors for meteorological parameter measurement at different heights; collection of air samples for gaseous concentration; high-volume air sampler; sound detection and ranging (SODAR) for planetary boundary layer studies; laser heterodyne and mm wave radio spectrometer to get ozone profiles; liquid nitrogen plant to produce liquid nitrogen needed for the laser heterodyne system; sun photometer; UV biometer; MICROTOP to estimate column ozone, water vapor, aerosol optical depth, and so on; gas chromatogram for regular monitoring of greenhouse gases such as CO_2, CH_4, and so on; and NonDispersive InfraRed (NDIR) CO analyzer for around-the-clock monitoring of CO. In the present chapter, the salient features of various experiments conducted by the NPL and results obtained are discussed in detail.

The Antarctic Ocean is the most efficiently coupled ice-air-ocean system, which has evolved biologically to be the most productive ocean around Antarctica (Parulekar et al., 1983). Therefore, the natural question is how to understand various parameters with the dynamics of the ice-air-ocean system of Antarctica and to predict what the role or fate of Antarctica will be in the next century facing global change. This will also answer our quest of restructuring the Antarctic program in such a way that it attracts and benefits most of the Antarctic laboratories. At the same time, it may also pose a challenge to apply introspection and mold some of our ongoing research and development (R&D) programs specifically to understand this mighty interactive system of Antarctica and to derive scientific and technological advantages.

The atmosphere of the southern hemisphere high latitudes has undergone pronounced changes over the past few decades. At the surface, the Antarctic region from the west coast of the Antarctic peninsula to Novolazarevskaya (065° W–011° E) has warmed by several degrees Kelvin over the past several decades. Total column ozone losses have exceeded 50% during October throughout the 1990s. Greenhouse gases are also found to be increasing in the Antarctic atmosphere, which may cause exceptional regional effect over the Antarctic continent. In short, since the Industrial Revolution, Antarctica has undergone pronounced changes, which in turn may alter the global climate exceptionally. The greenhouse gas concentrations predicted by the mathematical models are subject to large uncertainties in the effects of both natural processes and human activities. For this reason, a better understanding of their trend is crucial for the prediction of future concentration levels and atmospheric global warming and climatic changes. The most important truth of this icy continent is that over the past 200 million years, it has withstood and maintained a unique status of being the coldest, driest, windiest, and most difficult to approach and thus holds a pristine environment. Naturally, one is curious to know how this continent maintains the cold in an open storehouse of the world's largest piece of ice. The other related query is to know precisely the role or fate of Antarctica in controlling or affecting the global change with which mankind is concerned today. What are the key parameters in atmosphere, ice, ocean, and flora and fauna that must be monitored on a long-term basis to scientifically assess and predict global change? At the same time, it is important to note that initially, the magnitude of changes will have to be assessed through special computing techniques on long time-series field data (Woodward and Gray, 1993). This shows that only those countries that possess and study the long time-series data will have the ability to weigh in on the changing global scenario. In view of this, over the past 20 years, scientific programs from NPL, namely planetary boundary layer, trace gas measurements, ozonosphere, ionosphere, and UV-B radiation, have been undertaken on a long-term basis. A review of the results and conclusion of the work is given below.

Planetary Boundary Layer Program

Over the past 20 years, two programs from NPL, namely planetary boundary layer and UV-B radiation, were undertaken on a long-term basis. The planetary boundary layer program started in the year 1988, when NPL installed a 9-m instrumented tower at the Indian Antarctic station, Maitri, to compute heat and momentum fluxes (Pasricha et al., 1991). Later, a monostatic acoustic sounder and 28-m tower with four levels of meteorological sensors were installed at the Maitri site (Naithani and Dutta, 1995; Naithani et al., 1995). These instruments were supplemented with an automatic weather station, a three-axis sonic anemometer, UV-B measurements, and solar radiation measurements.

Probing the planetary boundary layer over Antarctica using modern techniques started with the International geophysical year (IGY), and in the last few decades, a variety of instruments have been deployed over this icy continent to study the short and long-term trends of meteorological parameters (Streten, 1980; Schwerdtfeger, 1984; Pasricha et al., 1991; Naithani et al., 1994, 1995; Naithani and Dutta, 1995). By now, we have installed both direct and remote sensing instruments at the Maitri station essential to understand atmospheric boundary layer (ABL) dynamics and to forecast weather to assist the day-to-day operations of convoys, helicopter services, field work, and so on. The instrumentation consists of a PC-based monostatic acoustic sounder, a 28-m instrumented tower, an automatic weather station, and a solar radiation measurement system. Figure 6.1 shows a block diagram of monostatic acoustic sound. The surface pressure charts received from Pretoria, South Africa, and data downloaded from various Internet sites supplement these measurements apart from the above India Meteorological Department, New Delhi, recording various meteorological parameters at Maitri.

Tower Based In-Situ Measurements

The 28-m tower was installed in the year 1991–92, while the sensors on it were mounted only in the year 1992–93. The sensors were installed at four levels measuring 1.8, 4.5, 11.3, and 28 m from the ground, respectively. The different heights were chosen so that the spacing between different level

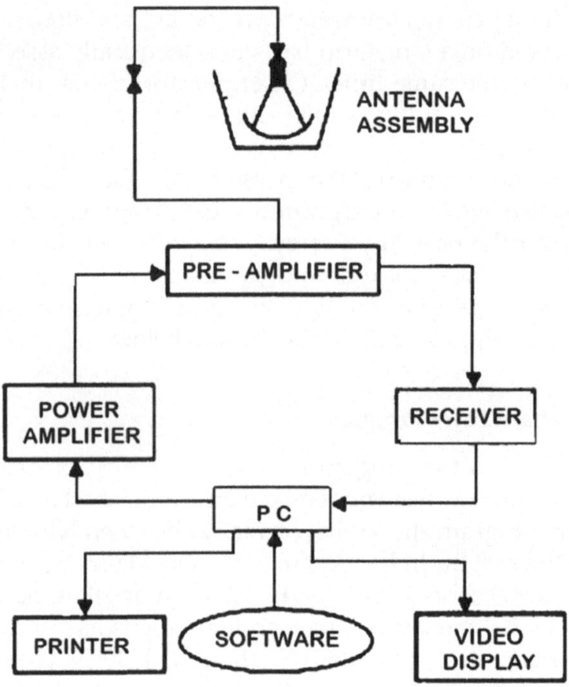

FIGURE 6.1
Block diagram of monostatic acoustic sounder.

sensors is nearly logarithmic. This is because the profiles of ABL parameters are height dependent and tend to decrease more steeply near the ground due to frictional drag, surface heating, evaporation, and so on. Above the surface layer, the spacing is linear because in this region turbulence scales with the ABL height and not with the height of measurement. The sensors could not be mounted exactly with logarithmic spacing due to structural problems in the tower. The meteorological sensors were of a slow response type and were mounted on 1-m booms. The air temperature and humidity were measured using a platinum resistance thermometer (PT 1000) and a sensitive humicap sensor, respectively. Wind speed and direction were measured using a cup anemometer and a sensitive rotating arm wind vane, respectively. The India Meteorological Department, Pune, calibrated all these sensors. A pyranometer for the measurement of solar insolation was mounted on a separate mast. An automatic weather station was placed inside the NPL hut with sensors mounted on a 3-m tower facing southeast.

Solar Ultra Violet-B (UV-B) Studies Program

Solar ultraviolet radiation studies have gained importance since this radiation has strong biological effects and its passage through the atmosphere is mainly dependent on the atmospheric ozone content, which was found to be changing very fast in some regions, particularly in the polar regions. The discovery of the ozone hole in Antarctica became a global concern, and steps were initiated to study the problem in detail. The World Meteorological Organization (WMO), after serious deliberations, recommended during the late 70s that a good network of ground-based measurements of solar UV-B radiation be established to monitor the magnitude of this radiation and its short- and long-term trends on a regional and global basis. Realizing the importance of these recommendations, the National Physical Laboratory, New Delhi, started regular measurements of UV-B radiation as early as 1979, and the program is continuing today. In continuation of this mainland program and keeping in view the sensitivity of the Antarctic ozone hole problem, the NPL took the first Indian initiative and extended the UV-B observation program at the Indian Antarctic station in 1986.

As soon as the role of atmospheric aerosols was established in affecting the radiative forcing of climate change, global aerosol climatology was undertaken as an international scientific venture, and the National Physical Laboratory, as a first Indian initiative, started the aerosol optical depth measurement program over Antarctica.

UV-B measurements by the NPL over Antarctica started with NPL UV-B filter photometers (Srivastava and Sharma, 1979). It is a narrow-band interference filter-based photometer operating at the 280, 290, 300, and 310-nm wavelengths. In addition to filter photometers, the NPL also procured (i) a monochromator-based spectroradiometer that operates in the 250 and 700-nm range with 1, 2, 5, and 10-nm resolution options; (ii) an erythemal probe attached to the UV spectroradiometer described above that measures the erythemal dose in absolute energy units; and (iii) a UV-B biometer that measures the total radiation in the UV-B region that is, 280–320 nm. The final output of the biometer is in the form of minimum erythemal dose per hour (med/hr). This is a measure of skin harmfulness of UV-B radiation. A single med/hr is defined as the energy equal to 5.83×10^{-6} w/cm^2 exposure for 1 hour and equals 21 m joule/cm^2.

Ultraviolet-Biometer

As UV-B radiation plays important role in the biotic components of the ecosystem, we focus our discussion mainly on living organisms. There are six biological action spectra: the erythema action spectrum, DNA to protein cross-links, DNA breaks, the polychromatic action spectrum for higher plants, phytoplankton photo-inhibition, and typhimurium killing. In our study, we consider only the erythema action spectrum (Morys and Berger, 1993).

Atmospheric Aerosol Study Program

The aerosol optical depth (AOD) measurement program over Antarctica by the NPL was started in 1986 with an interference filter-based sun photometer.

The instrument operates at the 368, 500, 675, and 778-nm wavelengths. The solar radiation intensities measured at these wavelengths are used to calculate the AOD at these wavelengths. Additionally, a spectroradiometer operating in the 350–1100-nm wavelength range with 5-nm resolution has also been used to derive the AOD. The calibration of these instruments is done at NPL, New Delhi.

Ozone Measurement Program

The Laser Heterodyne System

In the infrared region of the spectrum where the fundamental vibrational-rotational absorption bands of most of the atmospheric molecules are located, a study of window region (8–14 μm) with high-resolution instruments gives quantitative information and their concentrations at various altitudes. Infrared laser heterodyne spectroscopy provides a powerful tool for identification of weak molecular and atomic species. The advantages of the laser heterodyne system over other techniques are its ultra-high spectral resolution, high spatial resolution, high quantum detection efficiency, and very good signal-to-noise ratio. The high resolution makes the system very selective, as the interference problem due to overlapping lines or bands is minimized and the lines can be resolved completely. The laser heterodyne system with 1 GHz acousto-optic spectrometer (AOS) as a backend was set up at Maitri, an Indian Antarctic station (70° 46′ S, 11° 44′ E), during the 1993–94 and 1994–95 Antarctic summers to obtain ozone profiles and is first of its kind over the Antarctic region. However, the system was brought back that time due to non-availability of liquid nitrogen during winter, as the liquid nitrogen plant could not be operated during summer of 1993. Again, in the 16th expedition, the system was set up at Maitri and made operational in the summer as well as winter. The liquid nitrogen plant, which was brought during the 1993–94 Indian Scientific Antarctic Expedition for producing liquid nitrogen for the laser heterodyne system, was also made operational during 1997.

A block diagram of the laser heterodyne system designed, developed, and set up at NPL is depicted in Figure 6.2. In this, the CO_2 laser used as a local oscillator is tuned on a line corresponding to the absorption line of the minor constituent of interest. The solar tracker (heliostat) follows the sun and brings in solar radiation. The incoming chopped solar radiation is filtered out and infrared (IR) radiation (8–12 μm) and a CO_2 laser beam of moderately low power are combined via a zinc selenide beam splitter. The solar radiation and CO_2 laser beam are co-aligned and focused on the high-speed liquid nitrogen-cooled Hg Cd Te detector, which acts as a mixer as well as a narrow-band filter and detector. The detected signal is nothing but the difference frequency (IF). The IF signal is further amplified in low-noise wide-band (5–1200 MHz) RF amplifiers and passed through various RF filter channels (25–1200 MHz). The signal from various channels is square law-detected and fed to a lock-in amplifier where synchronous detection takes place. The signal

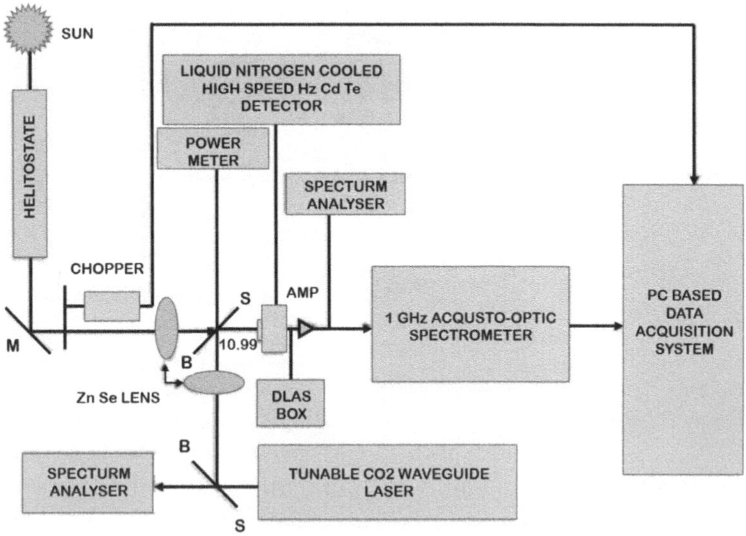

FIGURE 6.2
Block diagram of laser heterodyne system.

can be read directly from the lock-in-amplifier or recorded on a strip chart recorder. The observations taken at different channels in the wings of the absorption line of the minor constituents of interest would resolve the line completely, and a vertical profile can be obtained with these data using the inversion technique. The number of points in the wing of the line would decide the height resolution. This system was being manually operated by tuning various RF filters (25–1200 MHz), which was time consuming and subject to errors in the measurement. In view of the above, a wide band acousto-optic spectrometer has been developed.

Acousto-Optic Spectrometer

Acousto-optical techniques for power spectral measurements have been exploited for a variety of signal analysis applications (Turpin, 1981). A Bragg cell converts RF signal to ultrasonic traveling waves modulating the optical index of the cell. The cell is illuminated across its aperture by a laser beam. A fraction of light is diffracted by acoustic waves; the angle of diffraction is determined by the frequency, while the intensity of the diffracted light is proportional to the power of the input RF signal. The intensity distribution can be detected by a linear array of photo-detectors, which in turn represent the required RF power spectrum. An AOS with 1-GHz bandwidth was designed and developed in collaboration with Meudon observatory, France, and integrated with the LHS developed at the NPL, New Delhi. A block diagram of the AOS is shown in Figure 6.3. The acousto-optic spectrometer

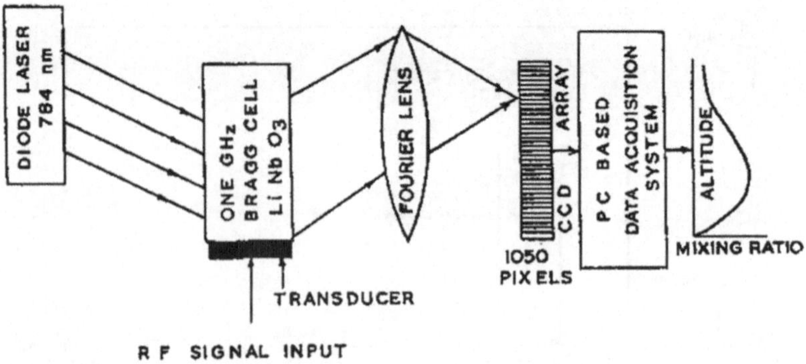

FIGURE 6.3
Block diagram of acoustic-optic spectrometer (AOS).

consists of a 1-GHz Bragg cell (Li Nb O_3), diode laser at 784 nm, Fourier lens, 1050-pixel CCD array, and PC-based data acquisition system. Proper software and hardware had to be developed to digitize the frequency power spectrum and stored for further analysis to get height profiles of various minor constituents.

The acousto-optic spectrometer developed was tested using the laser heterodyne system of Reims University, France. The frequency power spectra obtained using an empty cell of 200 cm length and 600 milli torr ozone in the cell were as expected. The P (24) CO_2 laser line in the 9.6 mm band was used. The spectra thus obtained show very well-defined ozone absorption lines. The use of AOS as the back end of the laser heterodyne system is timely and well suited to the detection of weak signal buried in noise. The main advantages of the AOS are wide bandwidth, high resolution, large number of channels, high detection sensitivity and compactness, light weight, and high energy efficiency. The system has been successfully operated during the 13th, 14th, and 16th Indian Scientific Antarctic Expeditions (ISAEs) and measured ozone height profiles in the harsh environmental conditions of Antarctica during both normal and ozone hole periods. A compact liquid nitrogen plant was also set up at Maitri to produce the liquid nitrogen needed for the cooling laser heterodyne system.

Inversion Technique

The composition profile or temperature profile can be obtained by analysis of individual spectral lines measured at ultra-high resolution through the inverse solution of radiative transfer equation (Kaplan, 1959). A software program was developed and tested to obtain the height profiles for ozone using the inversion technique (Jain, 1987). It was found that the retrieved profiles matched well with the model profiles and were independent of the initial guess (Jain, 1987). The inversion technique was also tested using actual profiles during normal and ozone hole conditions, and it was found that the retrieved profiles compared well with the original profiles. Different height zones were considered for Lorentz and Voigt line profiles depending on the constituent of interest. The various line parameters were computed using the AFGL HITRAN database (Rothman et al., 1987). A careful selection of the absorption line was made to get sharp contribution functions, hence the good height resolution. The contribution functions that depend on the absorption coefficient have a property of reaching maximum peak at different values of height for different values of frequencies. Computations were done for the line center and in the wing of the line. The inversion technique developed for ozone height profile is discussed in detail in Jain (1987).

Sunphotometer

A microprocessor-based sun photometer, MICROTOP II, has been used to measure total column ozone, water vapor, aerosol optical depth, and so on since 1997. The measurements were carried out at Maitri, and latitudinal measurements were also undertaken across the Indian Ocean during the sea voyage.

Greenhouse Gases Measurement Program

The Earth's environment is constantly changing. Scientific evidence indicates that these changes are result of complex interplay among several natural and human-related systems. Global human activities are changing the concentration of naturally existing greenhouse gases and adding additional greenhouse gases to the atmosphere, which in turn will be responsible for global warning and climate change. Therefore, in recent times, concern has grown about global change, which is related to natural and anthropogenic alteration of the Earth's environment.

Among the important issues, which are dangerous for the survival of life on Earth, are global greenhouse warming, urban and regional atmospheric pollution, regional increases in tropospheric ozone, the decrease in stratospheric ozone in general and the ozone hole over Antarctica in particular, acid rain, and so on. The monitoring of greenhouse gases has an important role in the understanding of global change. The concentrations of CO_2, CH_4, CO, water vapor, and total ozone fluctuate widely from place to place and even from day to day. For global warming studies, it is necessary to arrive at reliable estimates of true baseline values of these gases, which is possible only at the most pristine location possible, such as Antarctica. Keeping this in view, a project proposal entitled Monitoring of Green House Gases at Maitri, Antarctica was submitted to the National Centre for Polar and Ocean Research (NCPOR; Department of Ocean Development, Govt. of India), Goa, and was approved by NCAOR.

The necessary instruments were procured and were tested at NPL, New Delhi, before sending them to Maitri, Antarctica. The measurements of various greenhouse gases such as CO_2, CH_4, CO, water vapor, ozone, and so on were started during the 21st Indian Scientific Antarctic Expedition at Maitri on a regular basis, which in turn will go a long way to fill in the gaps and provide valuable data for modeling studies. All the measuring instruments were set up in a small hut, where all the experiments were carried out during the observational period. Observations made at a high latitude like Antarctica will also help in understanding the photochemical, heterogeneous, and dynamical processes that control the distribution of atmospheric trace gases. The studies at Maitri station were aimed toward the estimation of temporal variability and a complete annual variation in the near surface air over Antarctica.

A block diagram of the experimental system designed, developed, and set up for the online measurement of surface CO_2 and CH_4 using gas chromatograph at Maitri, Antarctica, is shown in Figure 6.4. At the same time, a CO gas analyzer has also been used for continuous measurement of concentrations at Maitri. The complete gas chromatogram (GC) system incorporates a nitrogen generator, hydrogen generator, computer-controlled sampling pump, gas chromatograph, and data processing unit.

Other Scientific Programs

In addition to the above major scientific equipment programs, earlier expeditions concentrated on VLF measurements. Cosmic noise absorption measurements with the help of solid state riometers were carried out at Antarctica. A global positioning system was used for the measurement of total electron content at Maitri in collaboration with Germany. Furthermore, in early 1986, air samples were collected and brought back to NPL to analyze for gas content in the Antarctic boundary layer. Later, a high-volume air sampler was used to collect samples en route to Antarctica and back for a

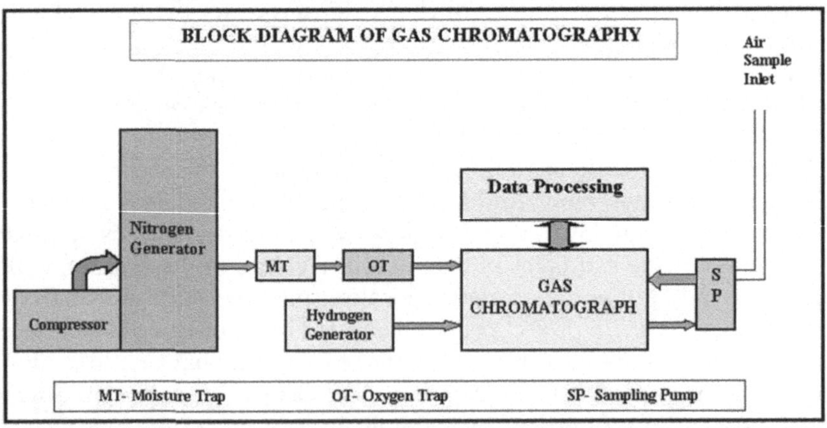

FIGURE 6.4
Block diagram for gas chromatographic measurements at Maitri.

few years. Sunphotometer observations have been taken since 1986 to study aerosols. Evaporation duct height measurements were also carried out in the second expedition. NPL has also worked in the field of oceanography.

Results

Planetary Boundary Layer: SODAR

During the period of the project, monostatic acoustic sounder data recorded at Maitri have been printed in a ready-to-use format and facsimile records have been compiled in book format (Dutta et al., 1999). Figure 6.5 shows a facsimile display of the passage of low-level clouds at Maitri. In this paragraph, we present general climatic features of the PBL, as observed by the monostatic acoustic sounder operating at Maitri. During local summer, there is a definite pattern observed in the facsimile charts, the daytime dominated by convection or thermal plumes. These thermal plumes originate due to the local heating of the rocky and dry surface, and this thermal convection can be seen in Figure 6.6. In the evening, as sun goes below the polar cap ice, the

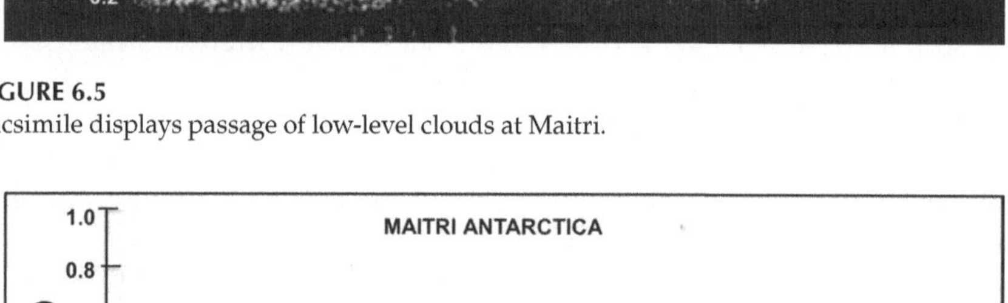

FIGURE 6.5
Facsimile displays passage of low-level clouds at Maitri.

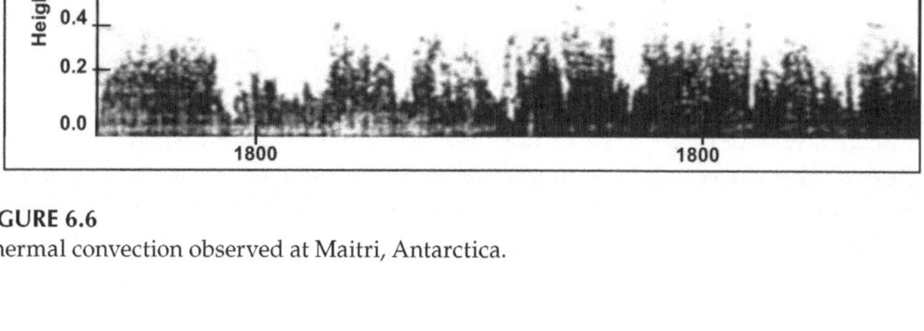

FIGURE 6.6
Thermal convection observed at Maitri, Antarctica.

rocky area becomes cooler, diminishing the presence of plumes, and thermal inversion is seen to prevail. Thus, there is a definite diurnal variation of the boundary layer, changing from convection to inversion. However, in the evening, the katabatic flow also starts, which might continue even at night, thus presenting diurnal variation both in the surface air temperature as well as in the observed wind speed.

During summer, if there is any snowfall, the snow in the rocky area gets melted during the daytime due to solar heating of the earth's surface. This melted snow water percolates into the pores of the rocks during the daytime and freezes at night, thereby forming ice due to the process of thawing. In this process, the volume of water increases and helps in developing cracks of the rocks. Further, the cracks break down and disintegrate into much smaller particles. Over time, all this disintegrated rock forms into sandy soil. Thus, with the help of erosion agents like wind, water, and so on, the weathered rock material is collected in an area where a lesser stress of erosion prevails. The small microorganisms of the region, on the other hand, inhabit these sandy soil materials. The microorganisms extract essential food material and nutrients from these soils.

The same process of thawing also occurs in the microorganism of the area. The diurnal variation of temperature affects the organisms of the region. Although the organisms release an "anti-freezing" chemical in their cells, sudden changes of temperature will be detrimental to them. In the austral summer when the temperature is above 0° C during daytime, the water melts and is absorbed by the cells of the organisms, and at night, the temperature decreases to less than 0°C and the water inside the cells freezes. This melting and freezing of water thus affects the metabolism of the organisms. The energy balance of the organisms will also be affected; therefore, reproduction will be minimal. Hence, the diversity of the area is limited to only a few hardy microorganisms such as lichen, algae, mosses, mites, midges, tardigrades, and so on. Organisms with protective covers have different strategies of survival in these conditions. Their bodies are kept warm by a protective layer and therefore no thawing takes place, for example, penguins. The erosion of soil is higher during the summer compared to any other season, and monostatic sounder data clearly provide the duration and the amount of thermal convection over the Schirmacher region.

During winter, once the inversion forms, it remains intact unless the katabatic winds weaken it. Thus, many stable layers are seen to be present. However, it is important to note that in Antarctica, between two successive depressions, there is a high-pressure zone, which, when observed, makes the sky clear and the surface air temperature immediately dip to form strong inversions. However, it is the formation of strong inversions that keeps Antarctica warmer due to re-radiation from the inversion toward the earth.

Due to the movement of depressions from east to west, pressure and temperature variations do show variation, with a periodicity of several days, and the daily variations are absolutely missing.

During the transition seasons, facsimile charts show highly turbulent structures and wave motions, indicating turbulent mixing in the atmosphere. In fact, during the transition seasons, the pressure over Antarctica is the minimum.

UV-B Studies

Comparison of UV-B Intensities at Antarctica and Indian Mainland

The first Indian measurements of UV-B intensities over Antarctica by the NPL established the absolute and relative magnitudes of these intensities (Singh et al., 1992; Hanjura et al., 1988). Through UV-B radiation measurements over Antarctica and the Indian mainland, it was established that the maximum noontime UV-B dose (a measure of total energy in total the UV-B radiation band) at Antarctica is less, even during the ozone hole period, than the maximum noontime dose obtained over the mainland, when the sun is overhead or nearly overhead (Sharma and Srivastava, 1992). This is attributed to the existence of much lower solar zenith angles over the Indian mainland than over Antarctica. The solar zenith angle never becomes less than 40° at our Antarctic station Maitri, while it remains zero or almost zero at many of the Indian mainland stations. Out of these conclusions, it was suggested that a virtual ozone hole exists over India as far as UV-B doses are concerned (Sharma and Srivastava, 1992; Singh, 1997). A comparison of UV-B doses over Antarctica and Delhi is shown in Figure 6.7. It is seen from this figure that the UV-B doses over Antarctica, which were maximum during November in that year over Maitri, are comparable to the July values over Delhi. In July, the noontime solar zenith angle over Maitri and for this solar zenith angle the

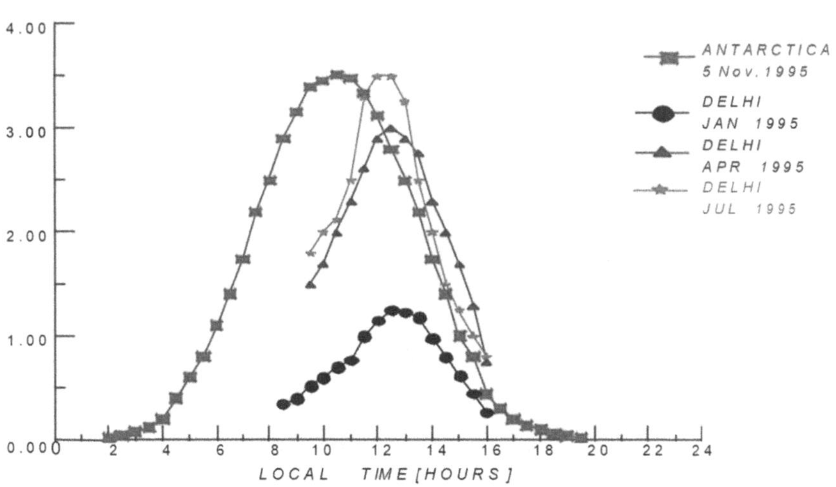

FIGURE 6.7
The first comparison of UV-B doses at Antarctica and over Delhi. The Antarctic values are a yearly maximum and are compared with Delhi values for different seasons.

noontime values over Delhi are comparable to the maximum Antarctic values; hence, UV-B doses at a place where the zenith angle becomes 0° will be higher.

Ultraviolet Radiation as Index of Total Ozone Variability

The variability in total ozone column content over Antarctica has been correlated with the variability in the UV-B radiation intensities at various wavelengths. It has been found that the variability in the radiation intensity at the 310-nm wavelength represents the best among all the wavelength (Singh et al., 1992) reference values.

Comparison of UV-B Time Integral over Antarctica and Indian Mainland

The first Indian effort from the NPL established (Figures 6.6 through 6.8) that due to the longer length of the day hours over Antarctica during the Antarctic summer season, the UV-B doses during its maximum epoch remain above the safe reference value, that is, 1.0 Med/hr continuously for 10 hours, for 7 hours above twice the safe reference value, and for 4 hours thrice above the safe reference value. The corresponding duration of these doses for the mainland (Delhi) in the month of July month are 8 hours, 5 hours, and 2 hours during the maximum epoch of these doses (Tables 6.1 and 6.2). This means that though the Antarctic values are comparable to the Delhi values, the Antarctic environment is exposed to UV-B doses for longer periods compared to the mainland environment. The diurnal time integrals of UV-B intensities at Antarctica are 1.5 times more than those for the mainland (Singh, 1997; Singh et al., 1997).

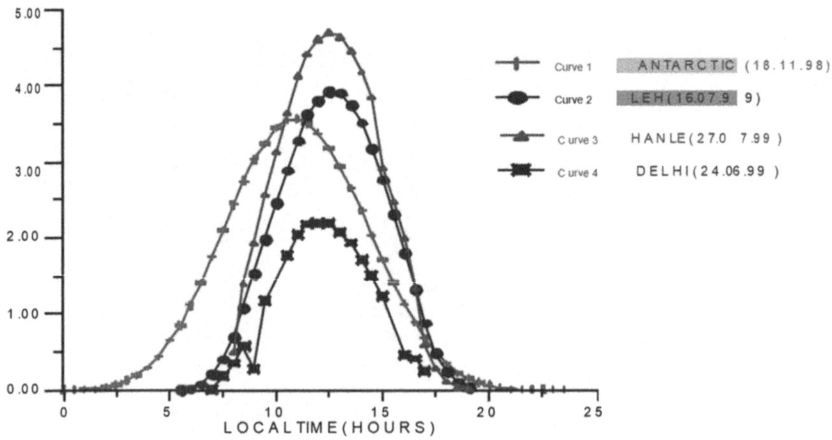

FIGURE 6.8
The first Indian comparison of the observed diurnal time integral of UV-B doses/intensities over Antarctica and over different sites over the Indian mainland. Delhi values are for a dusty day in June, while the values in the previous figure for July and April are for clear-sky conditions.

TABLE 6.1
Erythema Dose Variation (Per Hour) at Different Locations

Location	Dose Variation/Hr
Hanle (Ladakh, India)	0.9 MED
Antarctica (Maitri)	0.5 MED
Delhi (Summer)	0.7 MED
Delhi (Winter)	0.3 MED

TABLE 6.2
Time Duration (Continuous in a Day) for Different Erythemal Dose at Different Locations

Location	MED 1.20	MED 2.0	MED 3.0	MED 4.0
Antarctica (Maitri)	10.0	7.0	4.0	–
Hanle	9.0	7.0	6.0	5.0
Delhi (Summer)	8.0	5.0	2.0	–
Delhi (Winter)	3.0	–	–	–

Correlation between Column Ozone and UV-B Intensities

Generally anti-correlation between UV-B intensities and total ozone contents over Antarctica is observed, but under certain circumstances, this anti-correlation does not exist as shown, for example, in Figure 6.9. This is attributed to the special dynamical conditions during these changed synoptic weather conditions where marine/continental aerosols are pumped in or sucked out of the observation radiation field (Hanjura and Singh, 1995).

Forenoon-Afternoon Asymmetry in UV-B Intensities

Occasionally, it is found that the UV-B intensities are different in the forenoon and afternoon sectors of the day at the same solar zenith angles, as shown in Figure 6.10. The forenoon values are generally found to be higher than those for afternoon times. This is considered to be due to the moving nature of the ozone depletion regions over Antarctica (Hanjura and Singh, 1995).

Comparison of UV-B Intensities over Antarctica and Inside Mid-Troposphere

Comparison of UV-B intensities over Antarctica and at high-altitude station Leh/Hanle at 4.8 km height shows that the intensities at Hanle are more than those at Antarctica (Figure 6.8). This confirms the substantial role played by aerosol scattering on UV-B intensities (Singh et al., 1999; Nath et al., 2000).

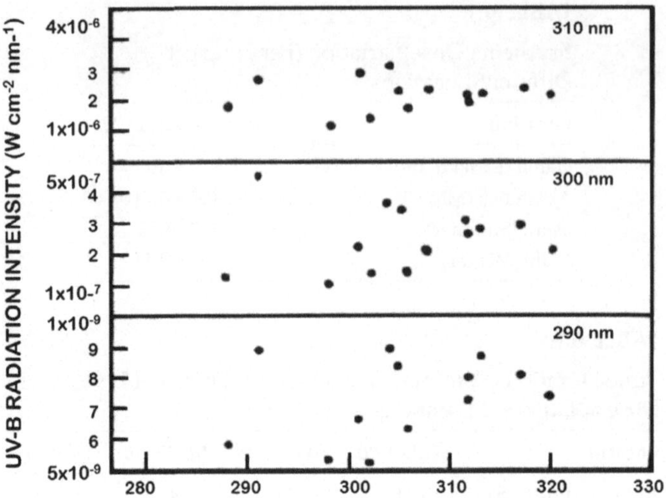

FIGURE 6.9
Variation of UV-B intensities over Maitri at 290, 300, and 310 nm at fixed zenith angle 56° for different total ozone contents. The inverse proportionality between UV-B intensities and ozone content seems to be broken here.

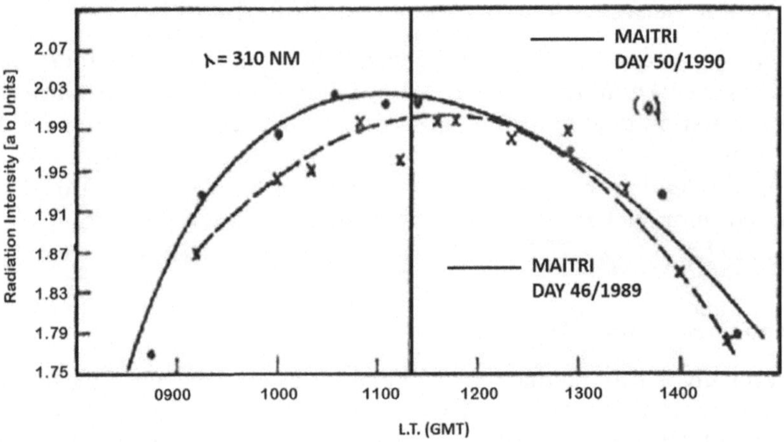

FIGURE 6.10
Diurnal variation of UV-B intensities showing the occasional morning—evening asymmetry with different values at the same zenith angles in the morning and evening sectors of the same day.

Aerosol Optical Depth Measurement in Pristine Environment

- The measurement of aerosol optical depth at Antarctica provided an opportunity to assess the AOD in a pristine environment and make comparisons with mainland values. It is found that AOD values are about 5–6 times less than the values at Delhi (Singh et al., 1999; Shambhu Nath et al., 2000).

- The signatures of the Pinatubo volcanic eruption were recorded in the AOD measurements over Antarctica. This provided an opportunity to the scientific community to assess the wind and global air circulation trajectories (Hanjura and Singh, 1995).
- The AOD measurements at Antarctica are compared with the in-situ measurements in the mid-troposphere at high-altitude station Leh/Hanle at 4.8 km height. It is found that the AOD values at Hanle are lower than even those at Antarctica (Figure 6.11). This provides a clue that in respect to aerosol content, the environment at these heights is more pristine than that even at Antarctica (Singh et al., 1999; Shambhu Nath et al., 2000). Aerosol optical depths were found to be increasing continuously during the last decade at Maitri, Antarctica (Shambhu Nath et al., 2000).
- The role of aerosols on UV-B intensities at Antarctica was identified during a unique phenomena of very heavy downpour during the night of February 2–3, 1996. The UV-B intensities increased by 20%–30%, while column ozone remained almost the same after the rains compared to the values before the rains. This has been attributed to the washout of aerosols during the downpour (Shambhu Nath et al., 2000).

In conclusion, the NPL has been continuously measuring UV-B and visible solar irradiance since 1987 over Antarctica and deriving erythemal doses and AODs from these observations. The analysis of the data for some important features has been presented here. More analysis is a continuing process out of this large amount of data collected in these parameters during the last 19 years.

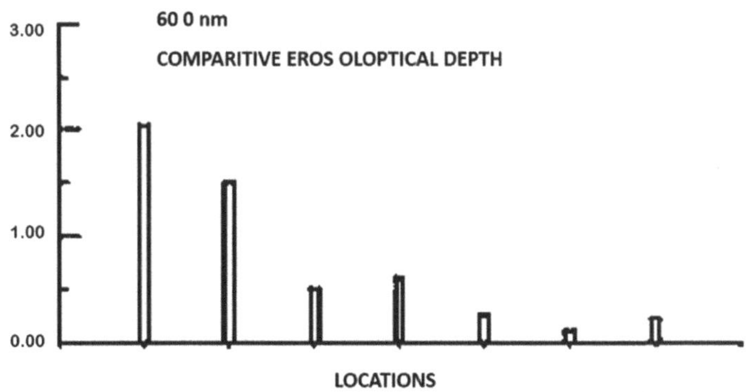

FIGURE 6.11
Comparative values of AOD at various sites showing minimum at Hanle and establishing the fact that Antarctic values are higher than those at Hanle and not the minimum at earth surface.

Ozonosphere

Laser Heterodyne System

Ozone depletion in general, and the ozone hole phenomenon over Antarctica during spring in particular, has generated a surge of interest in the measurement of ozone using various techniques. The laser heterodyne system (LHS) has been successfully operated at Maitri (70° 46′ S, 11° 44′ E) during the 13th, 14th, and 16th Indian Antarctic Expeditions to obtain ozone line spectra at 1043.1775 cm^{-1} with an ultra-high spectral resolution. The line spectra thus obtained with high spectral resolution in turn have been used to get vertical profiles of ozone in the atmosphere using the inversion technique developed at the NPL for Antarctic conditions. A guess profile of a uniform mixing ratio of 3 ppm was used for retrieval. The system was also operated during the ozone hole period. The retrieved profiles during the normal period and ozone hole period are shown in Figures 6.12a and b, respectively. The ozone was found to be depleting during the ozone hole period from 3% to 68% in the height range of 13–40 km (Jain, 1996). The laser heterodyne system needs liquid nitrogen for cooling the detector and therefore a liquid nitrogen plant was also installed at Maitri. However, the system has some limitations, as it can be operated only during a clear sky and on sunny days.

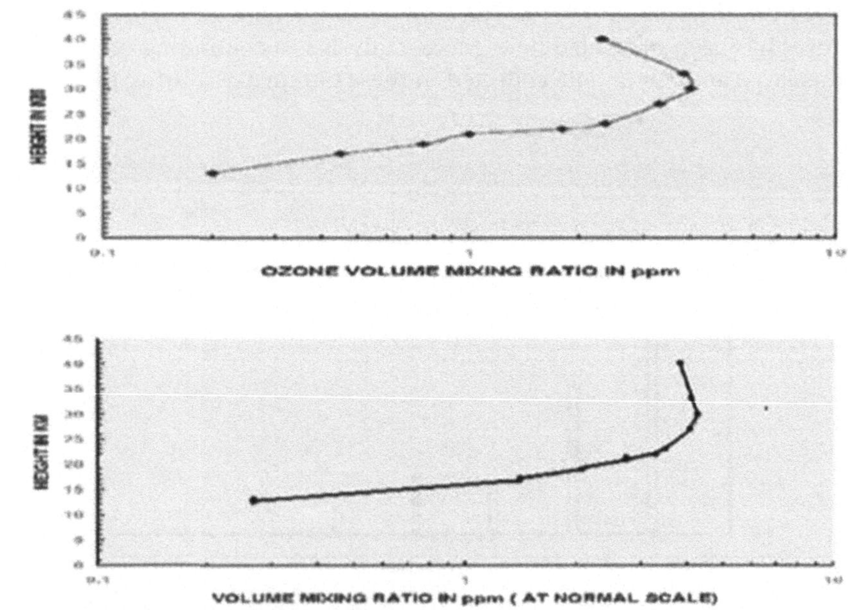

FIGURE 6.12
Vertical profiles of ozone (a) normal period; (b) ozone hole period.

MICROTOP Sun Photometer

The column ozone measurements were carried out during 1997 and again during 2002, 2003, and 2004 at Maitri on all clear days. Integrated column ozone observations during 1997, and 2002 and 2003 at Maitri show ozone concentration up to 320 DU in the months of January–February. The minimum value of column ozone observed during the spring of 1997 was about 135 DU in the first week of October, while during the spring of 2002, the minimum value of column ozone was found to be 185 DU, and 126 DU in the year 2003. Thus, it was found that the ozone hole during Antarctic spring of 2002 was not as deep as it was during 2003. Also, the ozone hole recovered earlier in 2002 as compared in spring 2003 (Jain et al., 2003; Ghude et al., 2005a,b). Figure 6.13 shows the daily averaged total column ozone over Maitri during 1997 and 2002–03–04. It was also found that year to year, the ozone hole depth as well as its duration is fluctuating depending on dynamics as well as meteorological conditions. The ozone hole again during 2003 was found to be quite severe and for a longer duration in contrast to years 1997 and 2004. The Maitri observations taken by the MICROTOP-II sun photometer were also compared with total ozone mapping spectrophotometer (TOMS) data, as well as nearby Russian station data, and were found to be in good agreement. The variability in the ozone hole in different years is attributed to the dynamics and meteorological conditions prevailing over Antarctica (Jain et al., 2003; Tripathi and Jain, 2003; Ghude et al., 2005a,b)

Greenhouse Gas Measurements

The study of greenhouse gases carried out at Maitri indicates the signature of regional climate change. Methane is the most abundant hydrocarbon in the atmosphere. Its troposphere chemistry affects hydroxyl (OH) radicals and CO

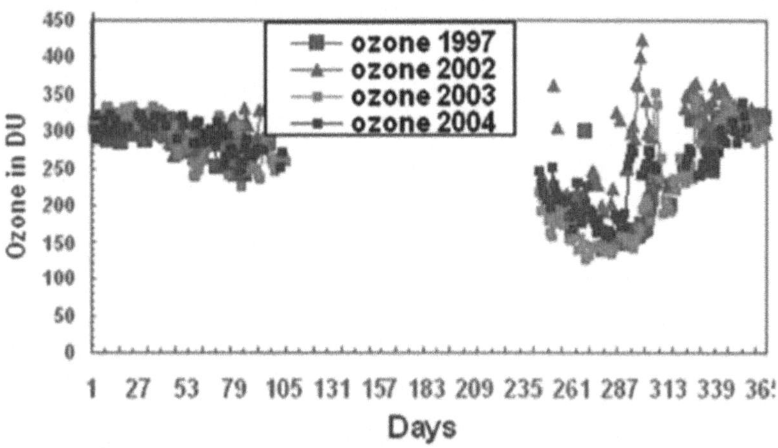

FIGURE 6.13
Daily averaged total column ozone at Maitri, Antarctica.

concentration. In the stratosphere, the oxidation of methane by OH radical is a major source of water vapor, and its reaction with chlorine atoms terminates the chlorine-catalyzed destruction of ozone. The measurements reveal that the level of carbon dioxide concentration is increasing in the atmosphere. On the other hand, the atmospheric concentration of CH_4 is found to be more or less stabilized (Jain et al., 2005a,b). The day-to-day variation in CO_2 and CH4 concentration is found to be of lesser magnitude, which illustrates that the air at Maitri is well mixed and far removed from the source strength characteristics of land area. The monthly mean CO_2 concentration is found to have increased from 367.8 ppm to 370.6 ppm, as seen in Figure 6.14 (the present value is 415 ppm). The observed trend of CO_2 concentration in near surface air illustrates a linear increase. Based on monthly mean CO_2 concentration, annual mean CO_2 concentration was found out to be 368.43 ppm in the year 2002 and 369.72 ppm in the year 2003. These results show that CO_2 concentration has increased by 1.3 ppm in 1 year (Jain et al., 2005a,b). No

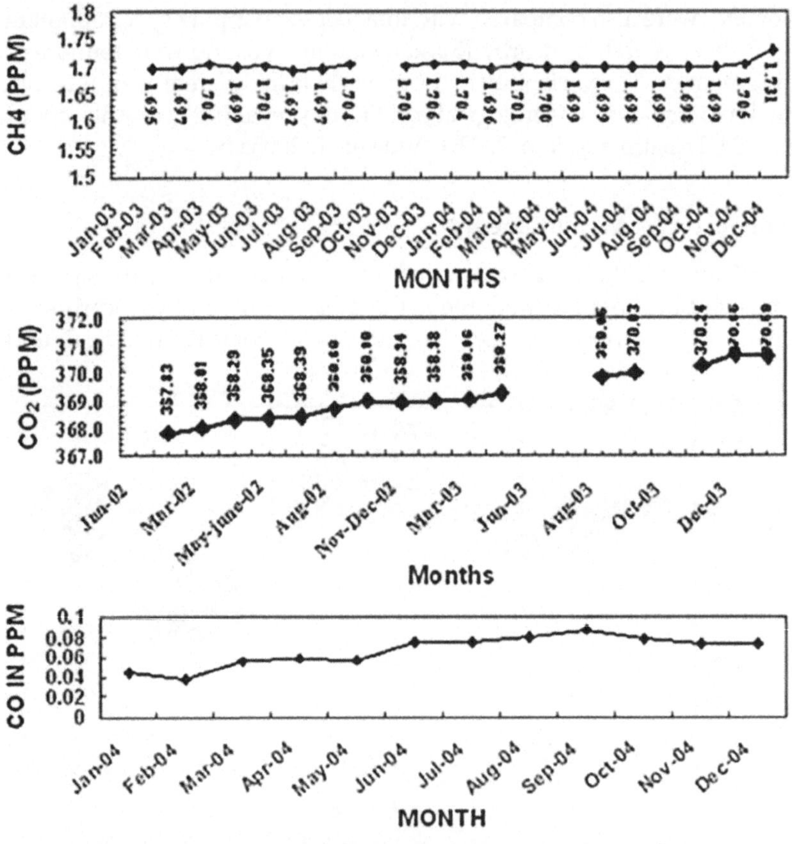

FIGURE 6.14
Monthly averaged CH_4, CO_2, and CO at Maitri, Antarctica.

seasonal variations were observed at Maitri during the observational period. The linearly increasing CO_2 concentration indicates that the global abundance of CO_2 in the atmosphere is increasing either due to the increase in CO_2 emission or imbalance between sources and sinks of CO_2. On the other hand, observed stable CH_4 concentration indicates a decrease in emissions, which may have accelerated the global methane budget toward a steady state (Jain et al., 2005a,b). The reason for the change is unclear. Some investigators believe that a major contributing factor was the break-up of the Soviet Union in 1991. Oil and gas production fell, the industry became more efficient at plugging gas leaks from pipes and wells, and methane emissions dropped significantly throughout the 1990s due to lower fossil fuel production. Based on current knowledge of the observed data, it is not possible to tell whether we are only observing a persistent, but temporary, pause in its increase or global levels of methane are likely to stabilize in the very near future. However, to derive acceleration and de-acceleration of global budget of these trace gases, it is necessary to have continuous measurement for a prolonged duration.

CO measurements made over Maitri showed large day-to-day variation and were found to be strongly correlated to changes in solar radiation. The study concluded that in the absence of human activity, CO concentration is controlled largely by the balance between photochemical production and extinction by reaction with OH radicals. It further indicates that solar radiation influences the atmospheric concentration and lifetime of carbon monoxide in the air. It was also concluded that photochemical reactions are involved and take place in the Antarctic air, which causes photochemical destruction, leading to large variation in CO values during the days when daylight hours are long. Higher concentration during polar night reveals termination of the reaction involved in the sink mechanism of CO, which eventually leads to a fairly long lifetime of CO in the dark. Observation indicates CO production processes in cold polar ice are activated by sunlight, initiating photochemical destruction of formaldehyde (HCHO). This shows that snow=covered regions may act as a source of CO, although they are not very important in the global CO budget. Figure 6.14 shows the monthly mean concentrations of CH_4, CO_2, and CO at Maitri for the observation period.

In the earlier expeditions, the NPL also measured concentration of trace gases like CH_4, CO, CO_2, and N_2O using a gas sampling technique during the V, VI, and VII expeditions. The samples were analyzed by gas chromatography using a Perkin Elmer (model Sigma 2000) gas chromatograph. The concentration of CH_4 was found to be 1.5 ± 0.02 ppm.

A signature of steady increase in total water vapor column over Maitri has been observed under this study. The correlation between surface temperature and water vapor also indicates that most of the variation in water vapor concentration probably took place in the lower troposphere due to variation in surface temperature (Arya and Jain, 1997; Jain et al., 2005a,b). Higher water vapor was observed during the ozone hole period in 2002 than in 1997, as shown in Figure 6.15. 2003 revealed anomalous strong tropospheric transport

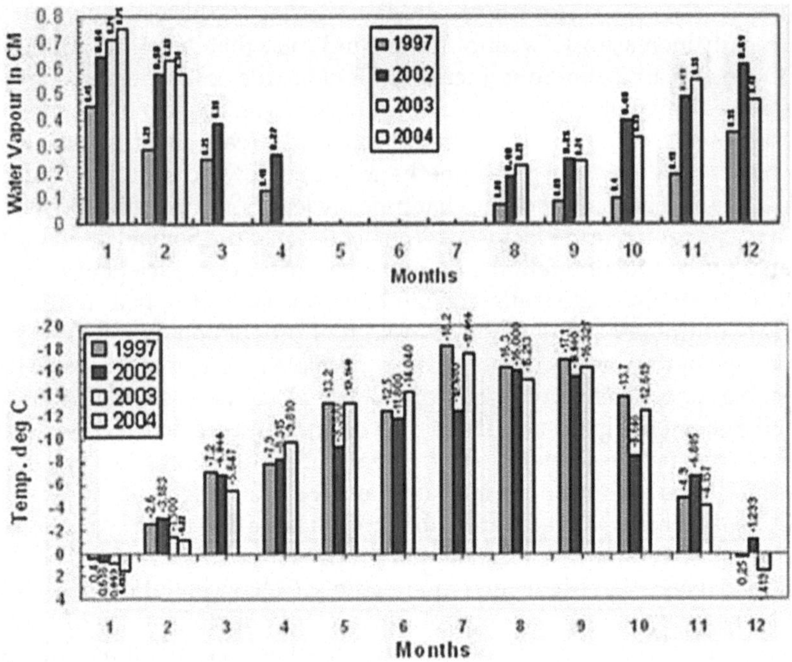

FIGURE 6.15
Monthly averaged total column water vapor and temperature at Maitri, Antarctica.

of air from the subtropic to the polar region, followed by coherent changes in surface temperature and water vapor. Measurements also showed a signature of increase in the surface temperature, found especially pronounced in the month of January at Maitri, as can be seen in Figure 6.15. The study of total column ozone (TOZ) showed a direct relationship between the stratospheric temperature anomaly and year-to-year variation in total ozone. The observations showed that chemical loss of ozone over Maitri during the ozone hole period in 2003 was increased by 18.6% compared to 1997 and 42% compared to 2002. The observations at Maitri also showed an event of major stratospheric warming, along with surface warming during 2002. A temporary sudden rise in the total column ozone before the recovery period (days 300–315) was also observed and found to be overlapping in all the observational years.

So, although few records are available at Maitri, the trend observed is a particularly important indicator of climate change. These changes in a small region of Antarctica demonstrate that the future of regional climate change may probably be one dominated by regional stimulation and feedback processes. The same may be true for more populated and polluted areas of the globe. The year-to-year variation of total column ozone and the early recovery of the ozone hole in 2002 can be attributed to major sudden stratospheric warming and decisions made by the international community under the Montreal Protocol and its subsequent adjustments and amendments to stop the use of ozone-depleting chemicals. These measurements indicate that the ozone hole is not worsening and may soon start to improve. However, this improvement is going to come very slowly. Further, changing atmospheric conditions, which modulate total ozone column over Antarctica, can complicate the task of detecting ozone hole recovery because a full explanation of meteorological anomalies observed over Antarctica is not yet available.

Water vapor measurement is based on a pair of radiometric measurements in the IR band. The 940-nm filter (10-nm full width at half maximum [FWHM]) is in a strong water vapor absorption band, while the 1020-nm filter (10-nm FWHM) is affected only by aerosol scattering. Day-to-day water vapor at Maitri was found to be highly variable in both years; however, the general trend showed a maximum during polar summer and minimum during polar winter. During polar winter, the minimum recorded water vapor was observed to be on the order of less than 0.1 cm. However, the maximum daily averaged total water vapor column was observed to be 0.51, 0.96, 0.90, and 0.95 cm in the January 1997, 2002, 2003, and 2004, respectively. A comparison of the monthly averaged water vapor and surface temperature during January 1997 to December 1997 and January 2002 to February 2004 is depicted in Figure 6.15 (Ghude et al., 2005a,b; Jain et al., 2005a,b). In 1997, the annual averaged water vapor was found to be 0.24 cm, while it was 0.42 cm in 2002 and 0.45 cm in 2003.

Aerosol Optical Depth Using Sunphotometer

Aerosols are of great significance as they play an important role in the understanding of global change. Columnar aerosol optical depth measurements were made over Maitri (70°,76′ S, 11°,73′ E and 140 amsl) during 1997, 2002, 2003, and 2004 at 1020 nm using a handheld portable sun photometer. The daily average values of AOD (1020 nm) for the measurement period are shown in Figure 6.16 and reveal a clear variation in month-to-month

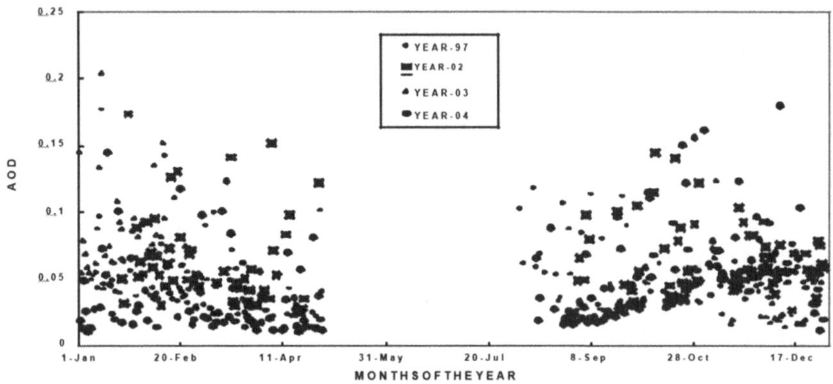

FIGURE 6.16
Daily mean values of aerosol optical depth (1020 nm) at Maitri, Antarctica.

values of AOD. The AOD values are high during January and decreasing up to April and attaining low AOD. Again, August-onward levels continuously increase until December. Because of the non-availability of the source (sun), measurement cannot be done during May to July. Diurnal and seasonal variation is almost negligible. As shown in Figure 6.16, most of the time, AOD values are less at Maitri due to the clean atmosphere.

During the VI and VII expeditions, measurements of turbidity and solar UV-B radiation were done. The AOD at 368 nm was found to be 0.04–0.16, and the atmosphere was clear at higher wavelengths. The consistent observations made it possible to see the Pinatubo aerosols just 6 months after their eruption in Philippines.

Ionosphere and Other Investigation Very Low Frequency (VLF)

In the earlier expeditions, the NPL concentrated on VLF communication. VLF measurements were carried out during the I, II, and VI Indian Antarctic Expeditions. The VLF signal received with the help of a tuned loop antenna is fed to the input of a phase-locked LF/VLF receiver, TRACOR model 599 K. The data were collected over a path from OMEGA/Argentina to the Indian base camp in Antarctica at 12.9 KHz carrier. The experiment was conducted for studies of the ionosphere D-region. This study was basically conducted in three parts: quiet time variation, disturbed time behavior, that is, the effect of solar flares on the sudden enhancement of ionization in the D-region, which recovers quickly due to recombination processes; and electron precipitation in the D-region. Wave guide modal analysis of daytime VLF data was carried out.

Riometers

Cosmic noise absorption measurements with the help of solid-state riometers were carried out at Antarctica. The frequencies used for these studies were 20, 25, and 30 MHz. It employs two half-wave dipoles with balloon transformers for impedance matching. It had a built-in calibration system, which provided both a secondary time standard and a monitor of the dynamical response to study rapid changes in signal. Riometer reading varies linearly with power. Absorption in decibels can be calculated using a standard formula. These riometers are characterized by an improved calibration sequence, accurate internal timing control, and a marked decrease in size and weight. They are suitable for long-term operations at remote locations where only limited quantities of power are available.

Global Positioning System

Global positioning system (GPS) observations for the measurement of total electron content (TEC) at Maitri were recorded in collaboration with Dr. Wilfried of Germany. These data collected at Maitri were then compared with the data collected at Bangalore, India. An analysis of the data showed that the diurnal range observed at Maitri was much less than that at Bangalore. This is due to the much lower variation of the solar zenith angle at Maitri as compared to Bangalore. TEC diurnal variation at Maitri is superposed by short-period small-amplitude variations. This is attributed to the stronger coupling of the high-latitude ionosphere with the magnetosphere as compared to low latitudes. Figures 6.17a and b show diurnal variation of TEC and solar zenith angle at Bangalore and Maitri, respectively.

The Evaporation-Duct Heights Computation

In the second IAE meteorological observations were carried out using an automatic weather station during the cruise for the computation of evaporation duct heights in the latitudinal range of 15° N to 70° S (Pasricha et al., 1991). Higher

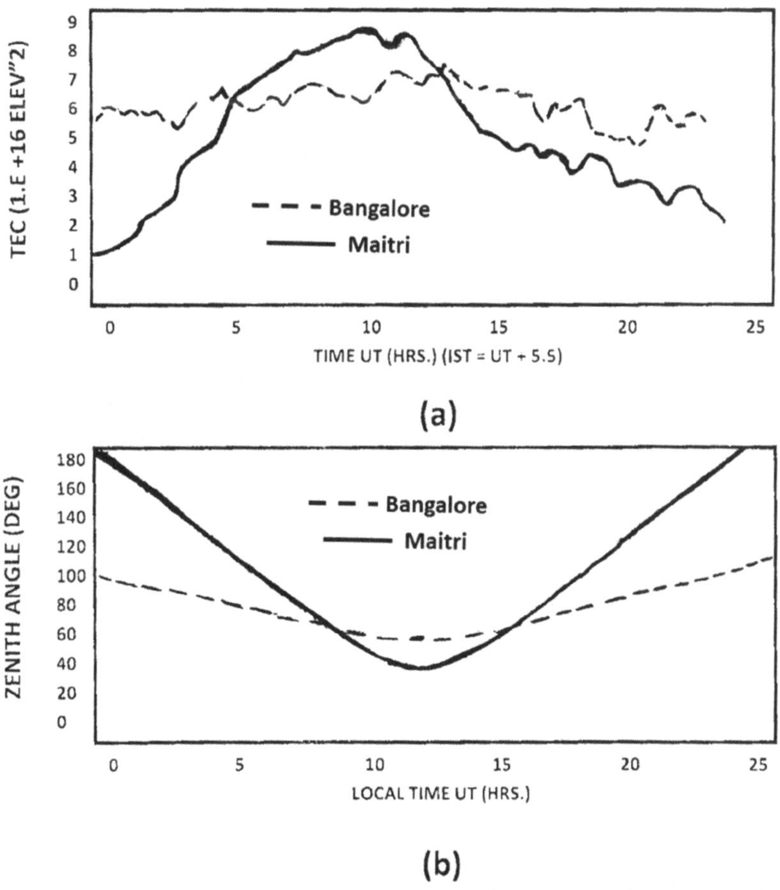

FIGURE 6.17
Diurnal variation (a) TEC; and (b) solar zenith angle at Bangalore and Maitri.

evaporation duct heights are observed at low latitudes. These are observed to reduce considerably in the cold-water region of the Southern Ocean.

Oceanography: Color of Antarctic Region Ocean-Water

The NPL has also worked in the field of oceanography. During the Sixth Antarctic Expedition, water samples were collected from different locations on the return journey, extending from lat 69° $58'$ S long 12° E to lat 60° S long 22° E. Optical studies, that is, color and refractive index, and physical analysis, that is, Sp.Gr., pH, and spectro-chemical analysis, were done on all the samples. This study confirms the low concentration of constituents in the Antarctic region ocean water.

Summary

The NPL has been participating in Indian Scientific Antarctic Expeditions right from its inception in 1981. During this period, about 35 personnel have participated in these expeditions, which includes about 10 winter-over participants. Some members undertook two to three expeditions. The NPL also contributed three leaders to these expeditions. A large variety of experiments have been undertaken during the period that mostly center around radio and atmospheric physics. Some significant findings are summarized below.

Earlier expeditions concentrated on VLF communication. It was found that 12.9-KHz signals from Omega station of Argentina at 4538 Km could easily be recorded at Maitri (70.77S, 11.73E). The day-to-day variation in the phase intensity depended upon variation in the number of sunspots. This gives an indirect idea about D-region electron density.

The natural radio waves coming toward Earth and passing through the ionosphere were monitored. It works on 10, 20, and 30 MHz. Data were collected for several years. Microbarograph records of extremely low-frequency noise (0.001–2 Hz) data were recorded for about 5 years. Meteorological parameters like temperature, wind speed, and wind direction were recorded by an automatic weather station. In 1992, a 28-m-height tower was erected, and sensors were installed at different heights to record meteorological parameters. Air samples were collected and brought back to the NPL to analyze for gas content in 1986. Later, a high-volume air sampler was used to collect samples en route, at Antarctica and back, for a few years. SODAR was established in 1991, and it worked for a few years and collected data about the planetary boundary layer to a height of about 1 km.

Laser heterodyne and mm wave radio spectrometer systems were installed to study the ozone profile of the atmosphere with an eye to decipher the ozone hole mystery discovered in 1985 in Antarctica. A laser heterodyne system, along with a 1-GHz acousto-optic spectrometer, was used to measure ozone height profiles during both normal and ozone hole periods. The ozone was found to be depleted during the ozone hole period from 3% to 68% in the height range of 13–40 km. Sometimes during the ozone hole period, high values of ozone were observed that revealed the influence of planetary waves on vortex dynamics. The increase and decrease of total ozone during the ozone hole period observed at Maitri showed the effect of planetary waves on dynamical behavior of the polar vortex. The study also revealed the early recovery of ozone hole during 2002 as compared to other years.

Sunphotometer observations have been taken since 1986 to study aerosols. The consistent observations made it possible to see the Pinatubo aerosols just 6 months after their eruption in Philippines. UV observations and erythemal observations (1995) have been undertaken to study the spectral intensities and intensity of integrated UV on human skin. These observations have led to interesting results for comparative study.

Hourly observations at 305, 312, 320, 940, and 1020 nm were carried out during clear sunny days to estimate column ozone, water vapor, aerosol optical depth, and so on. The regular measurements of water vapor from January 1997 to February 2004 showed a gradual increase of water vapor at Maitri. The annual averaged water vapor was found to be 0.24 cm in 1997, 0.42 cm in 2002, and 0.45 cm in 2003. The water vapor measurements were found to be in good co-relation with the increase of surface temperature at Maitri.

A gas chromatogram was set up for regular monitoring of greenhouse gases such as CO_2, CH_4, and so on. CO_2 was found to vary in the range 367–373 ppm during the period February 2002–January 2004. Extremely stable values of methane were observed during the observational period of February 2003–August 2004. The annual average of methane from February 2003 to February 2004 was found to be 1.69 ppm. Am NDIR CO analyzer was set up for around-the-clock monitoring of CO. Variability in concentration of carbon monoxide has been observed with hourly mean mixing ratios ranging from 30 to 75 ppb during the period February 2003–August 2004.

List of Publications on Antarctic Research by NPL Scientists

1. Jain, S.L. 1989. "Inversion Technique for Ozone Profiles Using Laser Heterodyne Radio Meter." *Indian Journal Radio & Space Physics*, v. 18, p. 103.
2. Jain, S.L. 1989. "Laser Heterodyne System and Measurement of Minor Constituents."*Indian J. Radio & Space Phys*, v. 18, pp.175–179.

3. Jain, S.L. 1992. "Retrieval of Ozone Profiles over Antarctica using Laser Heterodyne System." *Indian J. Radio and Space Physics*, v. 21, pp. 110-115.
4. Arya, B.C. and Jain, S.L. 1994. "Monitoring of Solar Radiation in 2.5–14.5 um Spectral Range." *Journal of Optics*, v. 24, pp. 87-95.
5. Arya, B.C. and Jain S.L. 1997. "The Measurement of Water Vapour at Maitri, Antarctica." *Indian J. Radio and Space Physics*, v. 26, p. 117.
6. Jain, S.L. 1996. "Laser Heterodyne System for Measurement of Minor Constituents of Atmosphere." *Review Paper in Indian J. Radio & Space Phys*, v. 25, pp. 309–317.
7. Tripathi, O.P. and Jain, S.L. 2001. "Measurement of Ozone at Maitri." *J. of Marine & Atmos Res*, v. 2, no. 1, pp. 37–39, Jan. 2001.
8. Jain, S.L. 2001. "Monitoring of Ozone, Water Vapour etc. During the Voyage to Antarctica." *Asian Journal of Physics*, v. 10, no. 3, pp. 315–321.
9. Tripathi, O.P. and Jain, S.L. 2002. "Abrupt Change in Total Ozone Column over Indian Antarctic Station." *Indian Journal of Radio & Space Physics*, v. 32, pp. 38–42, 2003.
10. Jain, S.L. 2001. "Monitoring of Ozone Using Laser Heterodyne System over Antarctica." *Bulletin Laser & Spectroscopy Society of India*, no. 11, pp. 39–53.
11. Jain, S.L., Arya B.C., Kumar, A. and Ghude, S.D. 2003. "Green House Gases Measurement at Maitri, Antarctica." *Asian Journal of Physics*, v. 12, no. 3, pp. 103–107.
12. Jain, S.L., Ghude, S.D. and Arya, B.C. 2003. "Signature of Early Ozone Hole Recovery during 2002." *Current Science*, v 86, no. 7, p. PP-963.
13. Sachin, D.G., Jain, S.L., Arya, B.C. and Bajaj, M.M. 2005. "Comparison of Total Column Water Vapour Measured at New Delhi, Leh, Hanle and Maitri (Antarctica)." *Indian Journal of Radio & Space Physics (2004)*, v. 34, pp. 264–268.
14. Sachin, D.G., Kumar, A., Jain, S.L., Arya, B.C. and Bajaj, M.M. 2005. "Comparative Study of Total Ozone Column (TOC) over Maitri, Antarctica during 1997, 2002 and 2003." *International Journal of Remote Sensing*, v. 26, no. 16, pp. 3413–3422
15. Jain, S.L., Ghude, S.D., Arya, B.C., Kumar, A., Bajaj, M.M. 2005. "Increase in Total Water Vapour Column and Surface Temperature at Maitri, Antarctica." *Current Science*, v. 89, no. 11, pp. 1917–1922.
16. Jain, S.L., Ghude, S.D., Kumar, A., Arya, B.C., Kulkarni, P.S., Bajaj, M.M. 2005. "Continuous Observations of Surface Air Concentration of Carbon Dioxide (CO_2) and Methane (CH_4) at Maitri, Antarctica." *Current Science*, v. 88, no. 12, pp. 1941–1948.
17. Sachin, D.G., Jain, S.L., Arya, B.C., Kulkarni, P.S., Kumar, A. and Ahammed, N. 2006. "Temporal and Spatial Variability of Surface Ozone at Delhi and Antarctica." *Int. J. of Climatol*, v. 26, pp. 2227–2242.

18. *Sachin, D.G., Jain, S.L., Kulkarni, P.S., Kumar, A. and Arya, B.C. 2005. "Year to Year Variation of Ozone Hole over Schirmacher Region of East Antarctica: A Synopsis of Four-Year Measurement." *Indian Journal of Radio & Space Physics*, v. 35, Aug 2006, pp. 253–258.
19. Naithani, J., Dutta, H.N. and Pasricha, P.K. 1995. "Evaluation of Turbulent Fluxes over Maitri, Antarctica.", *Boundary-Layer Met*, v. 74, pp. 195–208.
20. Naithani, J. and Dutta, H.N. 1995. "Acoustic Sounder Measurements of the Planetary Boundary Layer at Maitri, Antarctica." *Boundary layer Meteorol.*, v. 76, pp. 199–207.
21. Arora, R.S. and Nagar, V.C. 1997. "A Ground Based Millimeter-Wave Technique for Ozone Observations in Antarctica." *Ind. J. Radio & Space Phys*, v. 26, pp. 85–89.
22. Pasricha, P.K., Vijayakumar, P.N., Aggarwal, S. and Reddy, B.M. 1999. "Performance Analysis of an HF Radio Link between India and Antarctica." *J. Radio Space Physics*, v. 28, pp. 26–35.
23. Sharma, S.D. and Sharma, M.C. 1999. "Erythemal and Aerosol Studies at Maitri, Antarctica during Austral Spring of 1995." *Ind. J. Phys B*, v. 73B, pp. 463–472.
24. Deb, N.C., Chanda, A., Mazumdar, H.P. and Dutta, H.N. June 2001. "On the Structure of Horizontal Wind Flow in the Surface Layer of Maitri, Antarctica." *Indian J Phys B*, v. 75B, pp. 163–165.
25. Srivastava Manoj, K., Pasricha, P.K., Dutta, H.N. and Singh, R. 2002. "Summertime Turbulent Heat Fluxes at East Antarctica over Neighboring Rocky and Ice Shelf Sites." *Mausam*, v. 53, pp. 69–74.
26. Gajananda, Kh., Kaushik, A. and Dutta, H.N. 2004. "Thermal Convection over East Antarctica: Potential Microorganism Dispersal." *International Journal of Aerobiologia*, pp. 21–34.
27. Srivastava, A., Jain, V.K. and Dutta, H.N. Jan 2004. "Surface Wind Characterization at an Antarctic Coastal Station, Maitri." *Mausam*, v. 55, pp. 95–102.
28. Gajananda, Kh. and Dutta, H.N. 2005. "Terrestrial Vegetation Community Structure and Biomass of the Schirmacher Oasis Ecosystem, East Antarctica." *Int J. Ecology and Development*, v.. 3, no. Ma05, pp. 39–64
29. Pasricha, P.K. and Reddy, B.M. 1990. "Evaluation of the Structure Parameter Cn2 over the Sea-Surface." *IEE Proc. (Part F)*, v. 137, p. 384
30. Pasricha, P.K. et al. 1991. "Characteristics of Atmospheric Turbulence in the Surface Layer over Antarctica.", *Boundary Layer Meteorology*, v. 57, p. 207.
31. Singh, R., Pasricha, P.K., Sharma, M.C. and Srivastava, B.N. 1992. "Multi Wavelength Measurements of Atmospheric Turbidity And Determination of the Fluctuations in Total Ozone over Antarctica." *Atmospheric Environment*, v. 26a, no. 4, pp. 525–530.

32. Sharma, M.C. and Srivastava, B.N. 1992. "Ultraviolet Radiation Received in Antarctica in Comparison with the Indian Region." *Atmospheric Environment*, v. 26a, no.4, pp. 731–734.

Patent

"Detection of earthquake precursor by SODAR", by B S Gera, H N Dutta, Gurbir Singh and V K Ojha, patent filed on January 21, 2003. The patent has been filed by CSIR under No. 59 NF 04.

Acknowledgments

The authors wish to express their sincere gratitude to Head, Radio & Atmospheric Science Division; Dr. Vikram Kumar, Director NPL; and also the former NPL directors for their encouragement during the progress of the work. Thanks are also due to the NCAOR, Department of Ocean Development, and CSIR for logistic and financial assistance. Thanks are also due to all the leaders and team members of Indian Antarctic expeditions for their help and cooperation to carry out the R&D work successfully.

References

Arya, B.C. and Jain, S.L. 1997. "The Measurement of Water Vapour at Maitri, Antarctica." *Indian J. Radio and Space Physics*, v. 26, p. 117.

Ball, F.K. 1957. "The Katabatic Winds of Adelie Land and King George. V. Land." *Tellus*, v. 9, pp. 201–208.

Dutta, H.N., Srivastava Manoj, K., Pasricha, P.K., Gajananda, K.H., Naithani, J., Deb, N.C., Ojha, D. and Gupta, V.B. 1999. "Cyclone Induced Warming over East Antarctic Region during June 1994." Under communication.

Ghude Sachin, D., Ashok Kumar, Jain, S.L., Arya, B.C. and Bajaj, M.M. 2005a. "Comparative Study of Total Ozone Column (TOC) Over Maitri, Antarctica During 1997, 2002 and 2003." *Int. J. Remote Sensing*, v. 26, pp. 3413–3422.

Ghude Sachin, D., Jain, S.L., Arya, B.C. and Bajaj, M.M. 2005b. "Comparison of Total Column Water Vapour Measured at New Delhi, Leh, Hanle and Maitri (Antarctica)." *Indian J. Radiol & Space Physics (2004)*, v. 34, pp. 264–268.

Hanjura, A.K., Sharma, S.D., Shaarma, M.C. and Srivastava, B.N. 1988. "Turbidity and Ultra Violet Radiation Measurements at Antarctica during VI Indian Scientific Expedition NPL Res." Report No.82-02.

Hanjura, A.K. and Singh, R. 1995. "UV-B Radiation Intensity Measurements at Antarctica." *Department of Ocean Dev. Tech. Rep. No. 8*, pp. 17–34.

Jain, S.L. 1987. "Inversion Techniques to Get Vertical Profiles of Ozone in Atmosphere Using Laser Heterodyne System." *Indian J. Radiol Space Phys*, v. 16, p. 324.

Jain, S.L. 1996. "Laser Heterodyne System for Measurement of Minor Constituents of Atmosphere." *Review Paper in Indian J. Radiol Space Phys.*, v. 25, pp. 309-317.

Jain, S.L., Sachin, D.G. and Arya, B.C. 2003. "Signature of Early Ozone Hole Recovery during 2002." *Current Science*, v. 86, p. 963.

Jain, S.L., Sachin, D.G., Arya, B.C., Kumar, A. and Bajaj, M.M. 2005a. "Increase in Total Water Vapour Column and Surface Temperature at Maitri, Antarctica." *Current Sci*, v. 89, pp. 1917–1922.

Jain, S.L., Sachin, D.G., Kumar, A., Arya, B.C., Kulkarni, P.S. and Bajaj, M.M. 2005b. "Continuous Observations of Surface Air Concentration of Carbon Dioxide (CO_2) and Methane (CH_4) at Maitri, Antarctica." *Current Science*, v. 88, pp. 1941–1948.

Kaplan, L.D. 1959. "Inference of Atmospheric Structure from Remote Radiation Measurements." *J. Opt Soc Am*, v. 49, p. 1004.

Morys, M. and Berger, D. 1993. "The Accurate Measurement of Biological Effective Ultraviolet Radiation." *The International Symposium on High Latitude Optics*, Tromso, Norway.

Naithani, J. and Dutta, H.N. 1995. "Acoustic Sounder Measurements of The Planetary Boundary Layer at Maitri, Antarctica." *Boundary Layer Meteorol*, v. 76, pp. 199–207.

Naithani, J., Dutta, H.N. and Pasricha, P.K. 1995. "Evaluation of Turbulent Fluxes over Maitri, Antarctica." *Boundary-Layer Meterol*, v. 74, pp. 195–208.

Naithani, J., Dutta, H.N., Pasricha, P.K., Reddy, B.M. and Aggarwal, K.M. 1994. "Evaluation of Heat and Momentum Fluxes over Maitri, Antarctica." *Boundary Layer Meteorol*, v. 74, pp. 195–208.

Nath, S., Singh, R. and Srivastava, M. 2000. "Solar UV-B and Atmospheric Aerosol Studies at Antarctica, Scientific Report 2000." *Department of Ocean Dev. Tech. Publication No. 15*, pp. 20–42.

Parish, T.R. and Bromwich, D.H. 1998. "A Case Study of Antarctic Katabatic Wind Interaction with Large Scale Forcing." *Monthly Weather Rev.*, v. 126, pp. 199–209.

Parish, T.R. and Waight, III T.R. 1987. "The Forcing of Antarctic Katabatic Winds." *Mon. Wea. Rew.*, v. 115, pp. 2214–2226.

Parish, T.T., Petter, P. and Wendler, G. 1993. "A Numerical Study of the Diurnal Variation of the Adelie Land Katabatic Wind Regime." *J. Geophy. Res.*, v. 98, pp. 12933–12947.

Parulekar, A.H., Ansari, Z.A. and Harkantra, S.N. 1983. "Benthic Fauna of the Antarctic Ocean-Quantitative Aspects." *DOD Tech Rep No. v. 1*, p. 213.

Pasricha, P.K., Singh, R., Dutta, H.N. and Reddy, B.M. 1991. "Boundary Layer." *Meteorol.*, v. 57, pp. 207–217.

Philpot, H.R. and Zillman, J.W. 1970. "The Surface Temperature Inversion over the Antarctic Continent." *J. Geophys. Res.*, v. 75, 4161–4169.

Rothman, L.S., Camache, R.R., Goldman, A., Brown, L.R., Toth, R.A., Pickett, H.M., Poynter, R.L., Flaud, J.M., Camy-Peyret, C., Barbe, A., Husson, N., Rinsland, C.P. and Smith, M.A.H. 1987. "The HITRAN data base 1986 edition." *Appl. Opt.*, v. 26, pp. 40–58.

Schwerdtfeger, W. 1970. "The Climate of the Antarctic." In: Orvig, S., ed. *Climate of the Polar Region*. Amsterdam: Elsevier, pp. 253–355.

Schwertfeger, W. 1984. *Weather and Climate of the Antarctic.* Amsterdam: Elsevier, 261 pp.
Sharma, M.C. and Srivastava, B.N. 1992. "Ultra Violet Radiation Received in Antarctica in Comparison with the Indian Region." *Atmospheric Environment,* v. 26A, pp. 731–734.
Singh, R. 1997. "Standardisation of Solar UV-B and Atmospheric Aerosol Characteristic through Solar Radiation Measurements." *MAPAN J. Met. Soc. of India,* v. 12, pp. 149–156.
Singh, R., Jain, S.L., Arya, B.C. and Shambhu, N. 1999. *"Measurement of Atmospheric Aerosol, UV-B Radiation, Surface Ozone, Total Ozone, Water Vapour etc. at Hanle (Ladakh), N.P.L. Project, 'High Altitude Environmental Monitoring at Hanle.'"* A Campaign Completion Report.
Singh, R., Pasricha, P.K., Sharma, M.C. and Srivasta, B.N. 1992. "Multiwavelength Measurement of Atmospheric Turbidity and Determination of the Fluctuations in the Total Ozone over Antarctica." *Atmospheric Environ.,* v. 26-A, pp. 525–530.
Singh, N.P., Trivedi, K.I., Rathore, L.S. and Rizvi, S.R.H. 1986. "Meteorological Observations: 1983–84 Antarctic Summer." DOD Tech Pub No. 3, pp. 21–47.
Singh, R., Sharma, M.C. and Tanwar, R.S. 1997. "Characterisation of Response of the Atmospheric Environment on the Solar UV-B Induced Erythemal Action." *MAPAN, J. Met. Soc. of India,* v. 12, pp. 186–189.
Srivastava, B.N. and Sharma, M.C. 1979. "Design and Development of Solar UV-B Photometer." *RSD Tech. Rep. 101,* N.P.L., New Delhi.
Streten, N.A., 1980. "Some Synoptic Indices of the Southern Hemisphere Mean Sea Level Circulation 1972–77." *Mon. Wea. Rev.,* v. 108, 18–36.
Tripathi, O.P. and Jain, S.L. 2003. "Abrupt Change in Total Ozone Column over Indian Antarctic Station." *Indian Journal of Radio & Space Physics,* v. 32, pp. 38–42.
Turpin, T.M. 1981. "Spectrum Analysis Using Optical Processing." *Procs. IEEE,* v. 69, p. 79.
Van Loon, H. 1966. "On the Annual Temperature Range over the Southern Oceans." *Geogr Rev.,* v. 56, pp. 497–515.
Van Loon, H. 1967. "The Half Yearly Oscillations in Middle and High Southern Latitudes and Coreless Winter." *J. Atmosph Sci.,* v. 24, pp. 472–486.
Van Loon, H. 1984. "Climate of Oceans." In: *World Survey of Climatology,* Vol 15, Amsterdam: Elsevier, XVIII, p. 716.
Van Loon, H. and Roger, J.C. 1984. "Interannual Variations in the Half Yearly Cycle of Pressure Gradient and Zonal Wind at Sea Level on the Southern Hemisphere." *Tellus,* v. 36A, pp. 76–86.
Woodward, W.A. and Gray, H.L. 1993. "Global Warming and the Problem of Testing for Trend in a Time Series Data." *J. Climate,* v. 6, pp. 953–962.

7

Atmospheric Electricity and Aerosol Investigations over Antarctica

A. K. Kamra and C. G. Deshpande

CONTENTS

Introduction

The atmospheric electricity measurements made over land surface are often contaminated due to the presence of atmospheric aerosols and the radioactivity of the ground and its daughter products. The manifestation of the electrode effect and the occurrence of turbulence and convection currents are other sources of contamination for the atmospheric electrical measurements on the ground (Hoppel, 1967). Since the local effects produced by these contaminants mostly dominate in determining the values of electrical parameters, land measurements are generally not suitable to study global effects. On the other hand, the variations in atmospheric electrical parameters made at some "clean places" represent global effects and are therefore useful for studying the behavior of the global electric circuit.

The atmospheric electric conductivity measurements made over open oceans are important, as these are free from the presence of anthropogenic aerosols and the radioactivity of the ground. It is widely recognized, therefore, that the atmospheric electric conductivity over open oceans acts as an index of the background air pollution. The value of electric conductivity over oceans can also be influenced by some meteorological conditions such as fog, cloud, and strong winds causing wave-breaking on the sea surface.

Over oceans, part of the marine aerosol originates directly from the ocean, but a large fraction of it could be advected from the continents or formed within the atmosphere. Submicron-size aerosols play an important role in the formation of cloud droplets in the precipitation process and are of vital importance in assessing the background air pollution over oceans. The number, size distribution, and composition of the marine aerosol particles vary with time and space in response to several different processes such as the transport of continental air, the bursting of whitecaps, the oxidation of gases emitted by the oceans, and so on.

Background aerosols play an important role in the radiation budget, as they directly interact with solar terrestrial radiation. They influence the budget indirectly also by acting as cloud condensation/ice nuclei to form water droplets/ice crystals, thus involving them in the cloud formation processes. Sub-micrometer particles are suspended throughout the troposphere and lower stratosphere over the ice sheet of Antarctica. Their concentration shows a strong seasonal variation, being as low as a few cm^{-3} in winter to several thousand cm^{-3} in summer (Hogan, 1975).

We report here our measurements of the atmospheric electric field, conductivity, concentration, and size distribution of submicron aerosol particles made at Maitri from January 10 to February 24, 1997, during the 16th expedition from December 12, 1996, to April 5, 1997. Also reported are the measurements of the electric conductivity and the concentration and size distribution of submicron aerosol particles made onboard *Polar Bird* along the cruise track during the onward (December 15, 1996 to January 1997) and return (March 9 to April 4, 1997) journeys of the XVI expedition. Results are described in brief. The reader is advised to see the referenced research papers for detailed descriptions.

Scientific Importance of Antarctica

The remote glacial continent of Antarctica is almost free of anthropogenic aerosols and ground radioactivity. The weather remains clear 70% of the time.

High surface winds over the continent generate mechanical turbulence by friction with the rough bare ground and can make the electrode effect almost nonexistent. Deep temperature inversions suppress the local atmospheric convection. Being an area of large-scale subsidence, the convective turbulence is very small. Therefore, the continent of Antarctica provides a suitable site to investigate global changes in atmospheric electricity.

The background aerosols interact with the solar terrestrial radiation directly and indirectly and thus significantly influence the radiation budget. Slight changes in these properties and/or concentrations may cause large changes in the global atmospheric processes, leading to a global climatic change (Charlson et al., 1987). Therefore, continuous monitoring or at least occasional checking of the background level of aerosol concentrations needs to be made at sites free from anthropogenic pollutions. The Antarctic continent has no or very little activity, and air circulation over the continent prevents the direct transport of air from the lower latitudes through the lower atmosphere. Therefore, it provides an ideal site to study the background aerosols.

Research Projects

Investigations of atmospheric electricity and aerosols at Maitri, Antarctica.

Scientific Aims

To study the nature and behavior of atmospheric electric parameters and aerosols at Maitri, Antarctica.

Region of Investigation in Antarctica

Maitri

Diurnal Variation of the Atmospheric Electric Field and Conductivity at Maitri, Antarctica

Surface measurements of the atmospheric electric potential gradient and conductivity have been made at the Indian station, Maitri (70°45′ 52″ S, 11°44′3″ E, 117 m above mean sea level), Antarctica, from January 10 to February 24, 1997. The diurnal variation curve of potential gradient averaged for 20 fair-weather days is single periodic with a maximum at 1300 UT and another secondary but distinct maximum at 1900 UT and a minimum at 0100 UT. The mean value of the potential gradient is 83 V m^{-1}, and the maximum and minimum values are 1.55 and 0.73 times the mean value, respectively. The total conductivity is 2.1×10^{-14} S m^{-1} and does not show any significant diurnal variation. The diurnal variation of the potential gradient observed at

FIGURE 7.1
The field mill placed in a pit.

Maitri is like the one observed a few days later in the southern Indian ocean and the one reported by Kamra et al. (1994) with a few hours' difference in phase in the northern Indian Ocean. The results are discussed in terms of the seasonal variation of the longitudinal distribution of thunderstorm activity. It is proposed that the diurnal variation of flash activity when averaged over several days or weeks may cause a peak at an averaged time in the diurnal variation of ionospheric potential.

The experimental setup and the results are illustrated in Figures 7.1 through 7.9.

Physical Properties of Aerosols at Maitri

Measurements of submicron aerosol size distribution made at the Indian Antarctic station Maitri from January 10 to February 24, 1997, are reported. Total aerosol concentrations normally range from 800 to 1200 particles cm^{-3}, which are typical values for the coastal stations at Antarctica in summer. Aerosol size distributions are generally trimodal and open ended with a peak between 75 and 133 nm and two minima at 42 and 420 nm. Size distributions remain almost similar for several hours or even days in absence of any meteorological disturbance. Total aerosol concentration increases by approximately an order of magnitude whenever a low-pressure system passes over the station. Based on the evolution of aerosol size distributions during such aerosol enhancement periods, three types of cases have been identified. The nucleation mode in all three cases has been suggested to result from the photochemical conversion of the dimethyl sulfide (DMS) emissions transported either by the marine air or by the air from the ice-melt regions around Maitri. Subsidence of midtropospheric air during the weakening of radiative inversion is suggested as a possible source of the nucleation mode particles in the third case. Growth of the nucleation mode particles by

FIGURE 7.2
The Gerdien apparatus.

condensation, coagulation, and/or by cloud processes has been suggested to be responsible for other models in size distributions (Table 7.1).

The results are illustrated in Figures 7.10 through 7.21.

Atmospheric Electric Conductivity over the Indian Ocean during the Indian Antarctic Expedition in 1996–1997

The atmospheric electric conductivity measurements made over the Indian Ocean with a Gerdien apparatus mounted aboard MV *Polar Bird* during the XVI Indian Scientific Expedition in 1996–1997 are reported. Simultaneous 3-hourly measurements of aerosol concentrations of 13–1000 nm size and some meteorological parameters are also reported. Latitudinal variation of conductivity along the cruise route shows a minimum at 28°S. Further,

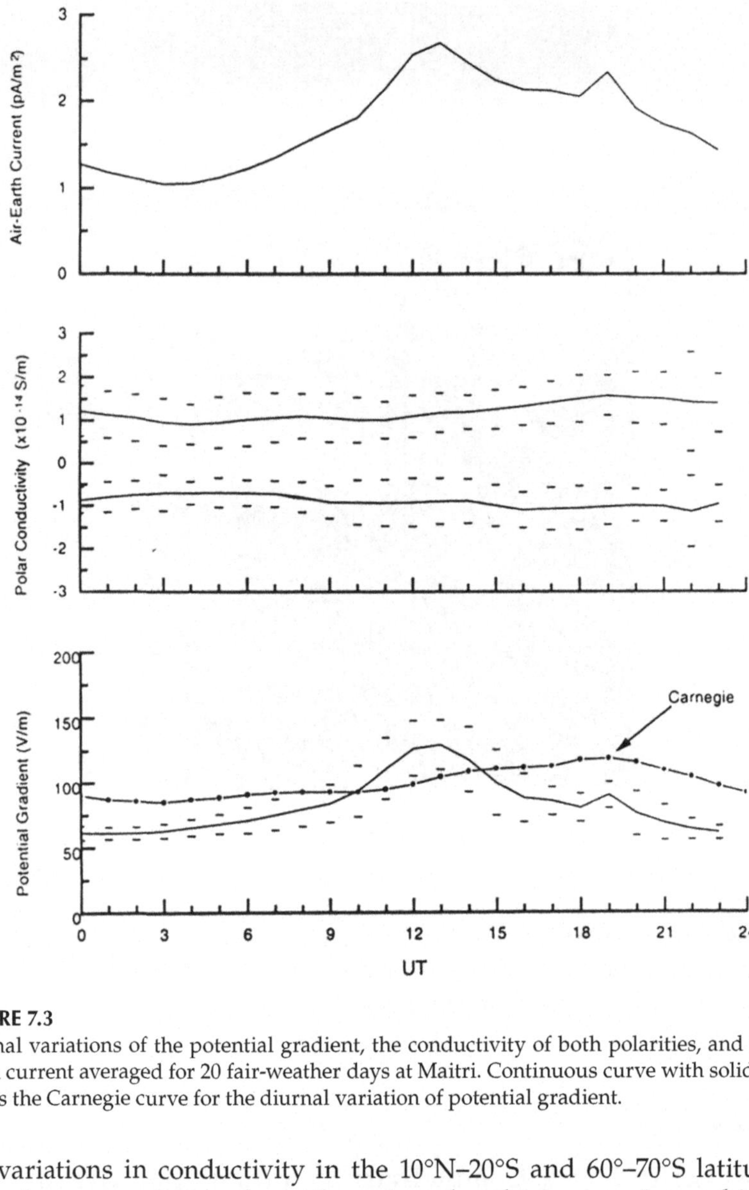

FIGURE 7.3
Diurnal variations of the potential gradient, the conductivity of both polarities, and the air-Earth current averaged for 20 fair-weather days at Maitri. Continuous curve with solid circles shows the Carnegie curve for the diurnal variation of potential gradient.

the variations in conductivity in the 10°N–20°S and 60°–70°S latitudinal belts show opposite trends on the outward and return cruises, which fall near the onset and withdrawal phases of the northeast monsoon season, respectively. The results are explained based on the well-known northward shift of the subsidence leg of the southern Hadley cell and of the position of the intertropical convergence zone during the months of March–April in this region and the observations of the cyclonic systems near the continent of Antarctica during the period of the outward cruise.

The results are illustrated in Figures 7.22 through 7.26.

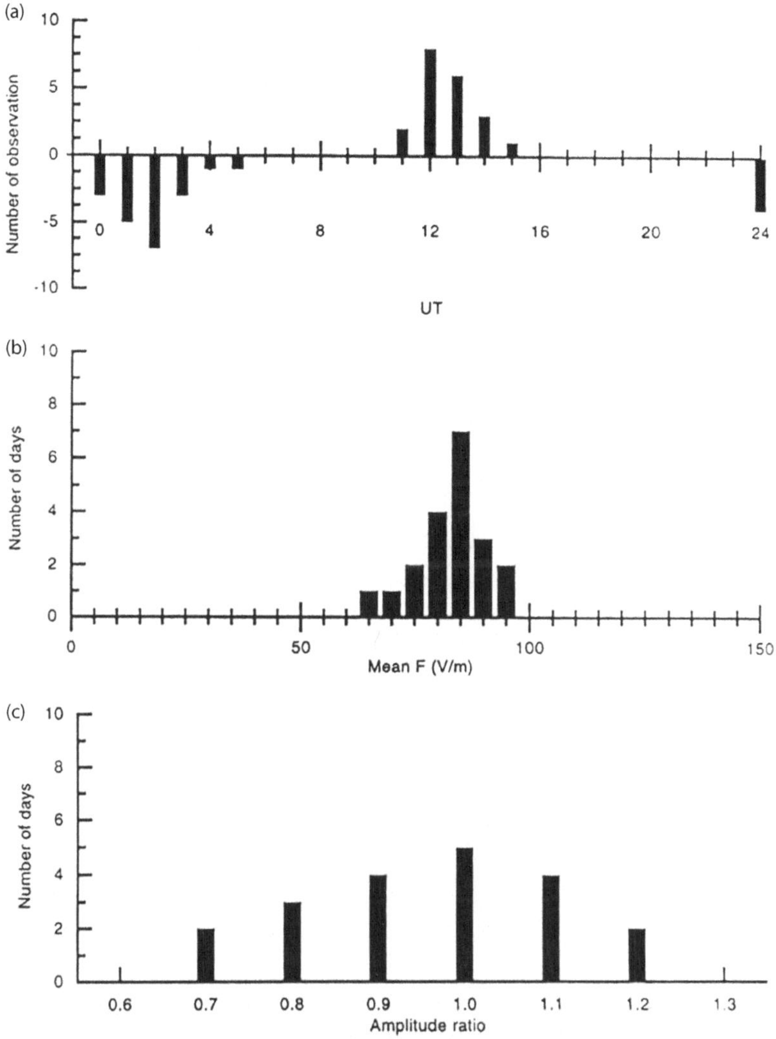

FIGURE 7.4
Statistics of surface potential gradient for 20 fair-weather days at Maitri: (a) times of maxima (represented by positive values) and minima (represented by negative values), (b) daily means, and (c) amplitude ratio (maximum–minimum m.

Aerosol Size Distributions over the Indian Ocean during the Northeast Monsoon Season

Measurements of aerosol size distributions (3–1000 nm diameter) were made over the Indian Ocean (15°N 75° + E to 70°S, 11°E) during the onward (December 12, 1996, to January 6, 1997) and return (March 9 to April 5, 1997) cruises of the

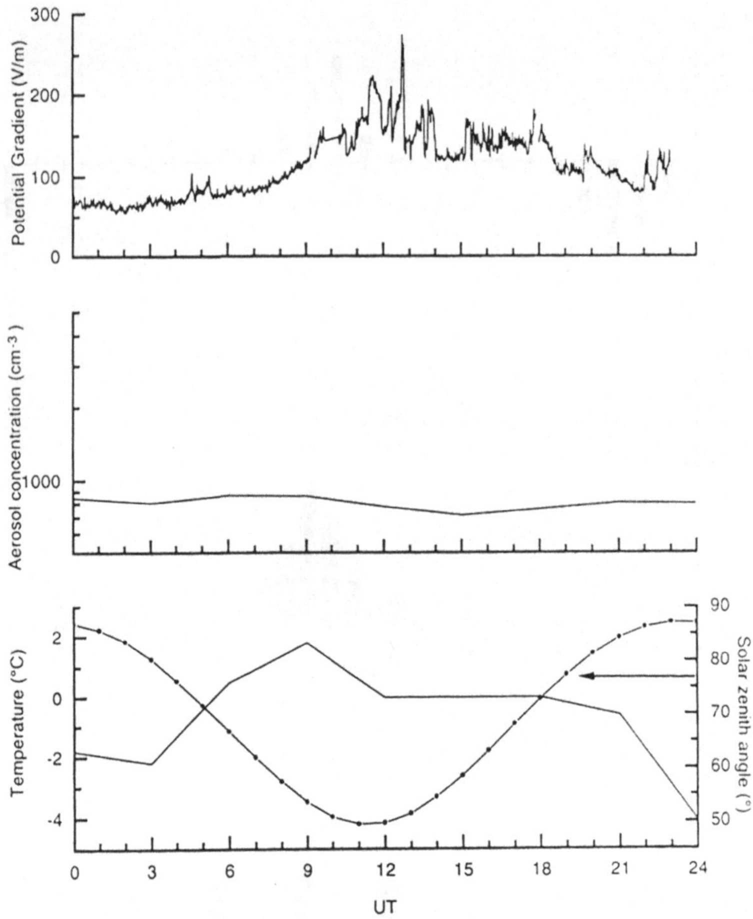

FIGURE 7.5
Diurnal variation of the 1-min averaged values of potential gradient, 1-hourly calculated values of the solar zenith angle, and 3-hourly observations of the aerosol concentrations and the atmospheric surface temperature on January 16, 1997.

XVI Indian Scientific Expedition to Antarctica. Observations show that during the January to April period, the environment over the Indian Ocean undergoes a transition from a relatively clean to a polluted one. Our observations of the large concentrations and the north-to-south positive gradient of aerosol concentration over the northern Indian Ocean strongly support the transport of aerosols and trace gases with the seasonal northeasterly winds from the Indian subcontinent to the Indian Ocean. The results also indicate the production of new particles over the oceanic region by the gas-to-particle conversion processes. During this season, because of the persistent northeasterly winds and the shift of the intertropical convergence zone (ITCZ) to the southern hemisphere in this region, the air pollutants over the northern Indian Ocean

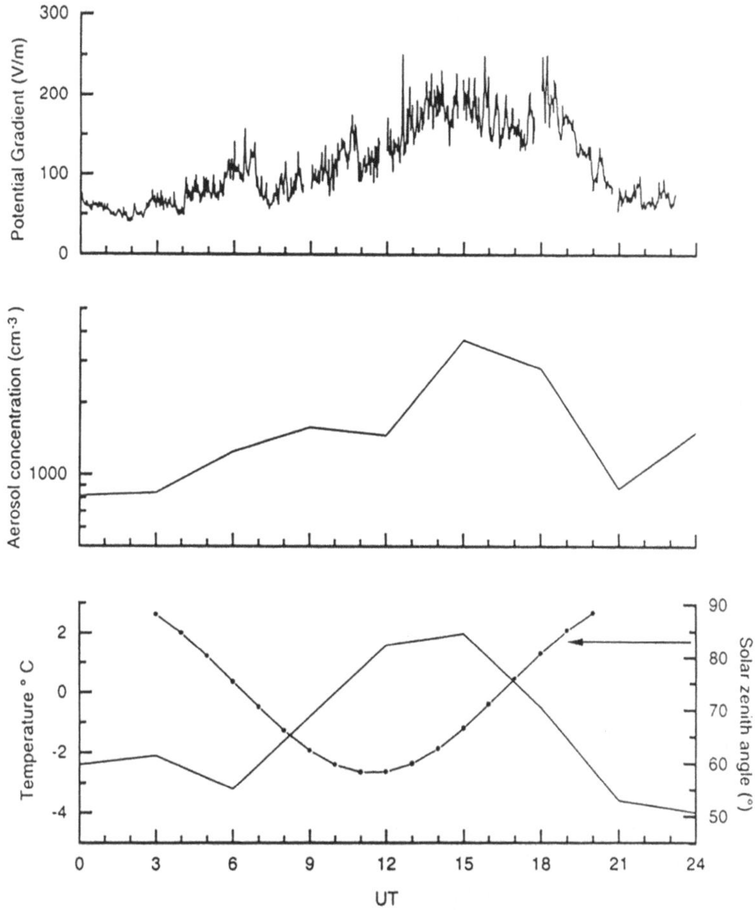

FIGURE 7.6
Diurnal variation of the 1-min averaged values of potential gradient, 1-hourly calculated values of the solar zenith angle, and 3-hourly observations of the aerosol concentrations and the atmospheric surface temperature on February 17, 1997.

are carried into the southern hemisphere with the cross-equatorial flow and reach up to the southern limit of the ITCZ. Some pockets of very high aerosol concentrations have been observed in and around the ITCZ.

Nucleation mode particles are observed in great abundance up to 30°S.

From the Indian coast to 5°N, aerosol particles are observed to have bimodal size distributions, with a maximum in accumulation mode at 133 nm diameter and a minimum for nucleation/Aitken mode particles of <74 nm in diameter. The size distributions gradually change to trimodal ones, with another minimum appearing for coarse mode particles >422 nm in the 5°N-equator belt and are generally trimodal in the southern hemisphere. Transport of nucleation mode particles from the

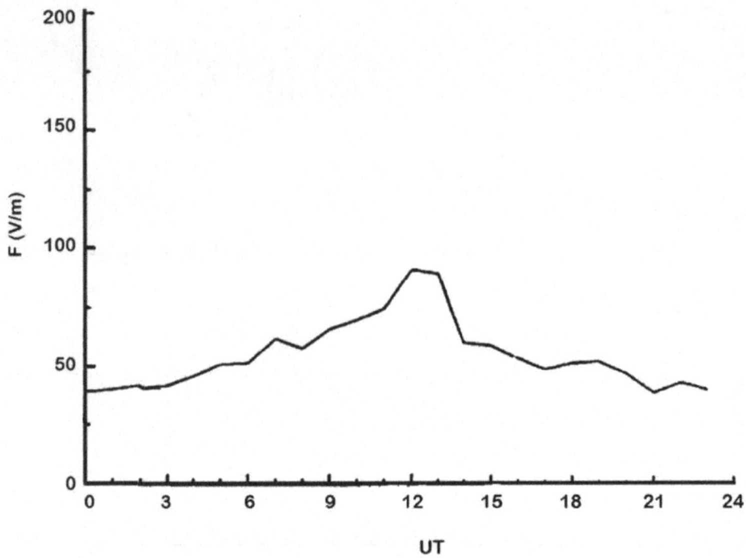

FIGURE 7.7
Diurnal variations of the potential gradient averaged for two fair-weather days over the Indian Ocean at 28°5′S, 41°38′E and 24°58′S, 50°42′E.

free troposphere to the marine boundary layer (MBL) associated with the large-scale subsidence and subsequent north-to-south transport of these and continental aerosols with the seasonal northeasterly surface winds over the Indian Ocean is proposed to explain the high concentration of nucleation mode particles observed up to the ITCZ. Size distributions change to Junge's power-law type in regions of high aerosol concentration inside the ITCZ. Comparatively small concentrations with almost uniform size distributions that are typical of the pristine air in the southern hemisphere are observed from 42°S to 56°S. Some high peaks in aerosol concentration that can be associated with the low-pressure systems that encircle Antarctica are observed in the 60–70°S latitudinal belt. Further, our observations strongly demonstrate the effect of wind direction on the land-to-ocean transport of the atmospheric aerosols.

The results are illustrated in Figures 7.27 through 7.35 and Tables 7.2 and 7.3.

Atmospheric Electric Conductivity and Aerosol Measurements during Fog over the Indian Ocean

The atmospheric electric conductivity and aerosol size-distribution measurements made over the Indian Ocean on two occasions of fog occurrence

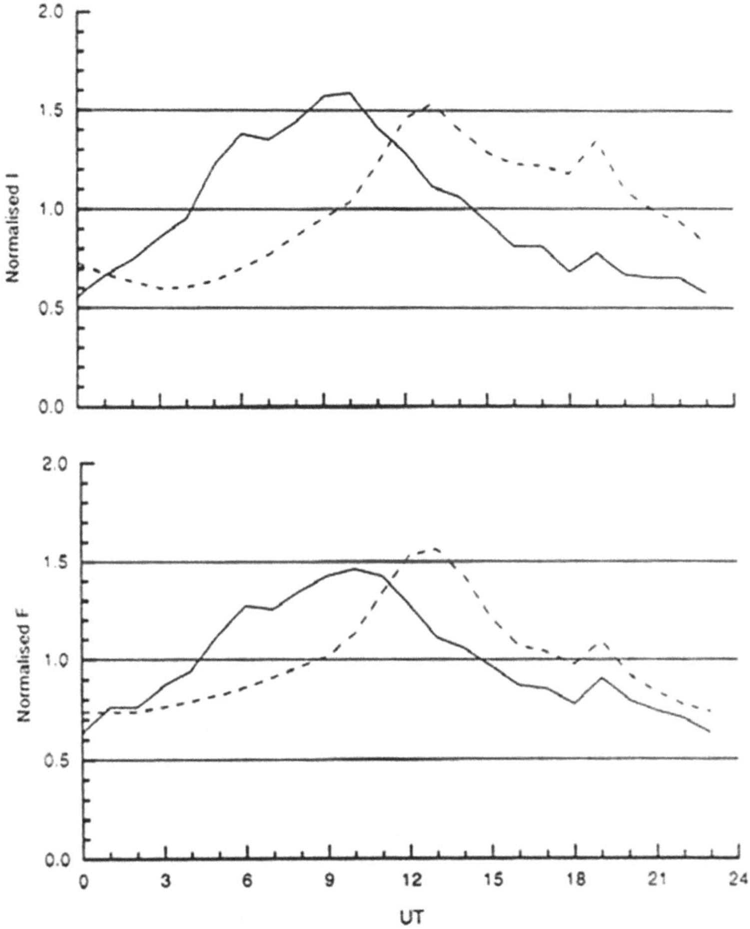

FIGURE 7.8
Diurnal variations or the normalized values of potential gradient and the air-Earth current. Continuous lines represent average of 40 fair-weather days in the Indian Ocean (Kamra et al., 1994), and dashed lines represent average of 20 fair-weather days at Maitri.

during the XVI Indian Scientific Expedition, 1996–1997, are reported. The locations and meteorological conditions associated with two fog episodes, one a warm fog and the other a cold fog, are widely different. Values of both polar conductivities, after showing a slight increase, start decreasing about 1 h before the onset of fog, and then start increasing about 0.5–2 h prior to the dissipation of fog. The conductivity values then attain their fair-weather values in case of cold fog but continue to increase and give a maximum ~ 2 h after the dissipation of fog in case of warm fog. The total number concentration of aerosol particles (between 13 and 1000 nm diameter) increases during the warm fog period but only just after the dissipation of cold fog. Aerosol size distributions show an increase in concentration of all particle sizes in warm

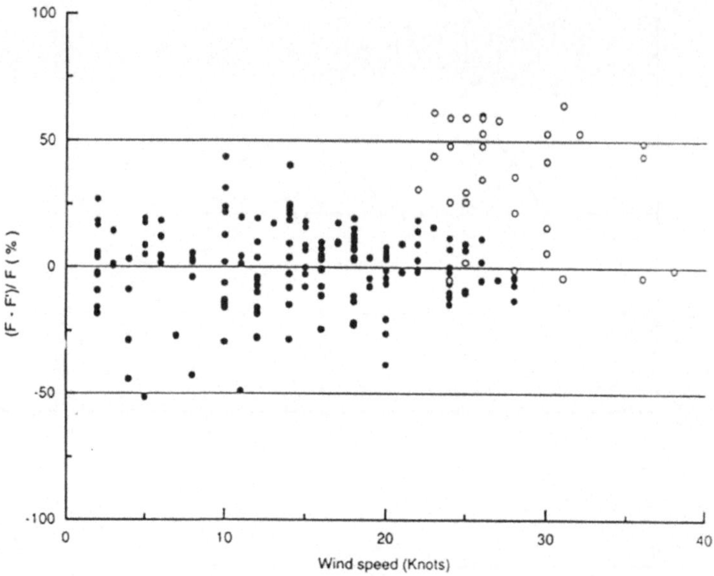

FIGURE 7.9
The percentage change in the average potential gradient for individual hours from the 20 fair-weather days' average values for the corresponding hour with respect to wind speed. Solid circles are computed from fair-weather days data and hollow circles from other days with wind speed >10 m s^{-1} (1 knot = 0.5 m s^{-1}).

TABLE 7.1

Average Modal Parameters for the Measurements Made at Maitri, Antarctica during January–February 1997

Particulars	Modal Parameters[a]	Aitken Number[b]	Accumulation Number[b]
Background Aerosol Average	N	427 ± 133	107 ± 39
	DgN	67.3 ± 18	153.2 ± 42
	σ	1.687 ± 0.007	1.706 ± 0.003
Case 1	N	750 ± 148	95 ± 25
	DgN	59.8 ± 10	175 ± 19
	σ	1.694 ± 0.002	1.706 ± 0.001
Case 2	N	3239 ± 1952	419 ± 337
	DgN	55.5 ± 19	242 ± 38
	σ	1.682 ± 0.007	1.706 ± 0.001
Case 3	N	616 ± 142	323 ± 100
	DgN	55.9 ± 7.6	278.1 ± 40
	σ	1.684 ± 0.004	1.703 ± 0.001

Note: ± refers to the atmospheric variability.

[a] N (cm^{-3}), D (geometric in nm), σ (standard deviation of the size distribution).

[b] Modal parameters are calculated from lognormal fits to the size distribution.

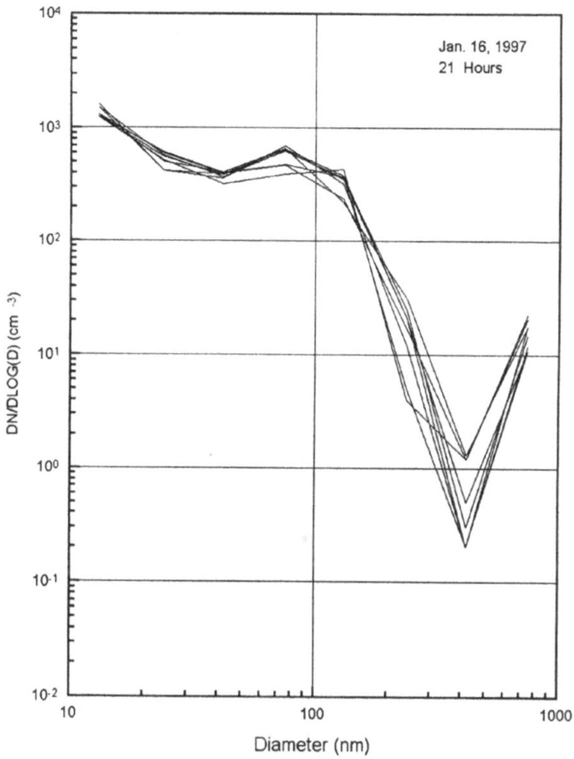

FIGURE 7.10
Typical aerosol size-distribution curves on a fair-weather day, January 16, 1997, at Maitri.

fog but mainly in concentration of the nucleation mode particles in cold fog. Possible causes of the changes observed in conductivity and aerosols and their utility in forecasting the onset and dissipation of fog are discussed.

The results are illustrated in Figures 7.36 through 7.39.

National Status

There have been practically no earlier measurements of atmospheric electric parameters and the submicron particle size distributions at Maitri. The only measurements relevant to this field are the small ion measurements carried out by the Bhabha Atomic Research Centre (Ramachandran and Balani, 1995). So, the present investigation is the first attempt in this direction.

Lal and Kapoor (1989, 1992) made measurements of the concentration of aerosol particles at Maitri and examined their characteristics associated with different surface meteorological features. Jayaraman et al. (1998) has

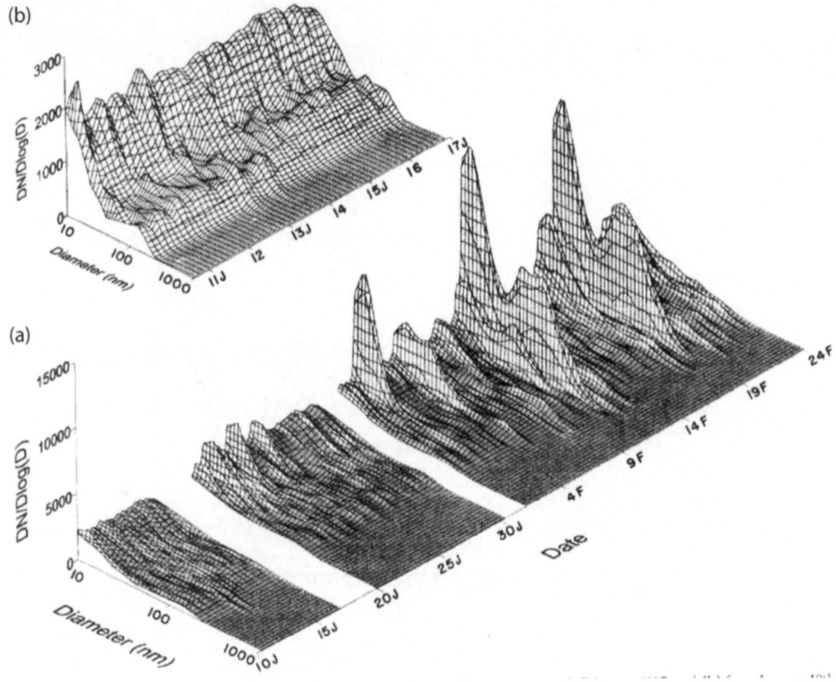

FIGURE 7.11
Three-dimensional plots of the number-size distributions of aerosols from (a) January 10th to 24th February, 1997, and (b) from January 10th to 16th, 1997.

examined the spatial and temporal distributions of the aerosol concentrations over the Indian Ocean

International Status

Atmospheric electricity measurements made over oceans (Mauchly, 1923; Ruttenberg and Holzer, 1955; Morita, 1971; Kamra et al. 1994), in the polar regions (Anderson, 1967; Kasemir, 1972; Takagi and Kanada, 1972; Park, 1976; Cobb, 1968, 1977; Byrne et al., 1993; Burns et al., 1995; Few et al., 1997; Bearing et al., 1998; Fuellekrug et al., 1999; also Frank-Kamenetsky, A. V., Burns, G. B., Troshichev, O. A., Papitashvili, V. O., Bering, E. A. and French, W. J. R. 1998. The geoelectric field at Vostok, Antarctica: Its relation to the interplanetary magnetic field and the polar cap potential, Personal communication), and over high mountains (Markson, 1977) have been used to study global effects. The unitary diurnal variation of the atmospheric electric field observed in these measurements is generally compared to and verified with respect to the Carnegie curve obtained all over world oceans

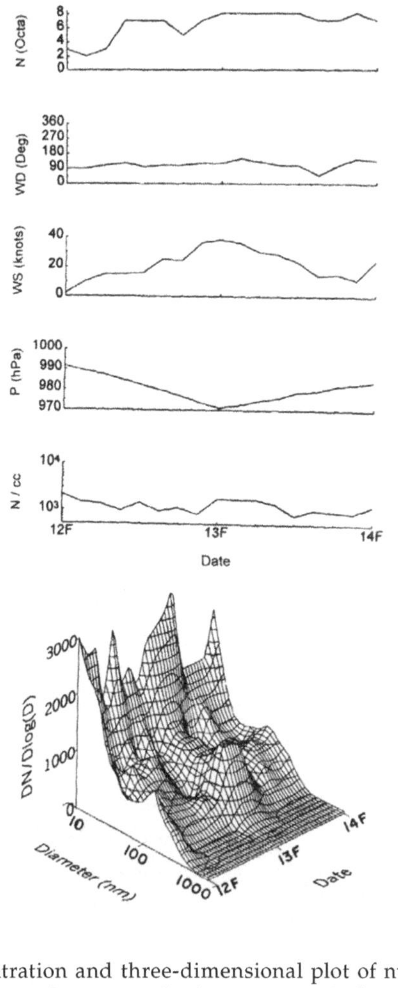

FIGURE 7.12
The total aerosol concentration and three-dimensional plot of number-size distribution of aerosols along with the surface atmospheric pressure, wind speed, wind direction, and cloud coverage during the passage of a low-pressure system over Maitri during February 1–4, 1997.

and which, as Hoppel et al. (1986, p. 155) put it, has long served as the "de facto standard with which to evaluate the validity of attempts to measure universal variations." However, significant deviations from the Carnegie curve have been reported in some of these measurements.

Fuellekrug et al. (1999) observed that the extremely low frequency (ELF) magnetic field at Arrival Heights, Antarctica, and the atmospheric electric field at the South Pole exhibit a remarkably similar diurnal variation. Park (1976), Tinsely et al. (1989), Tinsely and Heelis (1993), Burns et al. (1995), and Frank-Kamenetsky et al. (1998. The geoelectric field at Vostok, Antarctica: Its relation to the interplanetary magnetic field and the polar cap potential,

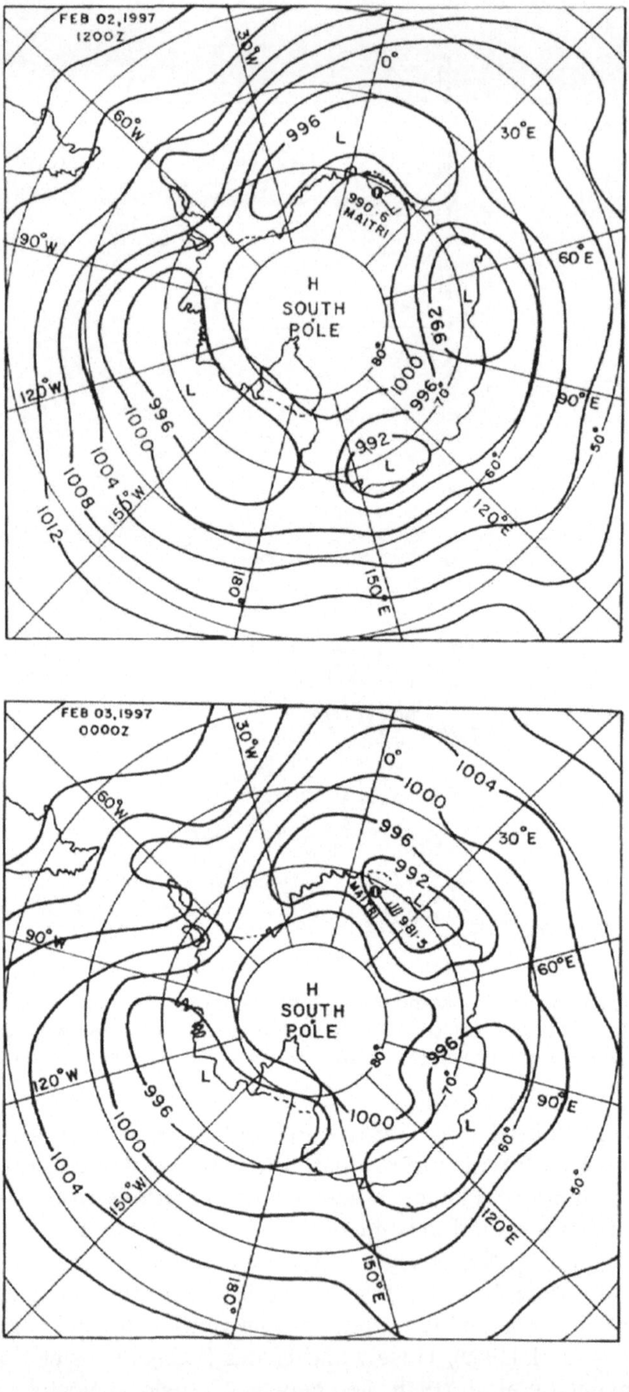

FIGURE 7.13
The six-hourly surface synoptic weather chart for 1200 hours on February 2nd, 1997 and 0000 hours on February 3rd, 1997.

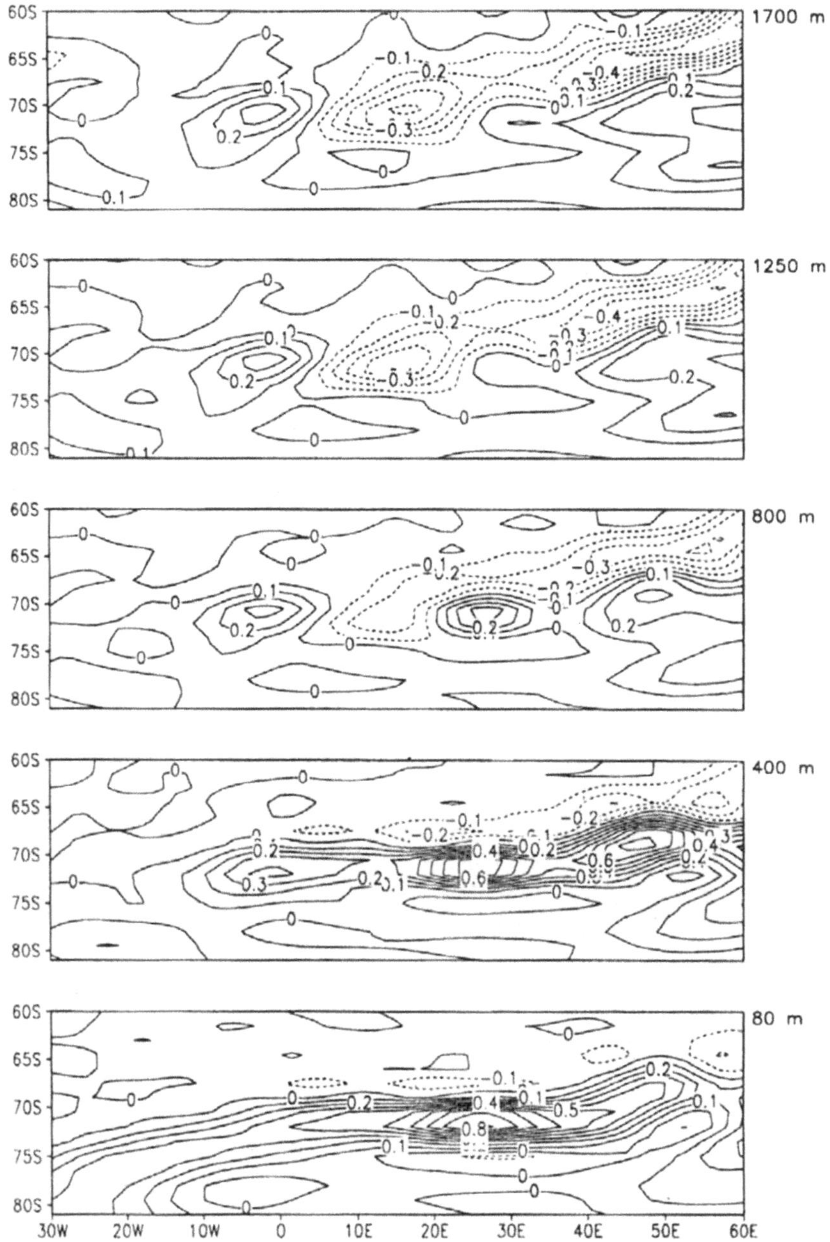

FIGURE 7.14
The vertical velocity analysis at 80, 400, 800, 1200 and 1700 levels at 1100 GMT on February 2, 1997. The vertical velocity data uses SI unit of Pascal per second. Continuous contours show upward and dotted contours downward velocities.

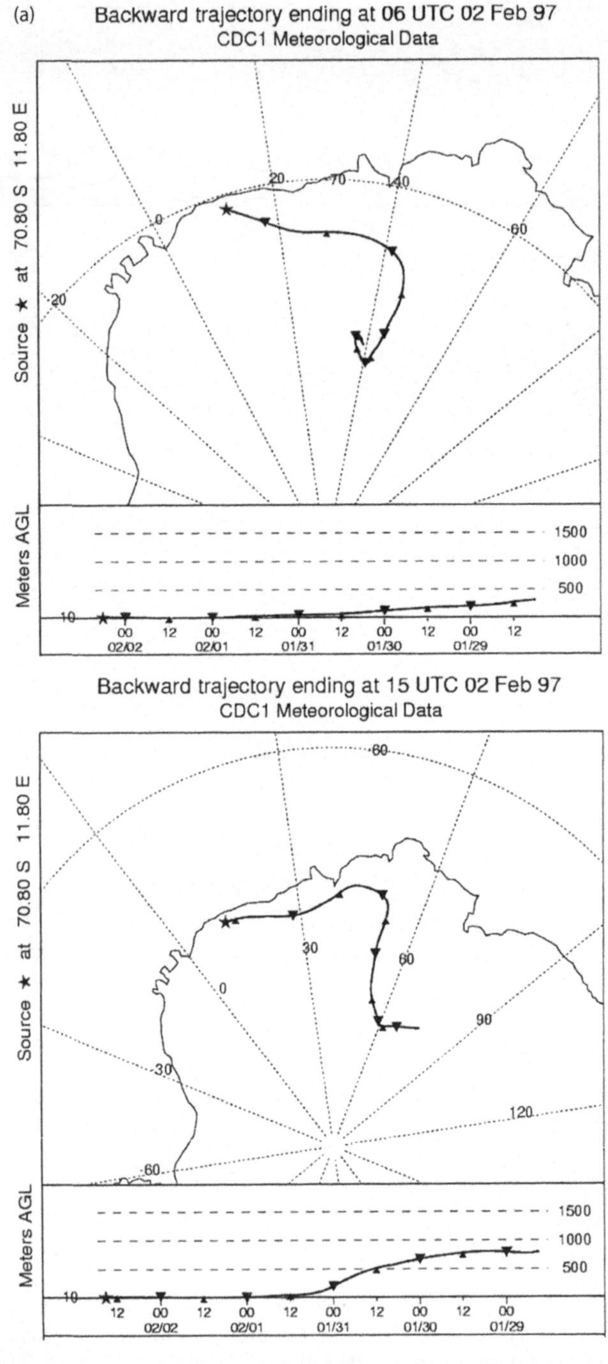

FIGURE 7.15
(a) The 5-day back trajectories at 0600 and 1500 GMT on February 2, 1997 and at 0300 GMT on February 3, 1997 at Maitri. *(Continued)*

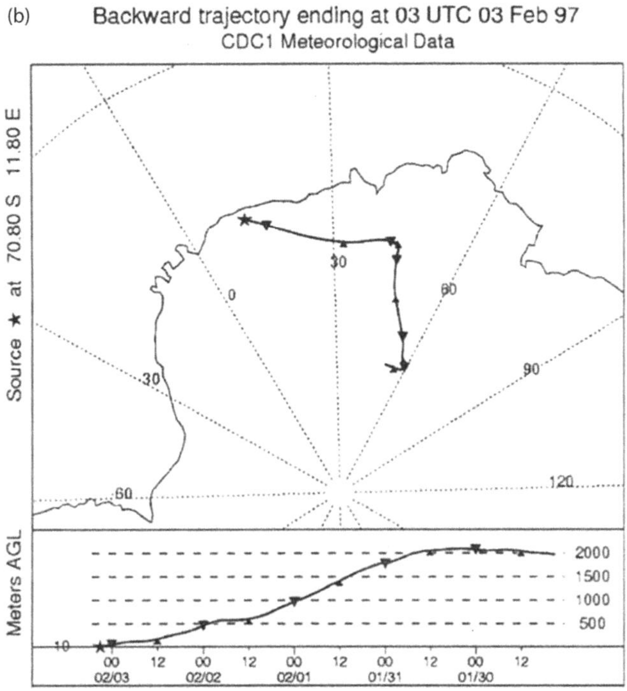

FIGURE 7.15 (Continued)
(b) The 5-day back trajectories at 0600 and 1500 GMT on February 2, 1997 and at 0300 GMT on February 3, 1997 at Maitri.

Personal communication) study solar-terrestrial relationships and the possible effects of ionospheric and magnetospheric generators on the global electric circuit from their observations at Antarctica.

Strong seasonality has been observed in condensation nuclei concentration at the coastal stations of Syowa (Iwai 1979; Ito, 1980, 1983, 1985; Ito and Iwai, 1981), Ross Island (Shaw, 1986), Georg-von-Neumayer (Jaenicke and Stingl, 1984; Jaenicke et al., 1992), and Mawson (Gras and Adriaansen, 1985). Most of the Antarctic aerosols have been found to be sulfate.

Cobb and Wells (1970) report a secular decrease in conductivity in the North Atlantic as compared to its value measured in the Carnegie cruises during 1909–1930, and they attribute it to a possible increase in the background aerosol pollution suspended in the atmosphere of the northern hemisphere. Several other oceanic measurements of atmospheric electric conductivity have been used to estimate the extension of air pollution from land to ocean (e.g., Misaki and Takeuti, 1970; Morita, 1971; Morita et al., 1973; Morita and Ishikawa, 1977; Kamra and Deshpande, 1995). Recent measurements of Kamra et al. (2001) conducted over the Indian Ocean in three consecutive northeast monsoon seasons during the Indian Ocean Experiment show that the conductivity values of the southern hemisphere are in general two to three times higher than that over the northern Indian Ocean. Further, the

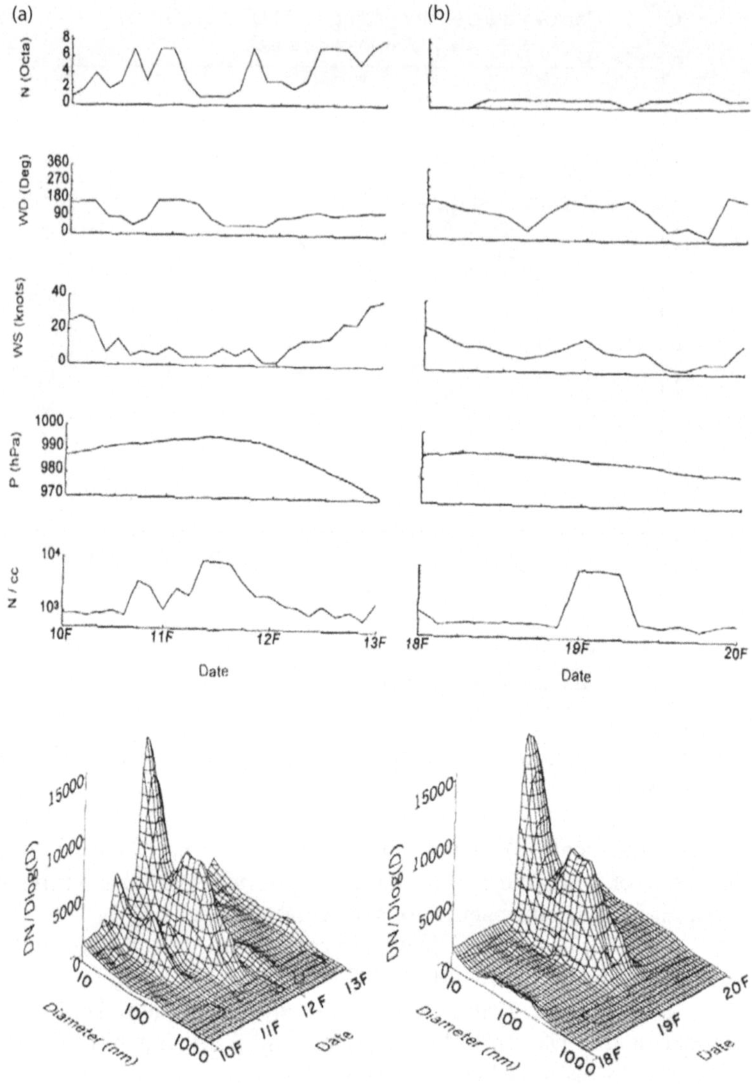

FIGURE 7.16
The total aerosol concentration and three-dimensional plot of number-size distribution of aerosols along with the surface atmospheric pressure, wind speed, wind direction and cloud coverage during the passage of a low pressure system over Maitri during (a) February 10–14, 1997 and (b) February 18–20, 1997.

north-to-south gradients of conductivity extend from the western coastlines of India to the intertropical convergence zone and have large inter- and intra-seasonal variations.

Several experiments have been conducted in the Pacific and Atlantic oceans to study the physical, chemical, and optical properties of submicron

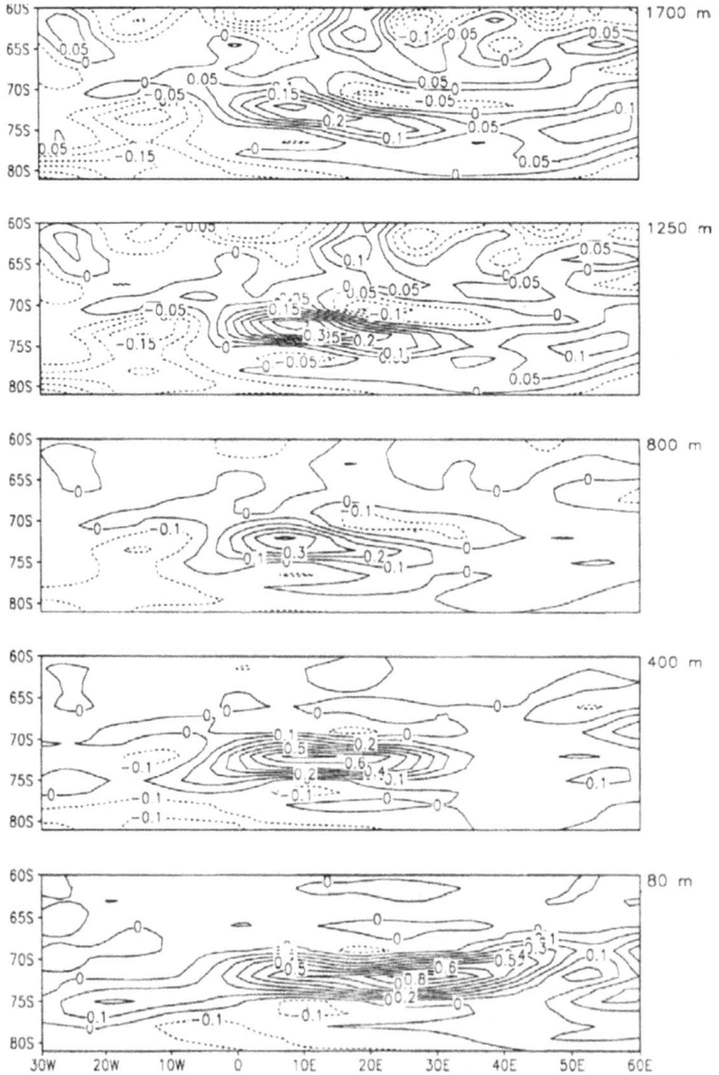

FIGURE 7.17
The vertical velocity analysis at 80, 400, 800, 1200 and 1700 levels at 2300 GMT on February 18, 1997. The vertical velocity data uses SI unit of Pascal per second. Continuous contours show upward and dotted contours downward velocities.

aerosols (e.g., Hoppel et al., 1985, 1994; Hoppel and Frick, 1990; Quinn et al., 1993, 1996; Covert et al., 1996; Bates et al., 1998, 2000; Kaufman et al., 1998; Russell et al., 1999; Raes et al., 2000). Because of these studies, several features in the unperturbed marine boundary layer size distributions and the processes that control them have been identified. However, the background

(a)

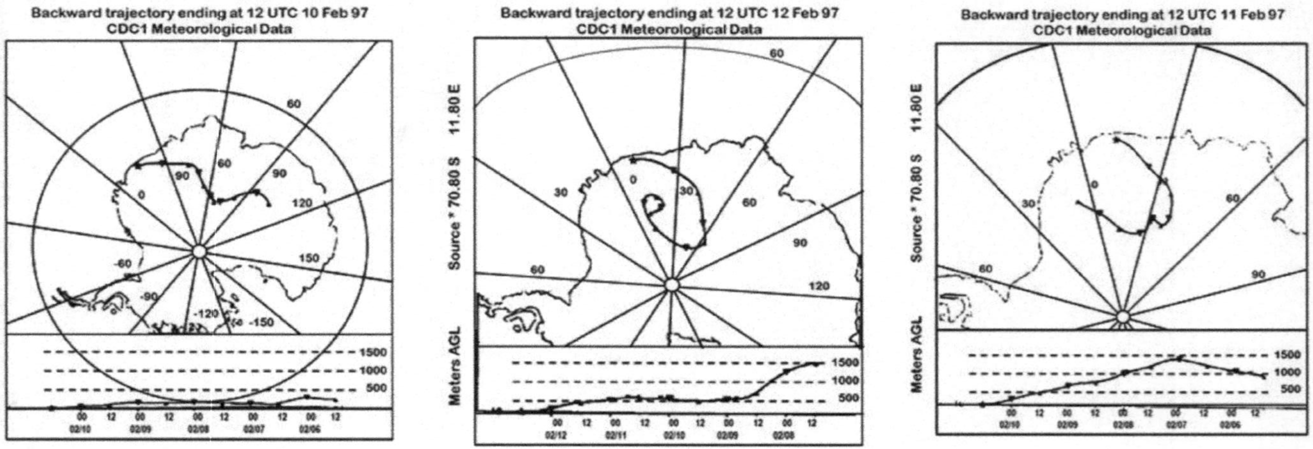

FIGURE 7.18
(a) The 5-day back trajectories at 1200 GMT on February 10, 11, and 12, 1997 at Maitri.

(*Continued*)

(a)

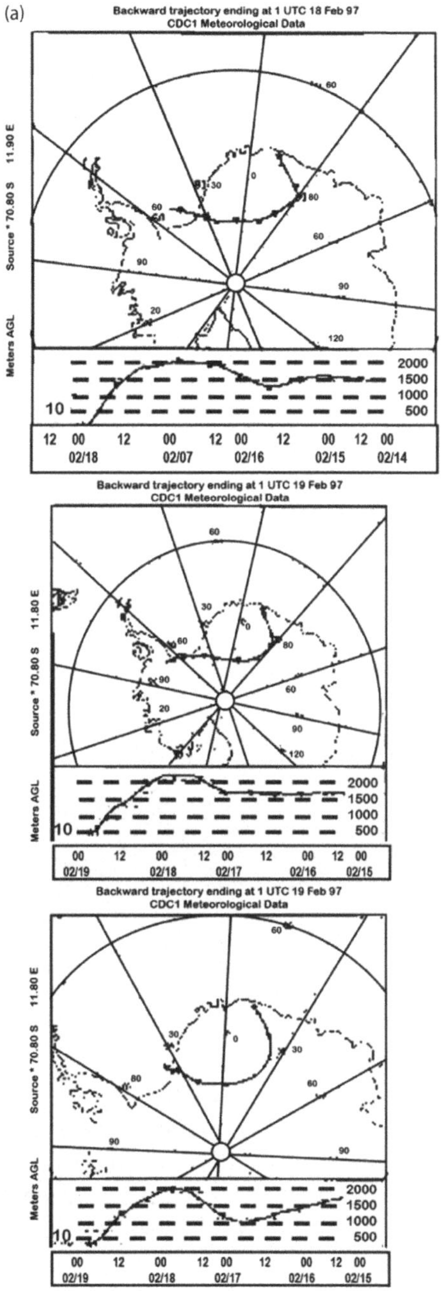

FIGURE 7.18 (Continued)
(b) The 5-day back trajectories at 1500 GMT on February 18, 0300 GMT and 1500 GMT on February 19, 1997 at Maitri.

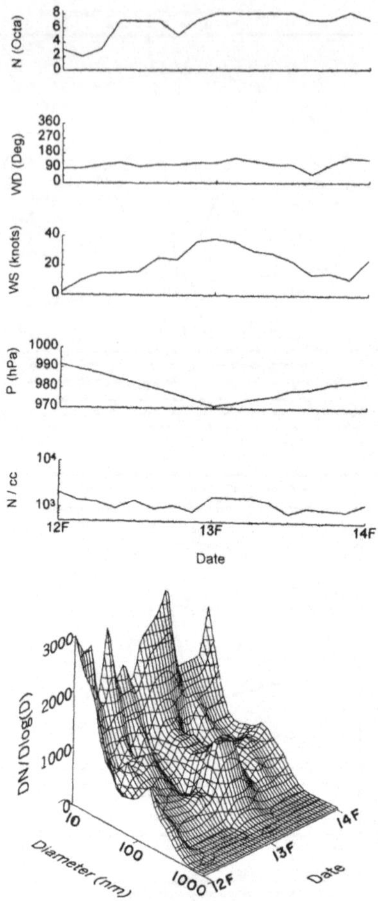

FIGURE 7.19
The total aerosol concentration and three-dimensional plot of number-size distribution of aerosols along with the surface atmospheric pressure, wind speed, wind direction and cloud coverage during the passage of a low pressure system over Maitri during February 12–14, 1997.

aerosol size distributions are substantially altered, and the radiative and cloud nucleating properties are modified if the air masses are influenced by some continent. The region of the Indian Ocean during the winter monsoon season is one such area. In spite of some measurements (Prodi et al., 1933; Savoie et al., 1987; Lal and Kapoor 1992; Rhoads et al., 1997; Jayaraman et al., 1998), detailed data on the spatial and temporal distributions of the aerosol concentrations over the Indian Ocean is lacking. The Indian Ocean south of the equator is virtually a data-void region. While compiling the available data all over the globe on the concentration and size distribution of marine aerosols, Heintzenberg et al. (2000) point out that major gaps occur in the Indian Ocean region.

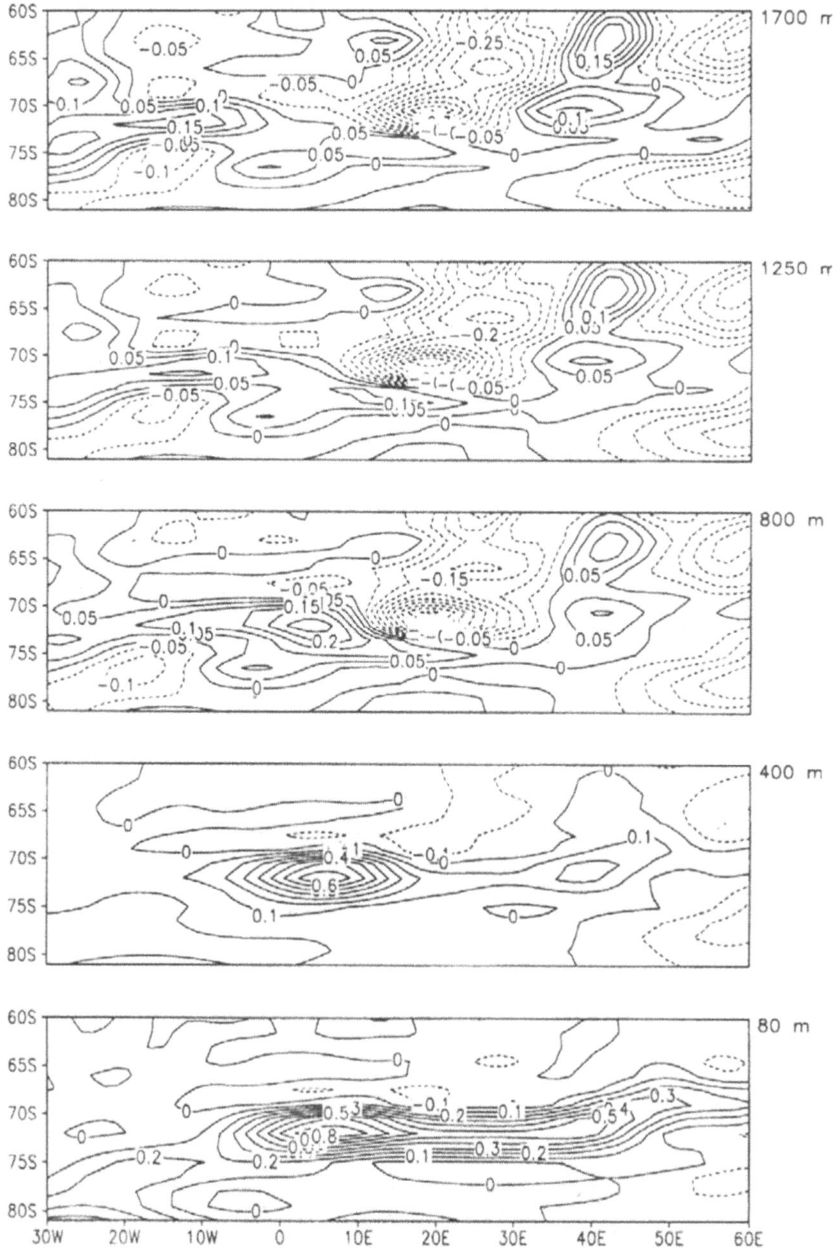

FIGURE 7.20
The vertical velocity analysis at 80, 400, 800, 1200 and 1700 levels at 2300 GMT on February 12, 1997. The vertical velocity data uses SI unit of Pascal per second. Continuous contours show upward and dottedcontours downward velocities.

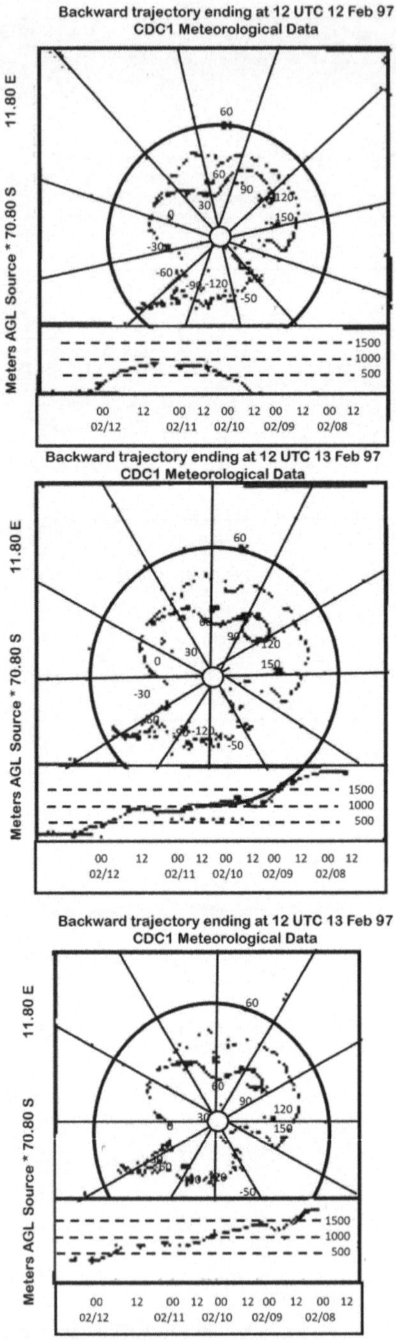

FIGURE 7.21
The 5-day back trajectories at 0900 GMT on February 12, 0000 GMT and 1200 GMT on February 13, 1997 at Maitri.

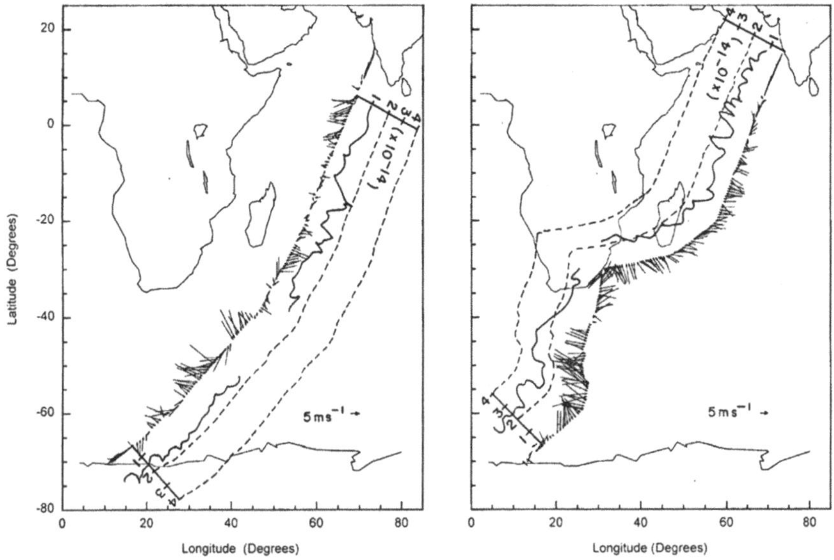

FIGURE 7.22
The 3-hourly averaged values of the total electric conductivity along the cruise tracks. A value of conductivity is plotted by taking its value as zero at the cruise track. Although dotted lines show the continued axis of conductivity, due care needs to be exercised for reading the conductivity absolute values whenever the cruise track changes its direction. The arrows show the 3-hourly values of the wind speed and direction measured on the ship.

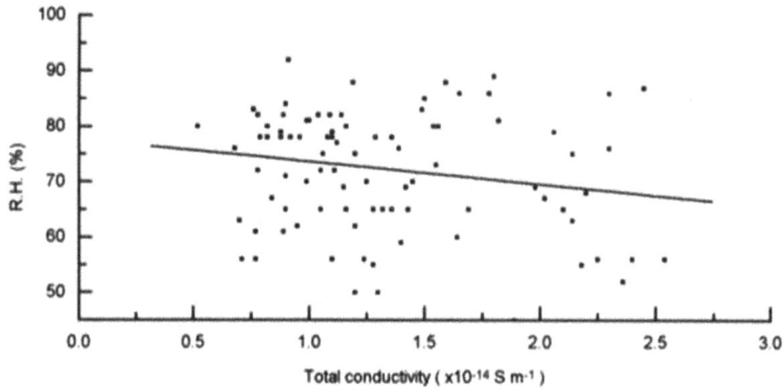

FIGURE 7.23
Variation of conductivity with relative humidity. Observations of the relative humidity made at every 3 hours are plotted against the corresponding conductivity values averaged for the preceding half an hour.

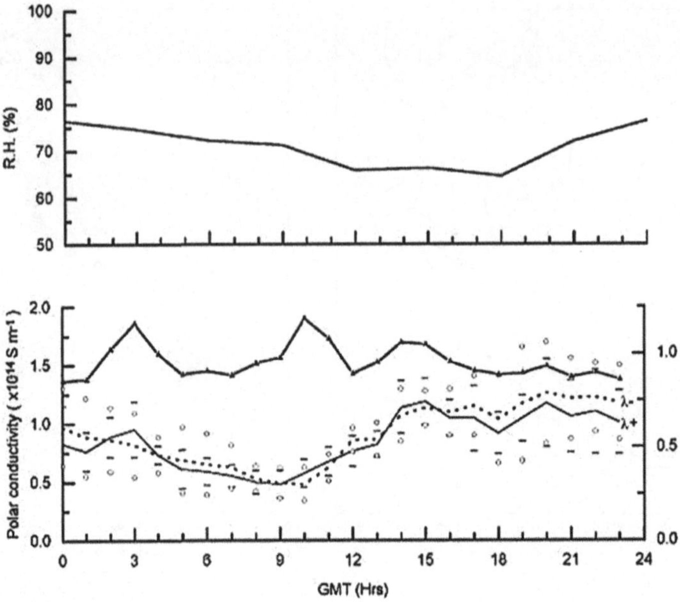

FIGURE 7.24
Diurnal variations of the polar conductivities (l_+ ______, $l-$ ....), ratio of the polar conductivities ($l_+/l-$, line with triangles), and the relative humidity (R.H.) averaged for eight fair-weather days during the cruise. Dashes and open circles show the standard deviations of the positive and negative conductivity, respectively, from their mean values.

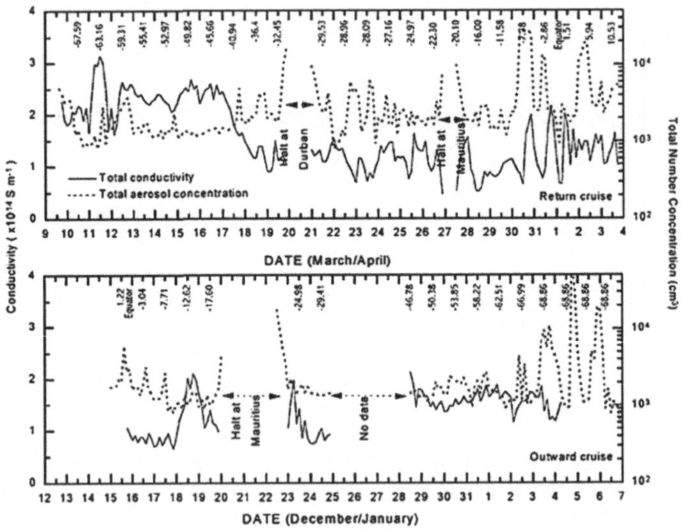

FIGURE 7.25
Variations of the total conductivity and aerosol concentrations with latitude for the outward and return cruises.

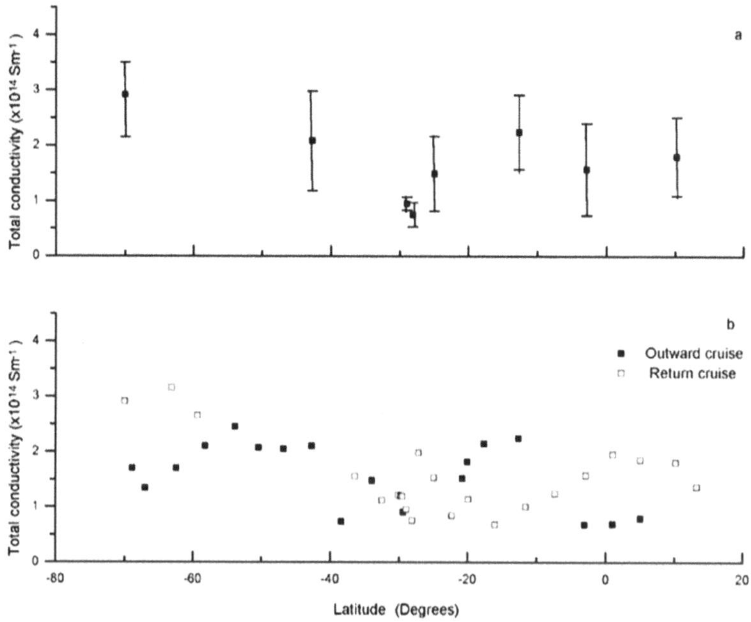

FIGURE 7.26
Latitudinal variation of the daily average total conductivity on (1) 8 fair-weather days, and (b) the fair-weather days with relaxed criteria as defined in the text.

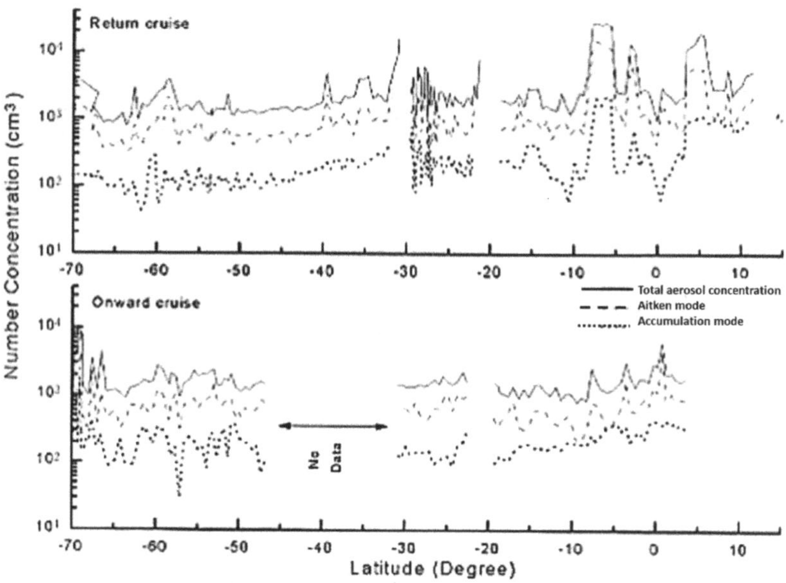

FIGURE 7.27
The total number concentration of the aerosol particles and concentrations in Aitken and accumulation modes along the onward and return cruises.

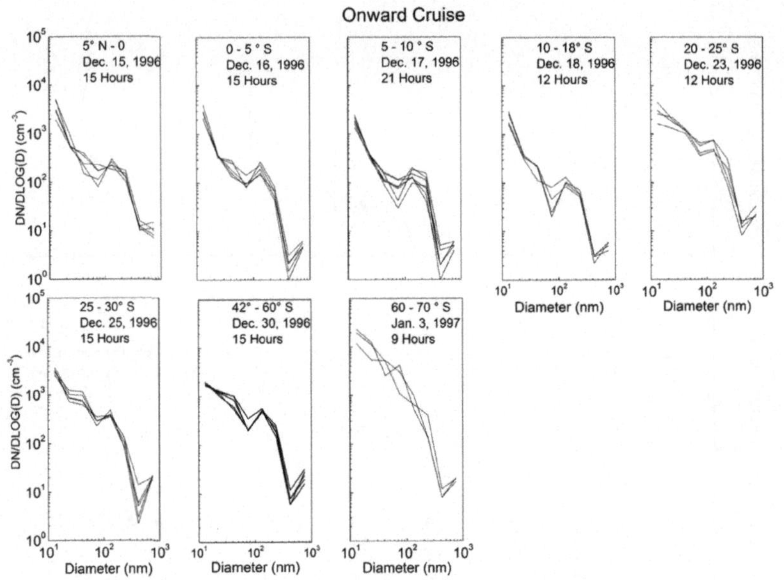

FIGURE 7.28
Aerosol size distributions that remained nearly similar continuously for several hours in different latitudinal belts during the onward cruise.

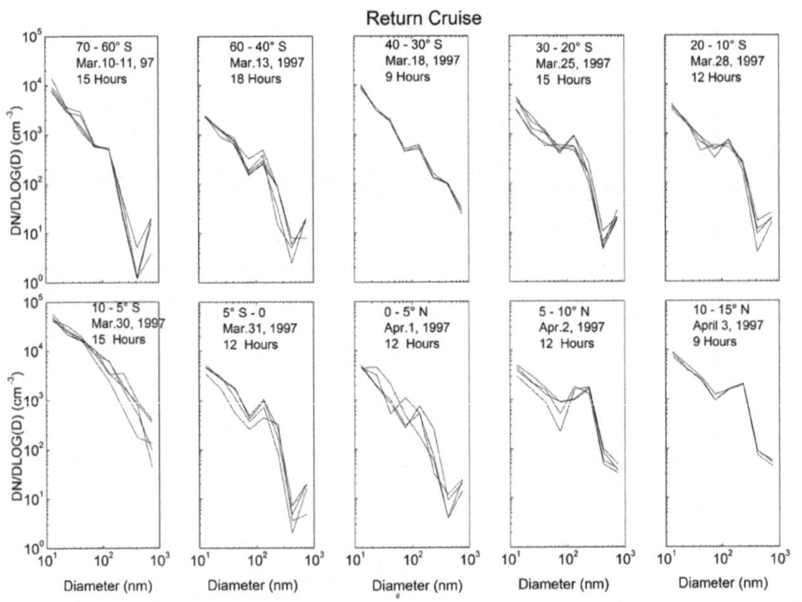

FIGURE 7.29
Aerosol size distributions which remained nearly similar continuously for several hours in different latitudinal belts during the return cruise.

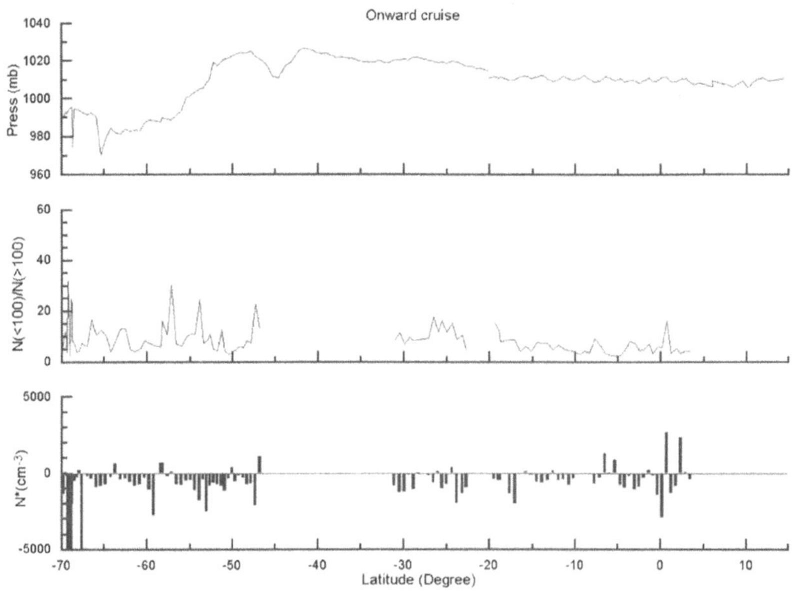

FIGURE 7.30
Plots of the difference in number concentrations of particles of diameter 42 and 24 nm, $N^* = dn(42)/d \log Dp - dn(24)/d \log Dp$, the ratio of the number concentrations of particles <100 nm to particles >100 nm, and the sea level pressure along the onward (a) and the return (b) cruises.

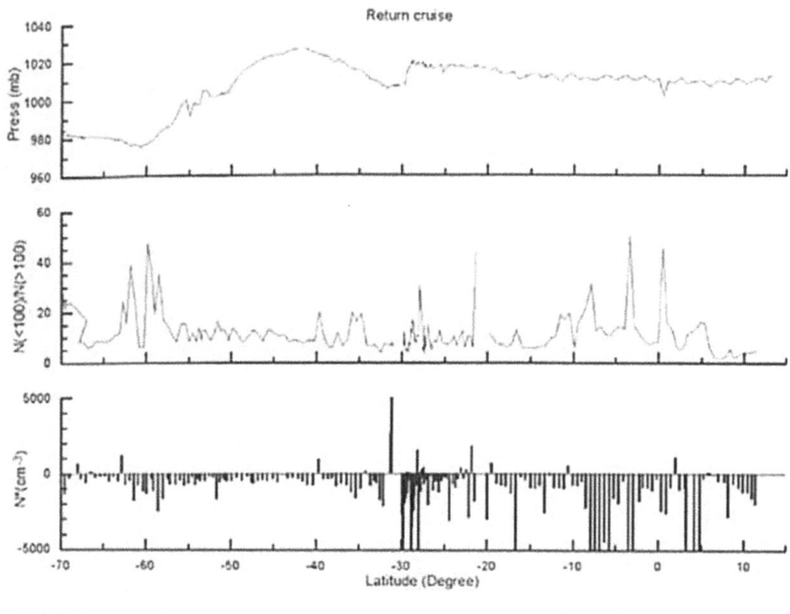

FIGURE 7.31
Plots of the difference in number concentrations of particles of diameter 42 nm and 24 nm, $N^* = dn(42)/d \log Dp - dn(24)/d \log Dp$, the ratio of the number concentrations of particles <100 nm to particles >100 nm, and the sea level pressure along the return cruises.

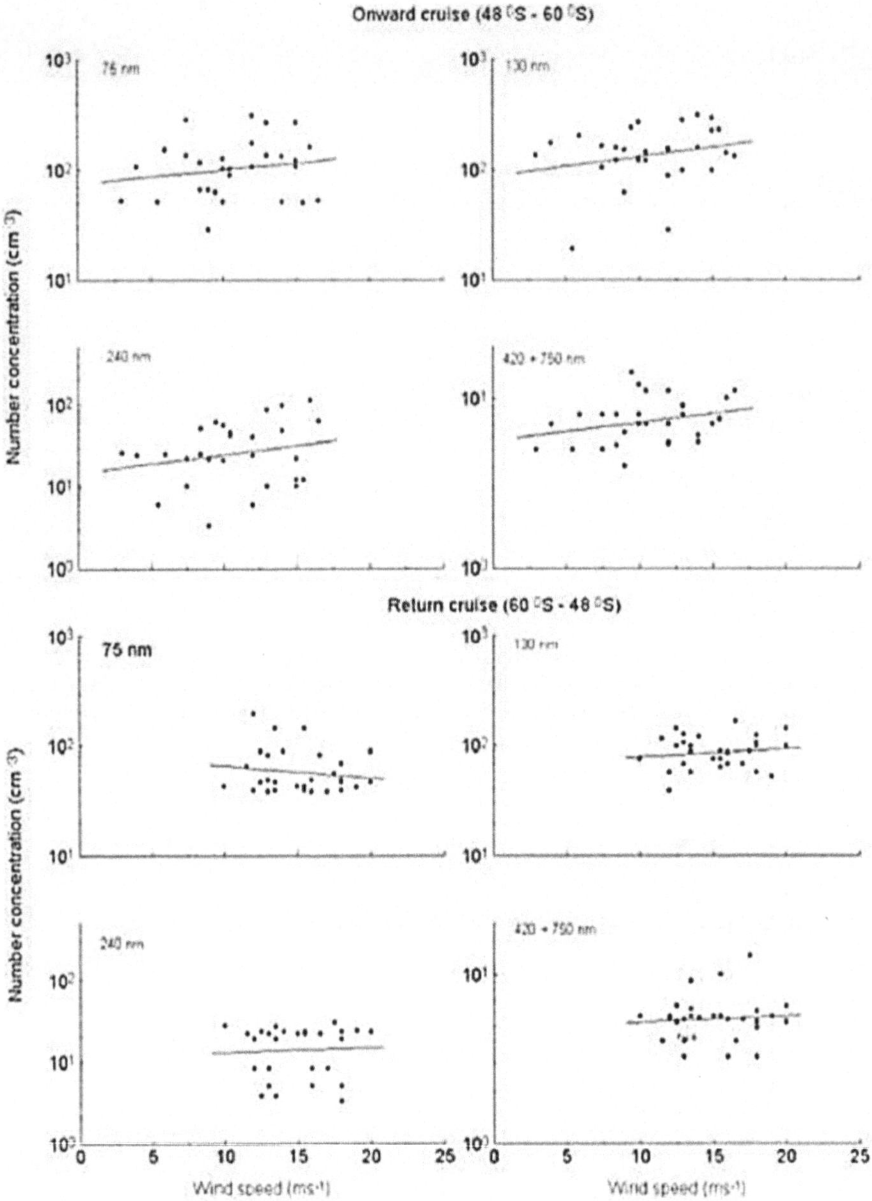

FIGURE 7.32
Total aerosol concentrations vs wind speed for different size ranges on the outward and return journeys.

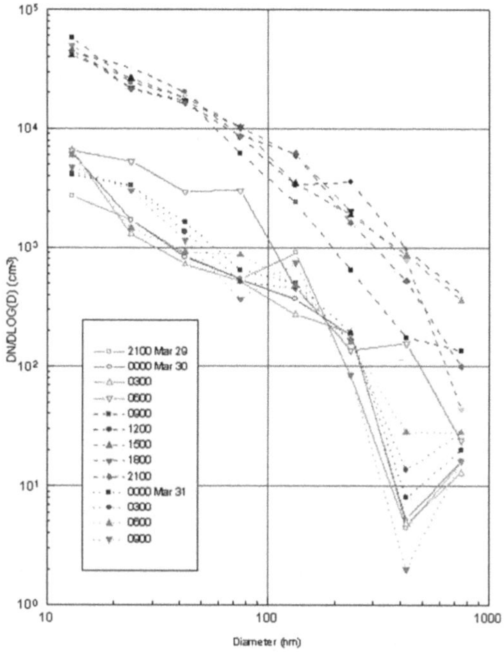

FIGURE 7.33
Aerosol size distributions on March 29–31, 1997 before, during, and after the passage of the ship through the ITCZ during the return cruise.

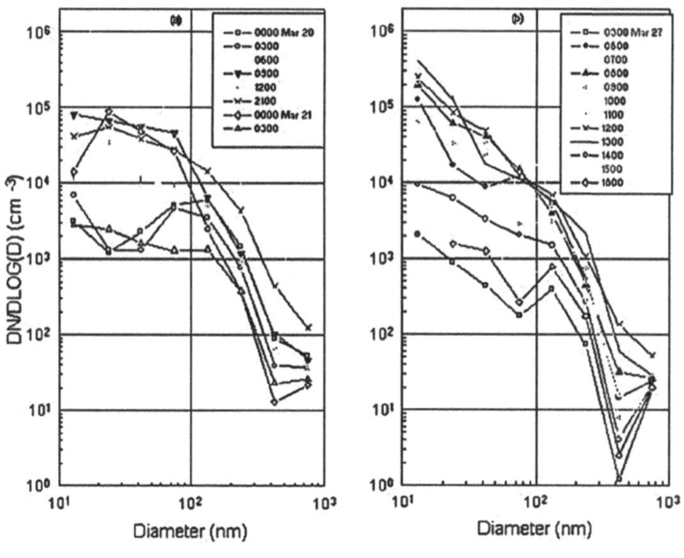

FIGURE 7.34
Aerosol size distributions while approaching, departing or berthed at (a) Durban during 0530 to 2205 UT on March 20, 1997, and (b) Mauritius during 0815 to 1315 UT on March 27, 1997.

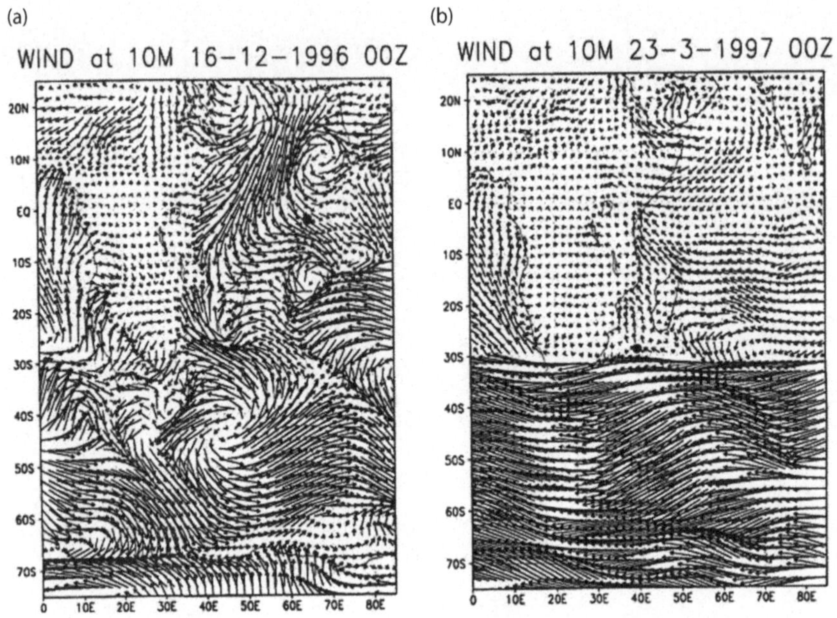

FIGURE 7.35
NCEP reanalysis plots of surface winds for 0000 UT on (a) December 16, 1996 and (b) March 23, 1997. The ship position is shown by the circle.

Major Findings

Our observations during this period lead to several major findings in the areas of atmospheric electricity and aerosol physics and are summarized below.

a. The diurnal variation curves of atmospheric electric field, averaged over the period of our observations, do not validate the classical concept of the global electric circuit. The observation indicates that other generators, viz. ionospheric and magnetospheric generators, may be significantly influencing the value of electric field in this region. Alternatively, the diurnal variation of flash activity when averaged over several days or weeks may cause a peak at an averaged time in the diurnal variation of ionospheric potential. Our observations also show that the value of the atmospheric electric conductivity at the ground surface does not undergo much diurnal variation.

b. Aerosol size distributions at Maitri are generally trimodal and remain almost similar for several hours or even days in absence of any significant change in meteorological conditions. Total aerosol concentrations significantly increase whenever a low-pressure system passes over the station. The enhancement in nucleation mode particles that are likely to form due to the photochemical conversion

TABLE 7.2

Average Modal Parameters for the Measurements Made during XVI Indian Antarctic Expedition 1996–1997

Position Latitude–Longitude	Modal Parameters[a]	Onward Journey		Onward Journey	
		Aitken Number[b]	Accumulation Number[b]	Aitken Number[b]	Accumulation Number[b]
15° N, 73° E–5° N, 69° E	N	–	–	2142 ± 1641	932 ± 163
	DgN	–	–	77.9 ± 15.6	88.9 ± 22.9
	σ	–	–	1.691 ± 0.004	1.686 ± 0.005
5° N, 69° E–Equator, 66° E	N	950 ± 126	378 ± 48	2145 ± 2103	391 ± 357
	DgN	85.2° 62.4	194.7 ± 111.8	103.2° 28.7	276.9° 208.4
	σ	1.684° 0.022	1.696 ± 0.006	1.694° 0.008	1.698° 0.004
Equator, 66° E–5° S, 64° E	N	864 ± 405	318 ± 81	2089 ± 2391	271 ± 174
	DgN	85.3 ± 25.4	177.9 ± 52.6	102.1 ± 23.4	248.9 ± 134.4
	σ	1.692 ± 0.004	1.696 ± 0.004	1.695 ± 0.003	1.698 ± 0.005
5° S, 64° E–10° S, 61° E	N	434 ± 214	244 ± 66	8691.5 ± 5995	1166 ± 932
	DgN	47.1 ± 24.8	131.5 ± 35.4	76.1 ± 10.1	157.5 ± 54.1
	σ	1.678 ± 0.017	1.693 ± 0.003	1.691 ± 0.002	1.694 ± 0.005
10° S, 61° E–30° S, 30° E	N	603 ± 224	150 ± 40	967 ± 482	248 ± 136
	DgN	88.8 ± 37.2	192.7 ± 110.1	92.6 ± 48.5	312.9 ± 172.6
	σ	1.692 ± 0.006	1.696 ± 0.005	1.692 ± 0.006	1.701 ± 0.003
30° S, 30° E–60° S, 25° E	N	767 ± 260	195 ± 87	806 ± 425	158 ± 76
	DgN	89.2 ± 45.6	392.6 ± 246.4	86.7 ± 18.7	386.1 ± 279.6
	σ	1.690 ± 0.009	1.701 ± 0.006	1.693 ± 0.003	1.701 ± 0.003
60° S, 25° E–70° S, 11° E	N	1194 ± 1230	287 ± 339	764 ± 425	126 ± 55
	DgN	89.2 ± 36.9	528.5 ± 533.4	87.9 ± 32.3	518.2 ± 340.4
	σ	1.691 ± 0.007	1.702 ± 0.004	1.691 ± 0.007	1.701 ± 0.009

Note: ± refers to the atmospheric variability.

[a] N (cm^{-3}), D (geometric in nm), σ (standard deviation of the size distribution).

[a] Modal Parameters Are Calculated from Lognormal Fits to the Size Distribution.

TABLE 7.3

Modal parameters are calculated from log normal fits to the size distribution. ± refers to the atmospheric variability

Onward Cruise	CC	SD
$\text{Log } N_{75} = 1.87 + 0.012\ U$	0.17	0.26
$\text{Log } N_{130} = 1.93 + 0.017\ U$	0.23	0.27
$\text{Log } N_{240} = 1.17 + 0.022\ U$	0.21	0.38
$\text{Log } N_{420+750} = 0.75 + 0.011\ U$	0.29	0.13
Return Cruise		
$\text{Log } N_{75} = 1.93–0.011\ U$	−0.15	0.20
$\text{Log } N_{130} = 1.83 + 0.006\ U$	0.12	0.15
$\text{Log } N_{240} = 1.06 + 0.005\ U$	0.049	0.31
$\text{Log } N_{420+750} = 0.68 + 0.003\ U$	0.073	0.13

Note: ± refers to the atmospheric variability.

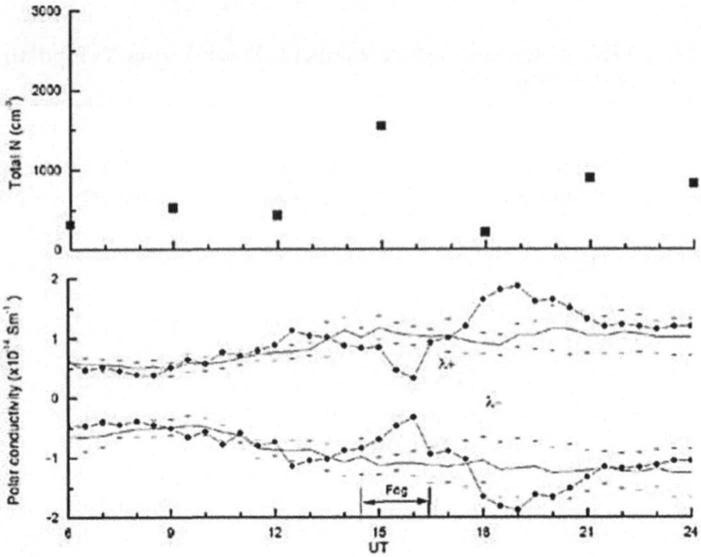

FIGURE 7.36
The variation of the electrical conductivity of both polarities and total aerosol concentration on December 25, 1996 (—•—). The continuous line (——) shows the average conductivity values along with their standard deviations shown by dashes for each half-hourly average values for 5 fair-weather days on the same cruise.

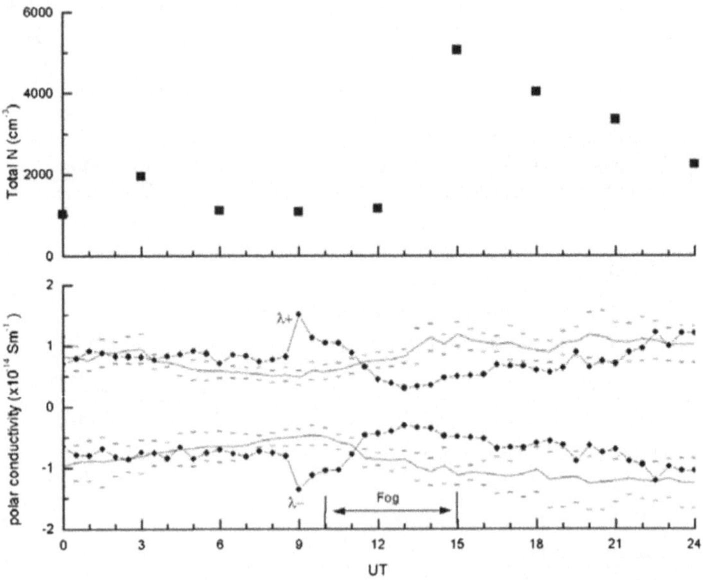

FIGURE 7.37
The variation of the electrical conductivity of both polarities and total aerosol concentration on January 5, 1997 (—•—). The continuous line (——) shows the average conductivity values along with their standard deviations shown by dashes for each half-hourly average values for 5 fair-weather days on the same cruise.

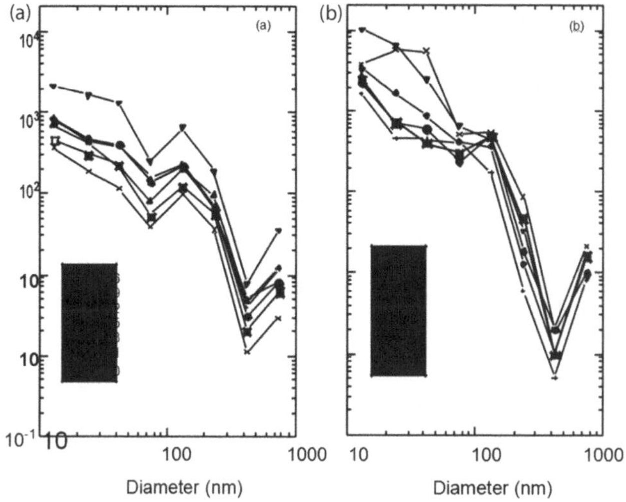

FIGURE 7.38
The aerosol size distributions during and around the fog episodes on (a) December 25, 1996 and (b) January 5, 1997. Each point is an average of five observations taken at 3-hourly intervals with an electrical aerosol analyzer system.

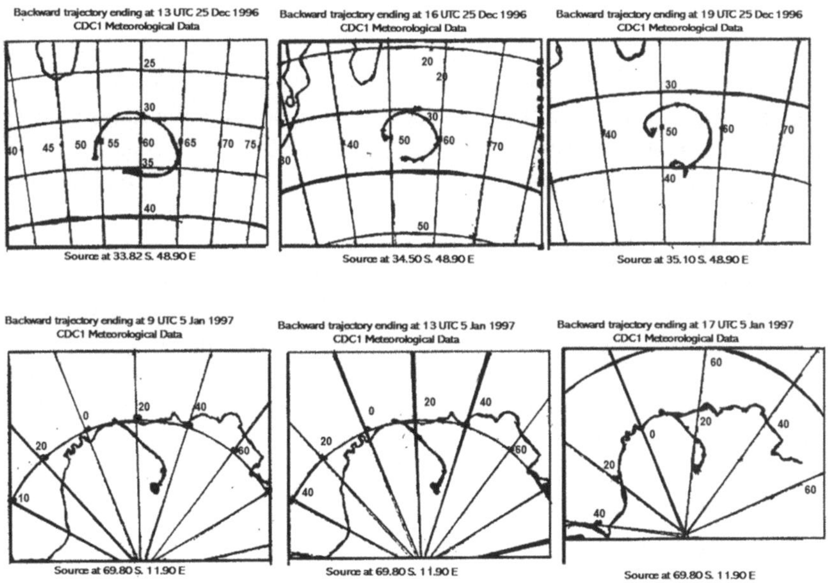

FIGURE 7.39
The 5-day backward trajectories at 1300, 1600, and 1900 UT on December 25, 1996 and at, 0900, 1300, and 1700 UT on January 5, 1997.

of the DMS emissions and/or the subsidence of midtropospheric air during the passage of low-pressure systems over Maitri have been proposed as possible causes for such observations.

c. Our observations over the Indian Ocean strongly support the transport of aerosols and trace gases with the northeasterly winds from the Indian continent to the Indian Ocean. Further, latitudinal variations of aerosols and conductivity suggest the downward transport of the nucleation mode of particles from the free troposphere to the marine boundary layer in the subsidence leg of the Hadley cell.
d. Values of electric conductivity change during warm and cold fog. The observed changes during warm and cold fog events can be explained in terms of the ion attachment and new particle formation processes, respectively.

Overall Contributions and Significant Achievements

The Institute's participation in the Summer Expedition of 1996–1997 immensely contributed to the advancement of our knowledge in the fields of atmospheric electricity and atmospheric aerosols. Several new findings were revealed. A large number of significant research papers brought out some novel and important findings in these areas. Above all, the results obtained during this participation confirmed the significance and importance of making such measurements at Maitri to study some problems of global importance.

The diurnal variation of the atmospheric electric field averaged over this period at Maitri does not validate the classical concept of the global electric circuit. The result illustrates that generators other than thunderstorm frequency and/or temporal variability of global thunderstorm frequency play significant roles in the global electric circuit.

Results of our investigations are published in national/international journals and contribute significantly to the advancement of our understanding of the subject. This way, our country is represented on the world map of research work in this area. Also, the aerosol data obtained over the Indian Oceans fills a gap in this data-void region on a global scale in understanding the changes in background air pollution and its role in climatic change.

References

Anderson, R. V. 1967. Measurement of worldwide diurnal atmospheric electricity variation. *Mon. Weather Rev.*, v. 95, pp. 899–904.

Bates, T. S., Kapustin, V. N., Quinn, P. K., Covert, D. S., Coffman, D. J., Mari, C., Durkee, P. A., DeBruyn, W. and Saltzman, E. 1998. Processes controlling the distribution of aerosol particles in the lower marine boundary layer during the first Aerosol Characterization Experiment (ACE-1). *J. Geophys. Res.*, v. 103, pp. 16369–16384.

Bates, T. S., Quinn, P. K., Covert, D. S., Coffman, D. J., Johnson, J. E. and Wiedensohler, A. 2000. Aerosol physical properties and processes in the lower marine boundary layer: A comparison of shipboard sub-micron data from ACE-1 and ACE-2. *Tellus*, v. 52B, pp. 258–272.

Bearing, E. A., Few, A. A. and Benbrook, J. R. 1998. The global electric circuit. *Phys. Today*, v. 51, pp. 24–30.

Burns, G. B., Hesse, M. H., Parcell, S. K., Malachowski, S. and Cole, K. D. 1995. The geoelectric field at Devis station, Antarctica. *J. Atmos. Terr. Phys.*, v. 57, pp. 1783–1797.

Byrne, G. J., Benbrook, J. R., Bering, E. A., Few, A. A., Morris, G. A., Trabucco, J. and Paschal, E. W. 1993. Ground-based instrumentation for measurements of atmospheric conduction current and electric field at the South Pole. *J. Geophys. Res.*, v. 98, pp. 2611–2618.

Charlson, R. J., Lovelock, J. E., Andreae, M. O. and Warren, S. G. 1987. Oceanic phytoplankton, atmospheric sulphur, cloud albedo and climate. *Nature*, v. 326, pp. 655–61.

Cobb, W. E. 1968. Atmospheric electric climate at Mauna Loa observatory. *J. Atmos. Sci.*, v. 25, pp. 470–480.

Cobb, W. E. 1977. Atmospheric electric measurements at the South Pole. In: Dolezalek, H. and R. Reiter, (Eds). *Electrical Processes in Atmosphere*. Steinkopff, Darmstadt, Germany, pp. 161–167.

Cobb, W. E. and Wells, H. J. 1970. The electrical conductivity of oceanic air and its correlation to global atmospheric pollution. *J. Atmos. Sci.*, v. 27, pp. 814–819.

Covert, D. S., Kapustin, V. N., Bates, T. S. and Quinn, P. K. 1996. Physical properties of marine boundary layer aerosol particles of the mid-Pacific in relation to sources and meteorological transport. *J. Geophys. Res.*, v. 101, pp. 6919–6930.

Few, A. A., Cleary, E. N., Bering, E. A. and Benbrook, J. R. 1997. Diurnal and seasonal observations of the global electric circuit from South Pole station—A proxy measure of the Earth's surface temperature. *Antarct. J. U. S.*, v. XXX(5), pp. 346–347.

Fuellekrug, M., Fraser-Smith, A. C., Bearing, E. A. and Few, A. A. 1999. On the hourly contribution of global cloud-to-ground lightning activity to the atmospheric electric field in the Antarctic during December 1992. *J. Atmos. Solar. Terr. Phys.*, v. 61, pp. 745–750.

Gras, J. L. and Adriaansen, A. 1985. Concentration and size variation of condensation nuclei at Mawson. *Antarctica. J. Atmos. Chem.*,v. 3, pp. 96–103.

Heintzenberg, J., Covert, D. C., Van Dingenen, R. 2000. Size distribution and chemical composition of marine aerosols: Compilation and review. *Tellus*, v. 52B, pp. 1104–1122.

Hogan, A. W. 1975. Antarctic aerosols. *J. App. Met.*, v. 14, pp. 550–559.

Hoppel, W. A. 1967. Theory of the electrode effect. *J. Atmos. Terr., Phys.*, v. 29, pp. 709–721.

Hoppel, W. A., Anderson, R. V. and Willet, J. C. 1986. Atmospheric electricity in the planetary boundary layer. In: *The Earth's Electrical Environment*. National Academy Press, Washington, D. C, pp. 149–165.

Hoppel, W. A., Fitzgerald, J. W. and Larson, R. E. 1985. Aerosol size distribution in air advecting off the east coast of the United States. *J. Geophys. Res.*, v. 90, pp. 2365–2379.

Hoppel, W. A. and Frick, G. M. 1990. Submicron aerosol size distribution measured over the tropical and south Pacific. *Atmos. Environ.*, v. 24A, pp. 645–659.

Hoppel, W. A., Frick, G. M., Fitzgerald, J. W. and Larson, R. E. 1994. Marine boundary layer measurements of new particle formation and the effect which non-precipitating clouds have on the aerosol size distribution. *J. Geophys. Res.*, v. 99, pp. 14,443–14,459.

Ito, T. 1980. Observations of atmospheric aerosols at Syowa Station, Antarctica (in Japanese). *Tenki Tokyo*, v. 27, pp. 13–24.

Ito, T. 1983. Study on properties and origin of aerosol particles in the Antarctic atmosphere (in Japanese, English abstract). *Pap. Meteorol. Geophys.*, v. 34(3), pp. 151–219.

Ito, T. 1985. Study of background aerosols in the Antarctic troposphere. *J. Atmos. Chem.*, v. 3, pp. 69–91.

Ito, T. and Iwai, K. 1981. On the sudden increase in the concentration of Aitken particles in the Antarctic atmosphere. *J. Meteorol. Soc., Jpn.*, v. 59, pp. 262–271.

Iwai, K. 1979. Concentration of Aitken particles observed at Syawo station, Antarctica (in Japanese). *Nankyoku Shiryo*, v. 67, pp. 172–179.

Jaenicke, R., Dreiling, V., Lehmann, E., Koutsenoguii, P. K. and Stingl, J. 1992. Condensation nuclei at the German Antarctic Station 'Georg von Neumayer'. *Tellus*, v. 44B, 311–317.

Jaenicke, R. and Stingl, J. 1984. Aitken particle size distribution in Antarctica. In: *Proceedings of International Conference on Aerosols and Ice Nuclei, Budapest, University of Budapest Press*, Budapest, Hungary.

Jayaraman, A., Lubin, D., Ramchandran, S., Ramnathan, V., Woodbridge, V., Collin, W. D., Zalpuri, K. S. 1998. Direct observations of aerosol radiative forcing over the tropical Indian Ocean during the January-February 1996 pre-INDOEX cruise. *J. Geophys. Res.*, v. 103, pp. 13827–12836.

Kamra, A. K., and Deshpande, C. G. 1995. Possible secular change and land-to-ocean extension of air pollution from measurements of atmospheric electrical conductivity over the Bay of Bengal. *J. Geophys. Res.*, v. 100, pp. 7105–7110.

Kamra, A. K., Deshpande, C. G. and Gopalakrishnan, V. 1994. Challenge to the assumption of the unitary diurnal variation of the atmospheric electric field based on observations in the Indian Ocean, Bay of Bengal and Arabian Sea. *J. Geophys. Res.*, v. 99, pp. 21,043–21,050.

Kamra, A. K., Murugavel, P., Pawar, S. D. and Gopalakrishnan, V. 2001. Background aerosol concentration derived from the atmospheric electric conductivity measurements made over the Indian Ocean during INDOEX. *J. Geophys. Res.*, v. 106, pp. 28643–28652.

Kasemir, H. 1972. Atmospheric electric measurements in the Arctic and Antarctic. *Pure Appl. Geophys.*, v. 100, pp. 70–80.

Kaufman, Y. J., Hobbs, P. V., Kirchoff, W., Artaxo, P., Remer, L. A., Holben, B. N., King, M. D., Ward, D. E., Prins, E. M., Longo, K. M., Mattos, L. F., Nobre, C. A., Spinhirne, J. D., Ji, Q., Thompson, A. M., Gleason, J. F., Christopher, S. A. and Tsay, S. C. 1998. Smoke, Clouds and Radiation-Brazil (SCAR-B) experiment. *J. Geophys. Res.*, v. 103, pp. 31,783–31,808.

Lal, M. and Kapoor, R. K. 1989. Certain meteorological features of submicron aerosols at Schirmacher oasis, East Antarctica. *Atmos. Environ.*, v. 23, pp. 803–808.

Lal, M. and Kapoor, R. K. 1992. Submicron aerosols over the Indian Ocean: Some meteorological characteristics atmos. *Environ.*, v. 27, pp. 291–303.

Markson, R. 1977. Airborne atmospheric electrical measurements of variation of ionospheric potential and electrical structure in the exchange layer over the ocean. In: Dolezalek, H. and Reiter, R. (Eds). *Electrical Processes in Atmosphere.* Steinkopff, Darmstadt, Germany, pp. 450–459.

Mauchly, S. J. 1923. On the diurnal variation of the potential gradient of atmospheric electricity. *Terr. Magn. Atmos. Electr.*, v. 28, pp. 61–81.

Misaki, M. and Takeuti, T. 1970. The extension of air pollution from land over ocean as related in the variation of atmospheric electric conductivity. *J. Meteorol. Soc., Jpn*, v. 48, pp. 263–269.

Morita, Y. 1971. The diurnal and latitudinal variation of electric field and electric conductivity in the atmosphere over the Pacific Ocean. *J. Meteor. Soc. Jpn*. Ser. II, v. 49(1), pp. 56–58.

Morita, Y. and Ishikawa, H. 1977. On recent measurements of electric parameters and aerosols in the oceanic atmosphere. In: Dolezalek, H. and Reiter, R. (Eds). *Electrical Processes in Atmosphere.* Steinkopff, Darmstadt, Germany, pp. 126–130. II, v. 49, pp. 56–58.

Morita, Y., Ishikawa, H., Nagasaka, J. and Kanada, M. 1973. The land-to-ocean transitional behaviour of the atmospheric electrical parameters and their relation to atmospheric pollution. *J. Meteorol. Soc. Jpn.*, v. 51, pp. 294–306.

Park, C. G., 1976. Solar magnetic sector effects on the vertical atmospheric electric field at Vostok. *Antarctica, Geophys. Res. Lett.*, v. 3(8), pp. 475–478.

Prodi, F., Santachiara, G. and Oloisi, F. 1993. Characterization of aerosols in marine environments (Mediterranean, Red Sea and Indian Ocean). *J. Geophys. Res.*, v. 88, pp. 10,957–10,968.

Quinn, P. K., Covert, D. S., Bates, T. S., Kapustin, V. N., Ramsey-Bell, D. C. and Mcinnes, L. M. 1993. Dimethylsulfide/cloud condensation nuclei/climate system: Relevant size-resolved measurements of the chemical and physical properties of atmospheric aerosols particles. *J. Geophys. Res.*, v. 98, pp. 10,411–10,427.

Quinn, P. K., Kapustin, V. N. and Bates, T. S. 1996. Chemical and optical properties of marine boundary layer aerosol particles of the mid-Pacific in relation to sources and meteorological transport. *J. Geophys. Res.*, v. 101, pp. 6931–6951.

Raes, F., Bates, T. S., Van Liedekerke, and Mcgovern, F. M. 2000. The second Aerosol Characterization Experiment (ACE-2): General overview and main results. *Tellus*, v. 52B, pp. 111–125.

Ramchandran, T. V. and Balani, M. C. 1995. *Report on the participation by the Bhabha Atomic Research Centre in the Tenth Indian Expedition to Antarctica.* Scientific Report Tenth Indian Expedition to Antarctica, Tech. Publ. No. 8, Department of Ocean Development, pp. 159–180.

Rhoads, K. P., Kelley, P., Dickerson, R., Carsey, T. P., Farmer, M., Savoie, D. L. and Prospero, J. M. 1997. The composition of the troposphere over the Indian Ocean during the monsoonal transition. *Geophys. Res.*, v. 102, pp. 18,981–18,995.

Russell, P. B., Hobbs, P. V. and Stowe, L. L. 1999. Aerosol properties and radiative effects in the US East Coast haze plume: An overview of the tropospheric aerosol radiative forcing observational experiment (TARFOX). *J. Geophys. Res.*, v. 104, pp. 2213–2222.

Ruttenberg, S. and Holzer, R. E. 1955. Atmospheric electric measurements in Pacific Ocean. *Geophys. Res.*, v. 42, pp. 101–108.

Savoie, D. L., Prospero, J. M. and Nees, R. T. 1987. Nitrate, non-sea-sulfate, and mineral aerosol over the northwestern Indian Ocean. *J. Geophys. Res.*, v. 92, pp. 933–942.

Shaw, G. E. 1986. On the physical properties of aerosol at Ross Island, Antarctica. *J. Aerosol. Sci.*, v. 17, pp. 937–945.

Takagi, M. and Kanada, M. 1972. Global variation in the atmospheric electric field. *Pure App. Geophys., v.* 100, pp. 44–53.

Tinsley, B. A. and Heelis, R. A. 1993. Correlations of atmospheric dynamics with solar activity. Evidence for a connection via the solar wind, atmospheric electricity, and cloud microphysics. *J. Geophys. Res., v.* 98, pp. 10375–10384.

Tinsley, B. A., Brown, G. M. and Scherrer, P. H. 1989. Solar variability influences on weather and climate: possible connections through cosmic ray fluxes and storm intensification. *J. Geophys. Res., v.* 94, pp. 14783–14792.

8

Observations of Unique Long-Duration Shadow Band Activity during the November 23, 2003, Total Solar Eclipse near Maitri, Antarctica

S. M. Bhandari

CONTENTS

Introduction

Total solar eclipses (TSEs) are periodic, though rare, short-duration celestial events, wherein the Moon perfectly overlaps the Sun in the sky and covers the visible disc of the Sun. Besides presenting a beautiful and awe-inspiring spectacle in the daytime sky for the enthusiastic observer, TSE events offer opportunities for investigating different aspects of the Sun, Moon, and Earth and the cosmos at large. Like all astronomical eclipses and occultations (of the Sun, Moon, planets, satellites, or stars), the location and timing of TSE events are precisely predictable years and decades in advance. This unique feature is exploited by scientists to conduct innovative well-thought-out and well-planned experiments. In the past, many fundamental discoveries have resulted from careful observations made during the short fleeting minutes of totality. Examples include the discovery of the element helium, confirmation of the gravitational bending of light, discovery of the extended solar atmosphere called solar corona, presence of sun-grazing comets, and so on (Brunier and Luminet, 2000, Mobberley, 2007).

TSE occurrences have also been exploited by the Earth scientists to study the Earth's surface and atmosphere/ionosphere by carefully investigating their responses to the stimuli provided by sudden short-duration removal and restoration of incoming solar radiation. Besides a study of changes in surface meteorological parameters, a number of studies have examined the effects of a TSE on surface and stratospheric ozone concentration and generation of gravity waves as a non-linear dynamic response to the sharp stimulus (Zerefos et al., 2007, Tzanis et al., 2008). Bhandari (1996) has focused on comprehensive high space-time resolution imaging the Earth from space simultaneously using the three geostationary meteorological satellites over India during the famous Diwali Day, Oct. 24, 1995, TSE. The Earth's biosphere is another arena where TSE events are exploited to study behavioral responses of plants and animals to the abnormal conditions of daylight darkness associated with TSE events.

In short, TSE occurrences, although rare and short-lived, provide unique unrealizable-in-lab opportunities to explore and study different aspects of nature.

The Phenomenon of Shadow Bands

During the process of movement of the Moon's disc over the Sun, a few minutes before and after totality, the Sun becomes a thin crescent and behaves like a linear array of star-like point sources incoherently illuminating the Earth's atmosphere. The turbulence present in the Earth's atmosphere introduces random phase perturbations on the incoming radiation, resulting in what may be termed "solar twinkling." This is observed on the ground in the form of a 2D pattern of quasi-linear dark and bright wave-like bands moving with the winds in atmosphere above. This phenomenon, referred to as the "shadow bands," is keenly sought after by the scientists during all TSEs. These shadow bands are not always produced with enough intensity and contrast, and their observations remain highly elusive due to short duration, feeble contrast, and fast propagation speed. So far, most of the shadow band experiments have either only visually observed or photographically recorded these enigmatic shadow bands. On a very few occasions, these have been recorded digitally for a careful analysis of their characteristics in detail (Jones, 1999).

Total Solar Eclipse over Antarctica on November 23, 2003

On November 23, 2003, a rare and unique total solar eclipse was predicted to occur over the continent of Antarctica. The uniqueness of this event lay in the fact that this was the first-ever TSE to occur over Antarctica after the discovery of the continent by humans nearly 100 years ago. Therefore, it presented an unprecedented opportunity to study the unique Antarctic surface and atmospheric characteristics during a TSE. For the astronomer, too, this TSE was special since it would allow studies of fine details of the solar corona, near the peak of the solar activity cycle, from most pristine Antarctic site, devoid of any atmospheric pollution. The atmospheric conditions of good visibility and poor seeing in the peripheral regions of Antarctica, along with low elevation of the Sun during the TSE, were considered favorable for strong shadow band activity (Codona, 1986).

Furthermore, the November 23, 2003, Antarctic TSE had special significance for the Indian scientific community since its track of totality passed over the permanent Indian Antarctic Research Station Maitri located at Lat. 70°45′ S, Long. 11°45′ E (Figure 8.1).

The short duration of totality, of 1 min. 17 sec, and the low horizon-grazing solar illumination geometry for the location of Maitri were areas of concern, primarily for experiments to study solar corona. Another area of concern was the high climatological probability of cloudiness in the month of November for the region of Maitri (Espenak and Anderson, 2002).

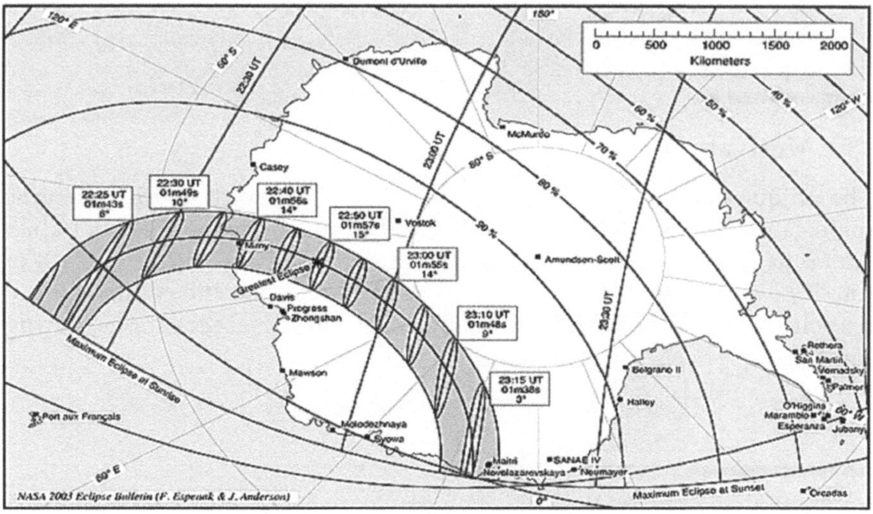

Indian Antarctic Station, Maitri **Latitude:** -70° 45' 34.19" S **Longitude:** 11° 44' 5.39" E

FIGURE 8.1
Path of Moon's shadow over the Southern Ocean and the Antarctic continent. Location of Maitri is indicated by an arrow at the bottom.

TABLE 8.1
Parameters of the November 23, 2003, TSE over Maitri, Antarctica

Beginning of the Eclipse	First Contact	22:28:56.6 UT
Beginning of Totality	Second Contact	23:16:40.3 UT
End of Totality	Third Contact	23:17:59.7 UT
End of Eclipse	Fourth Contact	00:05:42 UT (November 24)
Maximum Eclipse		23:17:20 UT
Duration of Totality		1 min. 17 sec.
Eclipse Magnitude		1.007
Solar Obscuration		1.067
Sun Elevation & Azimuth		$\sim 2°, 176°$
Umbral Shadow Dimension		$\sim$200 km EW $\times$ 1000 km NS
Umbral Velocity		3.5 km/sec
Location of Observation Site	Latitude	70° 52.37′ S,
	Longitude	11° 25.65′ E,
	Distance from Centerline	150 km

The November 23, 2003, TSE event over Maitri, Antarctica, nevertheless presented a unique and valuable celestial opportunity for the Indian scientific community to undertake well-planned experiments. It may be worthwhile to mention that the Antarctic continent would be experiencing its next TSE of the series only on December 4, 2021. Table 8.1 presents basic parameters associated with the November 23, 2003 Total Solar Eclipse for the location of Indian Antarctic Station Maitri.

The Motivation

The unique nature of the November 23, 2003, TSE event and its scientific importance, as well as the existence of a fully operational permanent Indian Antarctic Research Station at Maitri, Antarctica, lying within its path of totality, were factors that motivated the Indian scientific community to launch efforts to undertake meaningful scientific observations during the TSE.

The Challenge

Antarctica, located at the Earth's bottom, is known to be an icy white continent of extremes—extremes of temperatures, winds, and blizzards. All TSE-related experiments to be conducted in the open must address this, both in terms of technical and logistic challenges involved. To arrive at a meaningful set of experiments, a brainstorming discussion was arranged at NCAOR, Goa, where scientists from many interested organizations participated. The discussion resulted in finalization of two types of experiments:

- Digital imaging of shadow bands generated near second and third contacts to study the low-level atmospheric turbulence in the Antarctic atmosphere by exploiting the low horizon-grazing geometry of the TSE leading to a long path through the atmosphere
- Digital imaging of the fine structure of the solar corona during the active phase of the solar cycle using narrow-band filter spectrometry taking advantage of the extreme transparency of the Antarctic atmosphere

Both these experiments, of course, required clear sky conditions to be successful. The atmospheric science experiments, including the observations of shadow bands to study the Antarctic atmosphere, were proposed by a team of scientists from Physical Research Laboratory (PRL) and Space Application Centre (SAC), Ahmedabad. The astronomical experiments relating to observations of the solar corona were proposed by the team composed of scientists from Indian Institute of Astrophysics (IIA), Bangalore, and Uttar Pradesh Solar Observatory (UPSO), Nainital.

After narrowing down the potential scientific experiments, the following important aspects were addressed.

- Distance to the far-away Antarctic continent and logistic difficulties of the expedition in term of lead time available and nature of equipment.
- Environmental sub-zero temperature and high wind conditions prevailing at Maitri during the November period
- High climatological cloudiness probability, close to 50%, for the second half of the month of November (Espenak and Anderson, 2002)
- Location of Maitri in the Schirmacher Oasis in relation to the low-elevation TSE observation geometry

A look at the broad topographic details of the region of Maitri revealed that the TSE event would not be visible from the location of Maitri due to obscuration by the ice shelf towards the south. A reconnaissance survey would, therefore, be required to locate a suitable site for TSE observations.

In view of the short lead time at their disposal, NCAOR assumed the responsibility of logistics of the expedition and enlisted the requisite support from the 22nd IAE team (wintering-over-team) at Maitri.

Looking into the short lead time available and the requirements of procurements/fabrication and finalization of instruments, NCAOR decided to arrange the expedition travel using the air route as against the sea route between Cape Town and Antarctica. It was decided to charter the services of ALCI operating IL-76 flights between Cape Town and the Novo Russian Station close to Maitri.

The Indian Antarctic Eclipse Expedition

The decision to send the expedition to Maitri Antarctica, entirely by the air route restricted the participation to the absolute minimum number of scientists. After medical fitness certification, the following five-member expedition team was selected for undertaking the experiments during the November 23, 2003, TSE.

Atmospheric Studies

Dr. Satyendra M. Bhandari—SAC (ISRO), Ahmedabad

Solar Corona Studies

Dr. S. P. Bagare—IIA, Bangalore
Dr. S. S. Gupta—IIA, Kodaikanal
Dr. V. K. Varma and Shri Ramesh Chandra—UPSO, Nainital

Per the plan, the expedition team landed on the continent of Antarctica on November 10, 2003, and returned to Cape Town on November 27, 2003, after conducting planned experiments from a carefully selected open glacial site about 25 km from Maitri. In the following, the details of the shadow band experiment, the TSE observations, and the results are described.

The Experiments

Observation of Shadow Bands

The Shadow Band Experiment involved setting up a suitably designed and graduated 1.5 m × 1.5 m non-reflecting off-white screen to photograph and digitally image shadow band activity expected to occur prior to the second and after the third contacts. The horizon-grazing geometry of the TSE was expected to result in relatively high contrast shadow bands.

The following equipment was used to record shadow bands:

Digital and photographic cameras, digital video handycam

In addition, a five-channel EK-MS-120 digital sun-photometer was carried for measurement of aerosol optical depth.

Observations of the Solar Corona

A specially designed multi-wavelength low-noise CCD coronal imager with broad band and fine narrow-band spectral filters was mounted on a heliostat for tracking the Sun.

Digital camera and digital video handycam for photographic imaging of the event, including the large-scale structure of the solar corona and phenomena like diamond ring, Baily's Beads, and so on.

Pre-Eclipse Preparations

Prior to undertaking the experiments on TSE day on November 23, 2003, the pre-TSE period November 11–22, 2003, was utilized for a reconnaissance survey of the Maitri region with the help of wintering-over-team members to locate a site suitable for TSE observations providing a clear unobstructed view of the Sun during the event. This was necessary since Maitri station is in the Schirmacher Oasis, with the ice shelf blocking the view toward the southern side. During this exercise, the logistics of transporting the equipment to the selected site as well as other requirements of power and so on were also worked out. The selected site was on a high open glacial field near the Novo airstrip used by ALCI and about 25 km from Maitri. The geographical coordinates of the site, as determined using a GPS receiver, are: Lat. 70° 52.37′ S, Long.:11° 25.65′ E, Altitude: 500 m approx.

Figure 8.2 shows the location of Maitri, the NOVO airstrip, and the site selected for TSE observations. The pre-TSE period at Maitri was also utilized to get acclimatized to the sub-zero temperatures and windy conditions, rehearsing the setting up of the bulky Heliostat, the CCD imager, the CCD filter cooling assembly, and so on. This exercise was undertaken close to Maitri station near the Dodda Betta Hut located next to the Priyadarshini Lake.

This period was also used to make frequent sun-photometer observations and manually record solar radiation in five spectral bands for estimation of the aerosol optical depth. Except for November 16, 2003, on most other days, the Maitri area was clouded and most observations were thus contaminated. On November 16, 2003, the aerosol optical depth (AOD) estimates were in the range of 0.015–0.03, confirming the pristine nature of the Antarctic environment.

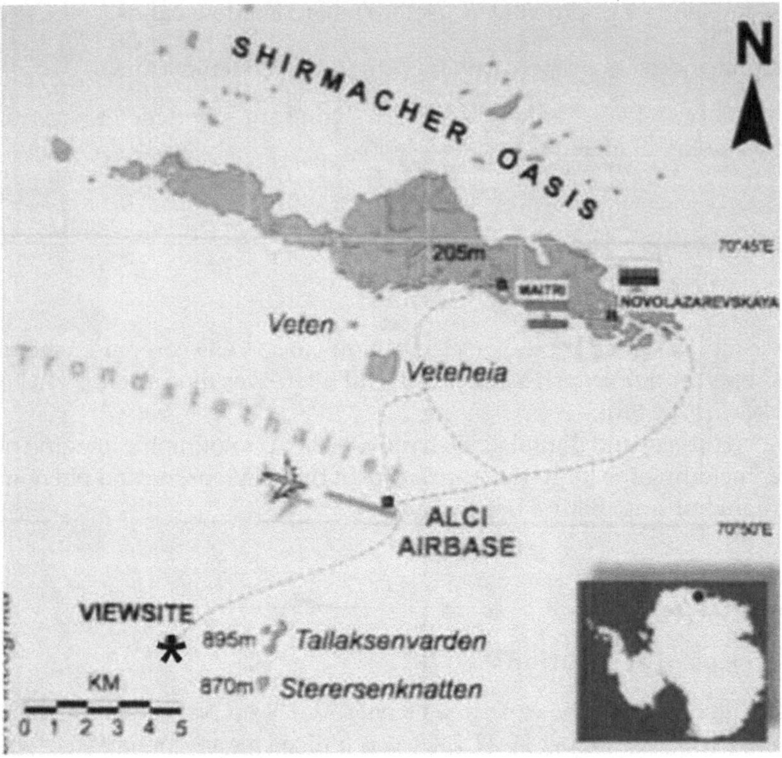

FIGURE 8.2
The location of the TSE observation site marked with *. The diagram shows Maitri and NOVO stations located in the Schirmacher Basin, the ALCI airstrip, and the route followed by the IAEE team from Maitri.

While the frequent cloudiness during the entire pre-eclipse period was a cause for concern, looking at the climatological statistics of nearly 40% cloud-free probability for the eclipse period in the month of November, it was reassuring that perhaps we were in for cloud-free days ahead. Compared to the location of Maitri, the weather conditions were also expected to be relatively favorable at the selected site on a high glacial plateau.

The Total Solar Eclipse Day

About 12 hours before the time of TSE occurrence, the expedition team, along with several enthusiastic members of the WOT, shifted to the pre-selected site, which was also occupied by several amateur eclipse watchers from many countries (USA, Japan, Europe), who were specially flown in for the event as tourists. A Sankalp trailer was used for housing the scientists and the equipment. The Jeevan-Jyoti generator trailer provided the necessary power for the CCD experiment.

With help from colleagues from WOT, the entire experimental setup, including the shadow band screen, was completed much before the beginning of the TSE event despite the low temperature (–15°C) and high wind chill resulting in effective temperature of –35°C. The shadow band screen was set-up on the Sun-facing side of the Sankalp trailer. There was great excitement as the weather cleared up to a brilliant blue Antarctic cloud-free sky, with excellent visibility extending tens of kilometers. A few small worrying patches of cloudiness in the sunward direction also melted before the TSE event.

Observations and Digital Recording of Shadow Bands

The TSE event proceeded per the predictions. It was a wonderful and fascinating sight to see the Sun getting eclipsed by the Moon low on the icy horizon. As we were approaching the time of second contact, expectations were high regarding the appearance of shadow bands. A close watch was maintained on the screen. Soon, about 5 min. before the second contact, we could clearly see the development of shadow bands—a clear wavy pattern with reasonably good contrast moving swiftly across the screen. The fast-moving shadow band pattern was observed to smoothly change its direction of movement. The shadow band activity continued for nearly 4 min. and began decaying near the time of totality. The entire 4-min sequence of the shadow band activity observed on the screen and the surrounding area was captured by the hand-held digital video handycam. Despite the extreme cold and wind chill (temp. –15°C and winds approaching 40–60 knots), the digital recording went very well with some jitter due to extreme cold. During the TSE, the light levels and temperatures first decreased gradually and then, as totality approached, darkness descended and there was a sudden drop in temperature.

With the cessation of the shadow band activity, the darkness descended very quickly as totality approached. The handycam was then turned to the sunward side to record the appearance of the diamond ring and the beautiful, nearly symmetric solar corona in all its glory. The bright diamond ring at the second contact and the corona showing broad coronal streamers at the 1 and 2 o'clock positions were remarkable. As expected, the lower portion of the solar disc and the corona were hidden below the horizon, as the Sun at this time was at a very low elevation.

The entire scenario from horizon to horizon was breathtakingly beautiful, with the totally eclipsed Sun surrounded by a golden aura of the bright orange-colored gleaming corona sitting on the icy white horizon. As the Moon's shadow moved swiftly across, the entire horizon was illuminated with an eerie hue.

FIGURE 8.3
Astronomical observations during the November 23, 2003, total solar eclipse from the observing site near Maitri, Antarctica.

Close to the time of the third contact, the diamond ring appeared on the other edge of the Moon's disc. This was soon followed by the reappearance of shadow bands, of similar nature. The shadow band activity continued for another unprecedented 7 min. and slowly faded away at the end. Except for the different durations of shadow band activity pre- and post-totality, the entire eclipse proceedings after totality were a repeat of pre-totality, in reverse order, of course.

Figure 8.3 shows images of the two diamond rings and the solar corona at the time of totality. Many such images were taken by different cameras by the expedition members as well as by other colleagues.

The CCD spectral imaging of the solar corona was adversely affected by strong reflections and scattering from the ice/snow surface due to the very low sun elevation angle.

Analysis of Shadow Band Digital Images

Figure 8.4 shows the shadow band screen setup against the Sankalp container. The screen was oriented in elevation and azimuth to directly face the sunward direction.

The entire 4 + 7 minutes of shadow band activity was captured by focusing on the screen using a high-resolution full-color digital video handycam. For analysis, a sequence of frames was extracted from the digital recording using the IrfanView software downloaded from the Web. Each extracted frame represents a small area within the screen and is of size 384 × 288 pixels. These are separated by 40 ms each. The entire digital recording represented several thousands of such 2D digital image frames. The spatial scale of the images was derived using the graduation marks on the screen.

Visual observations, as well as the visual analysis of the observed pattern in the contrast enhanced images, shown below in Figure 8.5, revealed the width of the dark and white bands to be about 30–50 cm, and their speed of movement nearly 4–5 m/sec.

FIGURE 8.4
The shadow band experimental setup at the observation site during November 23, 2003, total solar eclipse near Maitri, Antarctica.

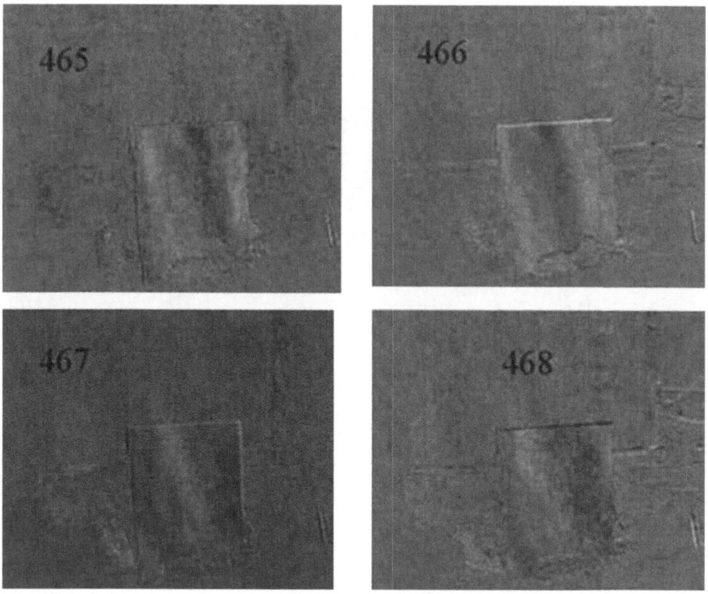

FIGURE 8.5
Sample sequence of contrast enhanced shadow band images (40 msec apart) showing shadow bands progressing across covering a small part of the picture.

The sequence of 2D images was subjected to time series analysis and 2D auto and cross-correlation and fast fourier transform (FFT) analyses to study the characteristics of the observed shadow bands.

A small sample of the type of intensity variations associated with the shadow and activity is presented in Figure 8.6. The random nature of fluctuations with an unusually high contrast (nearly 1%–2%) can be seen clearly.

Analysis of auto-correlation of the time series at a fixed location, and the cross-correlation analysis between time series at eight points separated by 4 cm each along the x-direction (Figure 8.7), shows that the auto-correlation falls to half at ~100 ms Further, from the spatial cross-correlation analysis, we see that the peak of maximum correlation shifts by ~100 ms over 35 cm. It is therefore estimated that the scale size of the turbulence irregularities in the atmosphere is on the order of 35 cm and these move with a speed of 3.5 m/sec in the x-direction.

The time series of intensity variations with a time sampling of 40 ms at each of the eight locations was subjected to power spectrum analysis. Figure 8.8 shows the average power spectra of the shadow band intensity fluctuations. After ~0.65 Hz, the spectrum shows a power law roll-off to 12.5 Hz, with a slope of –2, quite close to the Kolmogorov spectral slope of –5/3 for isotropic turbulence. The overall shape of the spectrum, along with the fine features, for example, the low frequency shoulder and the series of minima at regular intervals, agrees well with the theoretical expectations. Such spectra were suggested by Codona (1986) for intensity fluctuations caused by weak scattering due to atmospheric turbulence. The results also corroborate

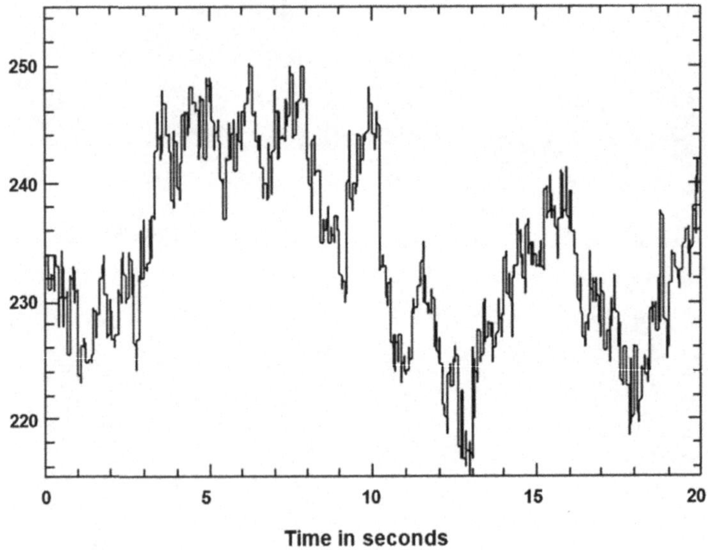

FIGURE 8.6
A sample record of intensity variations associated with shadow bands activity at the selected pixel (150, 110) on the screen.

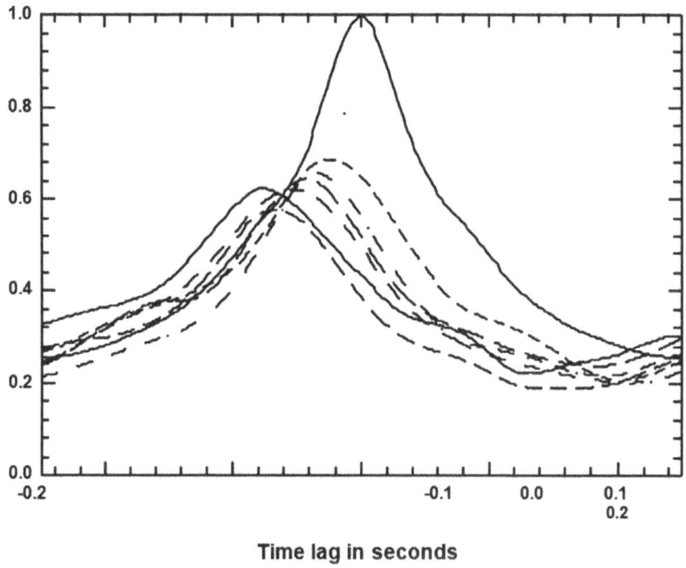

FIGURE 8.7
Auto-correlation and cross-correlations between time series of the intensity values from the individual frames. The correlograms are for time series separation in the range 0–28 cm in steps of 4 cm.

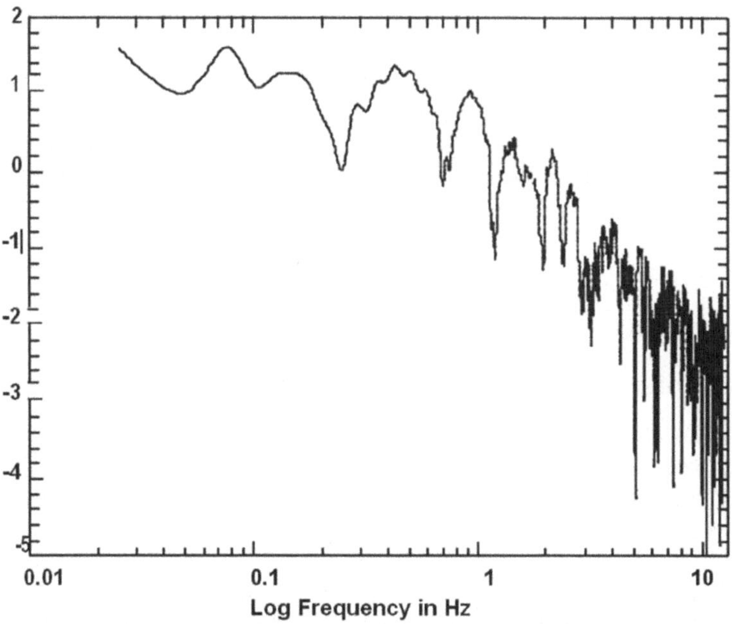

FIGURE 8.8
Averaged power spectra of shadow band intensity fluctuations.

TABLE 8.2

Characteristics of the Shadow Bands Observed during the November 23, 2003 Total Solar Eclipse near Maitri, Antarctica

Duration of shadow band activity	Before second contact	~ 4 min.
	After third contact	~ 7 min.
Spatial scale or wavelength		20–50 cm
Speed of movement		3–5 m/sec
Direction of movement		Variable
Image contrast		~ 1%–2%
Time scale of evolution		~ 100 msec
Angular size of solar crescent		2–3 min. arc
Effective distance to turbulent medium		About 350 m

theoretical results from Booker and Vats (1985) for horizontal propagation along a long atmospheric path. The November 23, 2003, TSE observations from near Maitri, Antarctica, represented horizontal path propagation at a very low elevation. Further details of the analysis and interpretation of shadow band digital data may be found in Vats et al. (2006).

Table 8.2 summarizes results of the shadow band experiment conducted during the November 23, 2003, total solar eclipse over Antarctica.

Recently, Gladysz et al. (2005) have reported spatial power spectra of shadow band intensity fluctuations using the 1 + 1 min. of digital data they were able to record during the December 04, 2002, TSE observed at a solar elevation angle of 34 deg. The similarity of their spatial spectra with the temporal spectra analyzed by us (see Figure 8.8) is striking.

The unique shadow band observations reported by us, as well as more recently by Gladysz et al. (2005), based on digital image data analysis recorded during two different TSEs, support interpretation of shadow band generation in terms of scintillation theory.

Conclusions

The Indian Antarctic Eclipse Expedition to Maitri, Antarctica, to undertake observations during the November 23, 2003, TSE was successful in observing and digitally recording unique long-duration shadow band activity. Such a long digital record of high-contrast shadow band activity had rarely been recorded in the past. The shadow band digital data was subjected to time series and FFT analysis to bring out the characteristics of the turbulence present in the lower region of the Antarctic atmosphere. The scale of turbulent patches was observed to be in the range of 20–50 cm, and the effective height of turbulence is estimated to be 350 m from the observation point. The coherence time of intensity fluctuations associated with shadow band activity were found to be about 0.1 sec. The pattern was observed to move with a speed of about 3–5 m/sec along the X-axis analyzed. The actual speed along the direction of their movement would be greater. The shape of the turbulence spectra showed a close match with the theoretical expectations for solar radiation propagation through a long horizontal atmospheric path close to Earth. The observations and analysis lend support to the scintillation theory of generation of shadow bands.

Further detailed analysis of the shadow band digital data is in progress to study various other parameters of interest, for example, the changes in the average intensity, wavelength and contrast, and the characteristics of atmospheric turbulence in relation to changing solar illumination conditions as the eclipse progressed. This analysis would throw more light on the origin and evolution of shadow bands observed during the November 23, 2003, total solar eclipse over Antarctica and the response of the Antarctic atmosphere to the stimulus generated by the total solar eclipse.

Shadow band studies continue to be an important component of total solar eclipse campaigns. In order to further investigate the finer details of the explanation for their origin and evolution, it would be necessary to continue to observe shadow bands in different TSE scenarios, at different locations and under different solar illumination conditions.

In this direction, an attempt was made by the author to explore shadow bands again over the polar region was the August 1, 2008, TSE over the Arctic. This observation campaign was undertaken based on the limitations of logistic support available at the Indian Arctic station in Svalbard, Norway (Bhandari, 2007). Unfortunately, the sky during the TSE event was totally clouded and shadow bands generated, if any, could not be observed.

The next crucial opportunity for a comprehensive shadow band experiment would be the forthcoming TSE of July 22, 2009, whose path of totality begins in western India, passes across the middle of the country, and moves further east towards China, and so on. This is an eclipse with a long duration of totality—of several minutes, favoring generation of distinct shadow bands. It would be interesting to set up shadow band imaging experiments at multiple locations along the path of totality stretching between Surat and Patna and passing over cities like Indore, Bhopal, Varanasi, and so on, by involving scientists and students from institutions and university departments. Observation from multiple sites would allow recording shadow bands at different solar elevation angles to constrain theoretical ideas. It would also help in increasing the probability of success of cloud-free observations, in view of the cloudiness associated with SW monsoon season during the time of TSE—not all sites would be clouded at the same time. Efforts are underway to make comprehensive shadow band observations during the July 22, 2009 total solar eclipse.

This would be in addition to the other international efforts to record shadow bands from various locations in China and elsewhere. It would be advisable to form an international shadow-band investigation group to look at the so-called enigmatic phenomenon of shadow bands in an integrated manner and put it on a firm scientific ground.

Acknowledgments

First and foremost, the author is grateful to then Director, NCAOR Dr. P.C. Pandey for extending the opportunity to lead the unique and challenging Indian Solar Eclipse Expedition to Antarctica, and Program Director (Science) Dr. N. Khare and Director (Logistics) Dr. Bhaskar Rao for their wholehearted support towards its realization. The author is also grateful to NCAOR (DOD) for technical, financial, and logistics support. The author also acknowledges the scientific collaboration and cooperation from all involved in making the expedition a success. Their contributions at different stages of conceptualization, planning, execution, analysis, and interpretation of data collected are gratefully acknowledged.

The expedition received the excellent support from the Station Commander and the enthusiastic members of the WOT. It is a pleasure to acknowledge the immense enthusiasm and deep involvement of the WTO Medical Officers- Dr. Pradip Malhotra and Dr. Ashok Aggarwal.

Special thanks are due to Dr. Fred Espenak of NASA/GSFC for providing copies of the NASA Eclipse Reports in a timely manner. The regular NASA Eclipse reports produced by Dr. Espenak and colleagues, much in advance of the events, are a comprehensive source of information for all eclipse scientists and amateurs alike.

Last but not least, the author is immensely appreciative of the support received from his family for undertaking scientific adventures like the IAEE expedition.

References

Bhandari, S. 2007. Forthcoming total solar eclipse over the arctic: Opportunities and prospects. *MEGHA*, v. 25, pp. 14–17.

Bhandari, S.M. 1996. Space-Based Observations of the Total Solar Eclipse of October 24, 1995 using INSAT-VHRRs. Invited Paper at NSSS "Special Session on TSE" 1996, Hyderabad, India.

Booker, H. G. and Vats, H. O. 1985. Application of refractive scintillation theory to laser transmission through the atmosphere near ground level. *Radio Sci.*, v. 20, pp. 833–841.

Brunier, S. and Luminet, J. 2000. *Glorious Eclipses: Their Past, Present and Future.* Cambridge Univ. Press.

Codona, C. 1986. The scintillation theory of eclipse shadow bands. *Astron. & Astrophys.*, v. 164, pp. 415–427.

Espenak, F. and Anderson, J. 2002. Annular and Total Solar Eclipses of 2003 NASA/ TP – 2002-211618.

Gladysz, S., Redfern, M. and Jones, B. W. 2005. Shadow bands observed during the total solar eclipse of 4 December 2002 by high resolution imaging Jl. *Atm. & Solar-Terr. Phys.*, v. 67, pp. 899–906.

Jones, B. W. 1999. Shadow bands during the total solar eclipse of 26 February 1998 Jl. *Atm. & Solar-Terr. Phys.*, v. 61, pp. 965–974.

Mobberley, M. 2007. *Total Solar Eclipses and How to Observe Them.* Springer-Verlag, New York.

Tzanis, C., Varotsos, C. and Viras, L. 2008. Impacts of the solar eclipse of 29 March 2006 on surface ozone concentration, the solar ultraviolet radiation and the meteorological.

Vats, H. O., Bagare, S. P. and Bhandari, S. M. 2006. Formation and observation of shadow bands during the total solar eclipse of November 23, 2003 near Maitri. *Antarctica Advances in Geosciences*, v. 2, pp. 361–368.

Zerefos, C. S. et al. 2007. Evidence of gravity waves into the atmosphere during the March 2006 total solar eclipse. *Atm. Chem. & Phys., Discussion*, v. 7, pp. 7603–7624.

9

Exploration of Antarctica as a Natural Test Site for Vicarious Calibration of Advanced Wide Field Sensor on Board Resourcesat-1

S. S. Manjul, P. Narayanbabu, D. R. M. Samudraiah, and S. M. Bhandari

CONTENTS

Introduction

The Indian Space Research Organization (ISRO) has launched a series of remote sensing satellites in polar sun-synchronous orbit incorporating state-of-the-art remote sensing sensors. They are being extensively used for monitoring and managing the earth's natural resources. Among these satellites, Resourcesat-1 has produced voluminous and useful data for various types of remote sensing applications.

Radiometric calibration of satellite sensors is a key to providing long-term continuous, reliable, and consistent data sets for a large range of remote sensing applications. Preflight calibration is carried out in the controlled environment of the laboratory against standard calibration sources with traceable international standards. However, most satellite sensors show a marked degradation with time due to launch stresses, out-gassing in the optics, bombardment by energetic particles when in space, variations in the filter transmittance and spectral response, slow deterioration of the electronic system, changes in the detector response, and so on. It has been pointed out that in some cases, for example, AVHRR and LANDSAT, these changes are on the order of 15%–25% in the first 2 years of operation (Guenther et al., 1997). Such a change not only invalidates the pre-launch calibration, but also leads to residual striping artifacts in the data due to inadequate calibration of the individual spectral channel. To deal with the problem of degradations of the sensor, various calibration mechanisms that include natural sources such as the Sun or artificial sources such as light-emitting diodes, black bodies, tungsten-halogen lamps, and so on are being used. But the addition of a built-in calibration mechanism involves extra cost and complexity. Additionally, estimation of top-of-atmosphere spectral radiance is a complex process, as there are several influencing parameters. Hence, relating digital count of a sensor signal with input spectral radiance/spectral reflectance of terrain surfaces becomes erroneous. To deal with these issues, vicarious calibration becomes necessary and essential as a calibration activity. To perform vicarious calibration, artificial surfaces could be fabricated, characterized, and used by illuminating with solar radiation. However, the feasible and cost-effective solution for undertaking vicarious radiometric calibration is to use natural test sites such as water bodies, sand dunes, and so on.

Antarctica is the seventh continent of the earth and is the stormiest, windiest, driest, coldest, and most inaccessible of all the continents. It contains more than 90% of Earth's permanent ice, with an average ice depth of 1600 m and maximum ice depth of 4500 m. It has an area of 14×10^6 km^2, of which about 94% is permanently covered with snow ice. Per international treaty, Antarctica is exclusively dedicated to conducting scientific research without disturbing its environment.

India has maintained its permanent presence in Antarctica for the last 25 years. The currently functional Indian Antarctic station is located at Maitri (Lat. 70°45′S, Long. 11°44′E) in the Schirmacher Basin of East Antarctica. India has been sending regular summer and winter expeditions to Antarctica for undertaking scientific research in a variety of disciplines. India is in the process of establishing a new Antarctic Research station in the Larsemann Hills region in the Indian longitudes and is ready to embark into the Arctic arena (http://nacor.nic.in, www.nacor.gov.in).

Although the environment of Antarctica is harsh and hostile, some of its unique characteristics make it a potential candidate as a natural test site for in-orbit radiometric calibration of optical sensors. The availability of a permanent Indian research base at Maitri, maintained by the National Centre for Antarctic and Ocean Research (NCAOR), Goa, provides the necessary support infrastructure and logistics help for the conduct of scientific experiments.

Keeping the abovementioned potential advantages of Antarctica in view, it was proposed to participate in a 26th Indian Scientific Expedition to Antarctica, organized by NCAOR, and to explore Antarctica as a natural test site for vicarious calibration of an advanced wide field sensor (AWiFS) on board Resourcesat-1 (Manjul and Bhandari, 2006).

This paper contains a short review of the scientific activities related to ground truth radiometric data collection undertaken during the 26th

Indian scientific expedition to Antarctica between November 2006 and April 2007.

Antarctica as a Test Site for Onboard Calibration

Antarctica could be a unique natural test site to perform vicarious radiometric calibration of a space-based optical sensor for several reasons given below:

- The snow/ice surface produces high spectral reflectance, as high as 100%. Hence, sensor performance is superior due to a higher S/N ratio.
- The snow surface acts as a very good Lambertian surface. Hence, radiometric uncertainties due to angular characteristics of terrain and sensor are less severe.
- The snow/ice surface covers a wide spectral reflectance range from visible to near infrared. Hence, it is useful to cover several spectral bands of satellite sensors to undertake the calibration.
- In general, the test site at Antarctica has distinct advantages with respect to that in the Himalayan region because of a large uniform and flat area that covers a greater number of pixels. Consequently, this improves the radiometric accuracy of the calibration.
- The Antarctic test site remains undisturbed, as there are no industrial activities and no atmospheric pollutants. Hence, atmospheric extinction is less and radiometric variability in measurements is predictable and accountable in temporal data sets for longer periods.
- The Antarctic test site gives near-constant solar illumination geometry during the summer period from November to February. This gives high radiometric stability in concurrent/non-concurrent ground truth measurements.
- In the summer season of Antarctica, it is possible to get several satellite sensor data sets and concurrent ground truth measurements that are free from clouds. Due to this, it is possible to estimate accurate calibration coefficients based on voluminous statistical data.

Aims and Objectives

The scientific research undertaken is complex, and it involved several specific tasks to be performed. Based on the individual task, aims and objectives can be listed as follows:

- To design and develop portable field-operated ground truth radiometers by incorporating spectral bands of an AWiFS of the Resourcesat-1 satellite and other useful spectral bands in the spectral range from 400 to 3000 nm.

Following are the types of radiometric measurement parameters:

- Spectral reflectance of snow/ice surfaces concurrent/non-concurrent to satellite passes. Direct and diffused components of incoming solar spectral irradiance concurrent/non-concurrent to satellite passes.
- To select a suitable natural test site at Antarctica and to carry out "in-situ" radiometric measurements of atmosphere and snow packs under different surface conditions/physical parameters of the snow such as age, depth, wetness, grain size, temperature profile, hardness of snow, and so on.
- To design and develop a database to store all the field data collected during the expedition.
- To establish the methodology of onboard radiometric calibration and validation of the visible channels of AWiFS data with snow as a standard terrain.
- To develop a natural calibration test site in Antarctica for future space-borne optical sensors.
- To gain more hands-on experience that can be used as feedback for the development of a new generation of ground truth instrumentation that addresses specific requirements and work under the harsh environment of Antarctica.

Work Accomplished during the Expedition

The ground truth data collection experiments were performed in the field from December 1, 2006, to March 17, 2007. During the expedition period, there were 45 cloud-free days and concurrent/non-concurrent measurements were undertaken on all these days. Among these, there were 27 days where concurrent measurements with respect to Resourcesat-1 satellite passes were possible.

Selection of Test Site

The Sankalp area close to the Indian station Maitri was selected as a natural test site. The central co-ordinates of Sankalp are: Lat.: 70° 47.914′ S, Long.: 11° 44.277′ E. The glacial test site has a large uniform coverage area, more than 20 × 20 km. It contains permanent continental hard ice with an average depth of 500 m. The area remains covered with uniform ice throughout the year. During December and January, ice starts melting a little at some spatial points,

FIGURE 9.1
Hard continental ice of the Sankalp area.

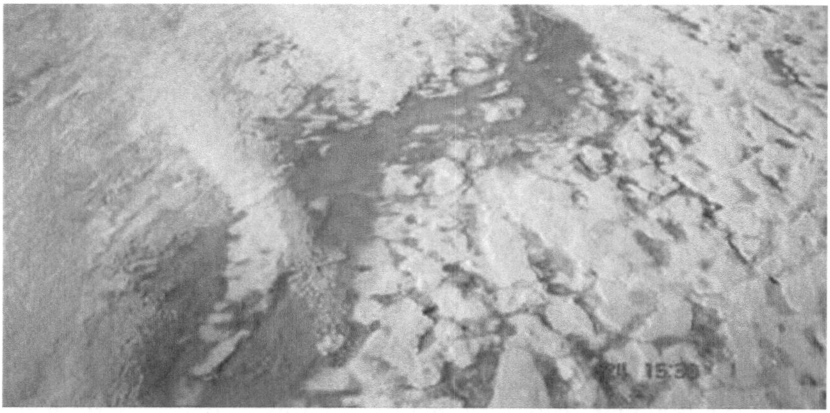

FIGURE 9.2
Fresh snowfall over hard continental ice of the Sankalp area.

and the melt water accumulates in the nearby lakes/catchment areas. At the same time, there is accumulation of fresh snow in this period. To undertake field measurements at Sankalp, logistic facilities including transportation and so on were made available by the WOT at Maitri. Photographs of the test site are shown in Figures 9.1 through 9.3.

Instrumentation

The expedition work started from December 1, 2006, and ended on March 17, 2007. During the expedition, the following types of indigenously designed and developed ground truth radiometers were operated:

FIGURE 9.3
Snow/ice features of the test site of the Sankalp area showing uniform hard ice.

- **Narrow Band Ground Truth Radiometer (NBGTR)** with 12 spectral bands in the spectral range from 400 to 1100 nm. The spectral bandwidth of each band is 10 nm.
- **Broad Band Ground Truth Radiometer (BBGTR)** with 12 spectral bands in the spectral range from 800 to 1800 nm. The band characteristics for 4 bands are same as that of the AWiFS. For the remaining bands, the spectral bandwidth varies from 30 to 70 nm.
- **Near Infrared Ground Truth Radiometer (NIGTR)** with 6 bands in the spectral range from 1000 to 3000 nm.
- **Ratio Radiometer Modules (RRM)**, working in the spectral range from 400 to 1800 nm.

Some of the important features of these instruments are:

- Useful for radiometric measurements, evaluation, characterization, and calibration in absolute values of spectral radiance (w/cm^2-sr-μ).
- Incorporate silicon, germanium, and lead sulfide detectors that cover the spectral range from 400 to 3000 nm with high D* values.
- Flexibility to incorporate any desired spectral band.
- Measurements in four different field-of-view values: 1, 3, 15, 20.
- High calibration accuracy, calibration uncertainty of the module ±1.5% for 3 s.
- 12-bit radiometric resolution.
- Software for corrections of offsets and temperature coefficients.
- Compact, portable and battery operated.

Field Experiments

Field experiments were conducted for the collection of concurrent/non-concurrent ground truth data with respect to Resourcesat-1 satellite overpasses. Ground truth measurements in terms of the following parameters were conducted:

- Reflectance measurement of snow surfaces around Maitri station.
- Reflectance measurement of ice and fresh snow surfaces around Maitri, Sankalp, Vatiayah Hill, and the Russian DROMLAN airstrip.
- Measurement of terrestrial solar spectral irradiance: direct and diffused components.

The experimental test setups are shown in Figures 9.4 through 9.10. The measurements were carried out by fixing the instruments on a tripod, Piston Bully snow vehicle, and helicopter. The ground height for the tripod and Piston Bully were typically between 1 and 2 m. In the case of the helicopter, measurements were carried out at various ground altitudes ranging from 200 to 1200 m. To fully characterize the test site along with its atmosphere, it was planned to carry out measurements at variable heights from 200 to 3000 m by mounting the instruments on the helicopter platform. Due to practical constraints, the helicopter was made available on February 21, 2007, for the Sankalp area and on March 14, 2007, for the Larsemann Hill area. Attempts were made to collect data over the Sankalp and Larsemann Hill areas on these dates

FIGURE 9.4
Experimental test setup for the measurement of direct components of solar spectral irradiance with narrow band ground truth radiometer (NBGTR) near Maitri station.

FIGURE 9.5
Experimental test setup for the measurement of spectral reflectance of a fresh snow patch near Maitri station.

FIGURE 9.6
Experimental test setup for the measurement of direct plus diffused components of solar spectral irradiance with near-infrared ground truth radiometer (NIGTR).

only. The instrument, a narrow band ground truth radiometer (NBGTR), was mounted over the corner-side window of the cockpit. Two other instruments, ratio radiometer modules: voltage type (VRRM) and current type (IRRM), were mounted outside the helicopter. All the instruments were positioned so that they looked vertically downward. To cover the full field of view of the instruments, a snow-covered terrain surface with area not less than 4 × 4 km

FIGURE 9.7
Experimental test setup for the measurement of direct plus diffused components of solar spectral irradiance with ratio radiometer module (RRM).

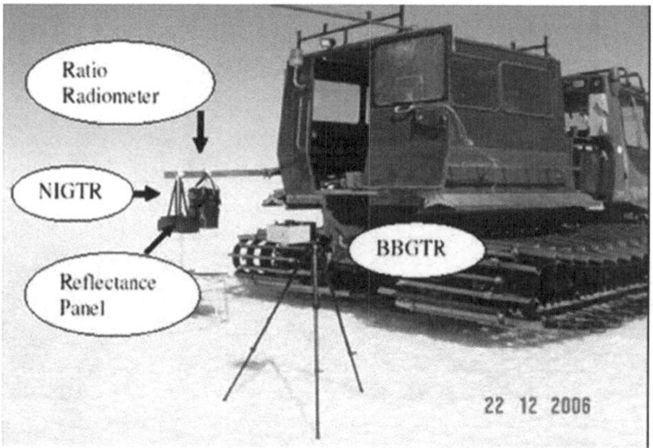

FIGURE 9.8
Experimental test setup for the measurement of spectral reflectance of snow/ice at the test site with ground truth radiometers that were mounted on Piston Bully.

was ensured. The helicopter was kept stationary during the measurements. Spectral radiance data at different ground heights was successfully recorded.

Test Results

Data processing and analysis of ground truth measurements are in progress. Typical values of direct components of solar spectral irradiance

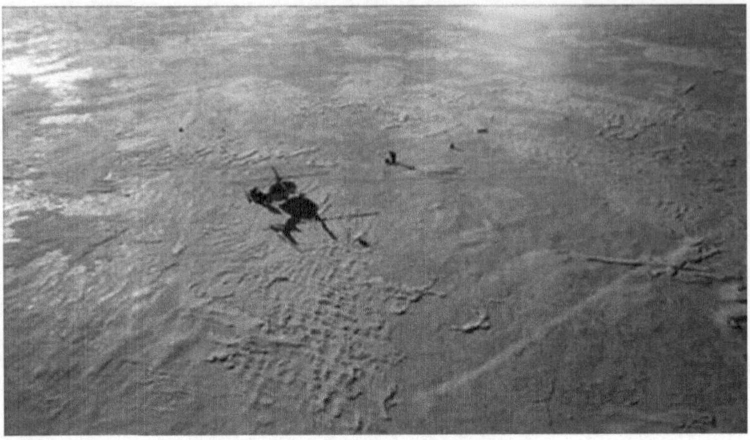

FIGURE 9.9
Photograph of snow/ice surface and helicopter that was used to take aerial radiometric measurements for estimating atmospheric effects.

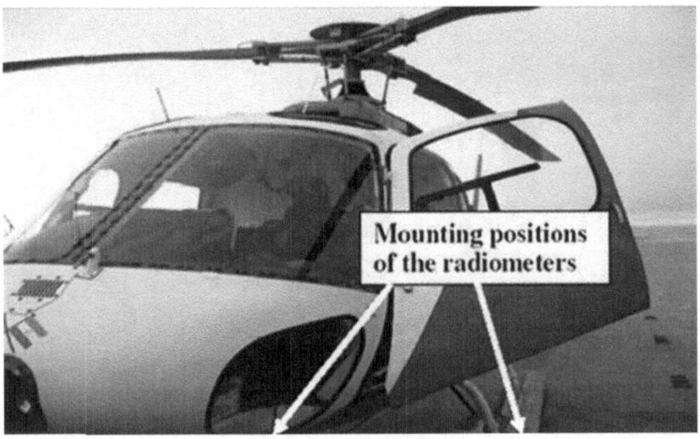

FIGURE 9.10
Photograph showing mounting positions of the radiometers for use during aerial radiometric measurements for estimating the atmospheric effects.

at band B1 of NBGTR are plotted in Figure 9.11. The figure shows how the illumination conditions remain constant from 10.00 to 13.00 hours (UT). It has been observed in the months of December and January that the solar zenith angle remains nearly constant and consequently the illumination conditions remain constant. Hence, this period is suitable to conduct ground truth experiments. The measured spectral reflectance values for fresh snow at the Sankalp area are shown in Figure 9.12. The measured values of reflected spectral radiance from the standard reflectance panel with the narrow band ground truth radiometer mounted at a ground height of 0.5 m are shown in Figure 9.13. These measurements were taken on March 14, 2007, under clear sky conditions at a sun zenith angle of 70°. Measured values of reflected

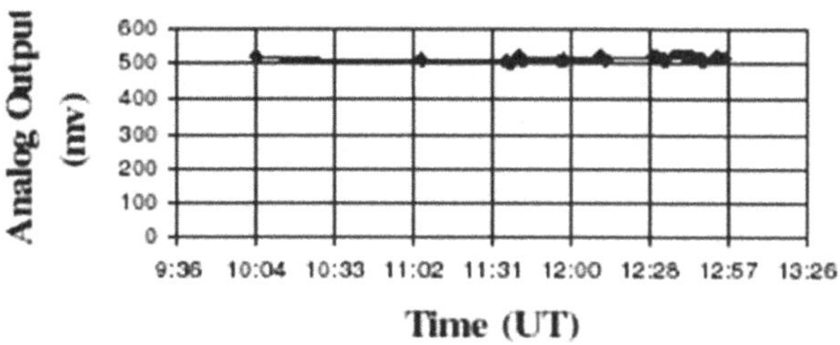

FIGURE 9.11
Typical values of direct component of solar spectral irradiance at band B1 of NBGTR as observed on December 3, 2006.

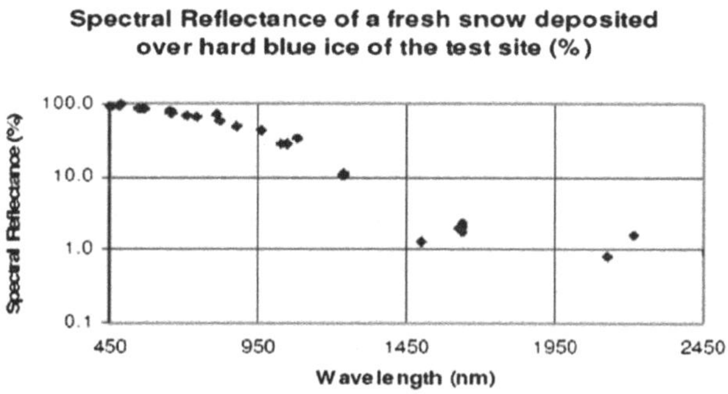

FIGURE 9.12
Measured spectral reflectance for fresh snow at Sankalp area of Antarctica.

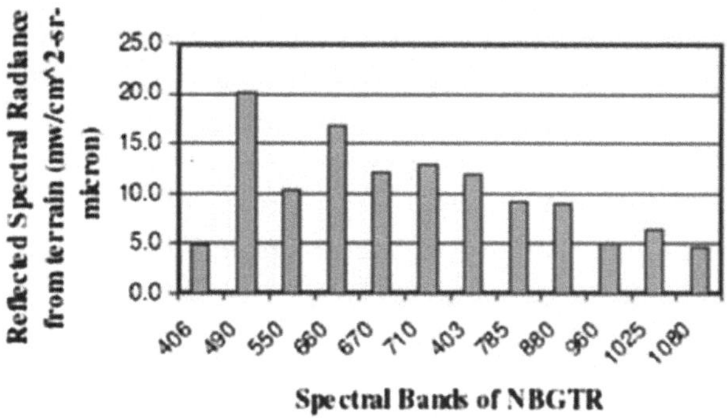

FIGURE 9.13
Measured values of reflected spectral radiance from the standard reflectance panel with narrow band ground truth radiometer mounted at ground height = 0.5 m.

spectral radiance from snow/ice-bound terrain surface with a narrow band ground truth radiometer mounted at a ground height of 0.5 m (series 1 data) and mounted at a ground height of 3000 m (series 2 data) are shown in Figure 9.14. These measurements were taken on March 4, 2007, under clear sky conditions at a sun zenith angle from 66.5° to 67.8°. The observed path

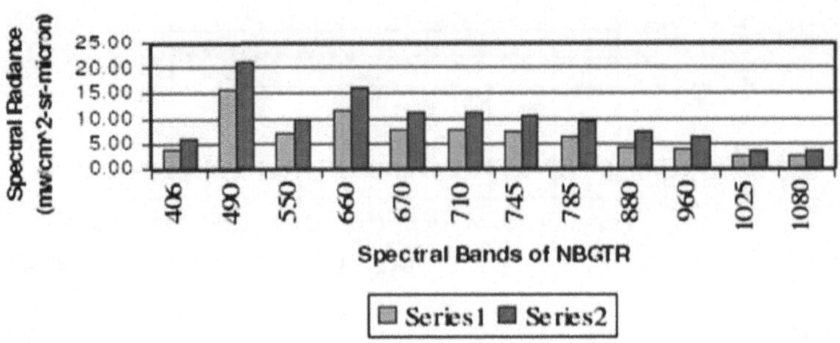

FIGURE 9.14
Measured values of reflected spectral radiance from snow/ice-bound terrain surface with narrow band ground truth radiometer mounted at ground height = 0.5 m (series 1 data) and mounted at ground height = 3000 m (series 2 data).

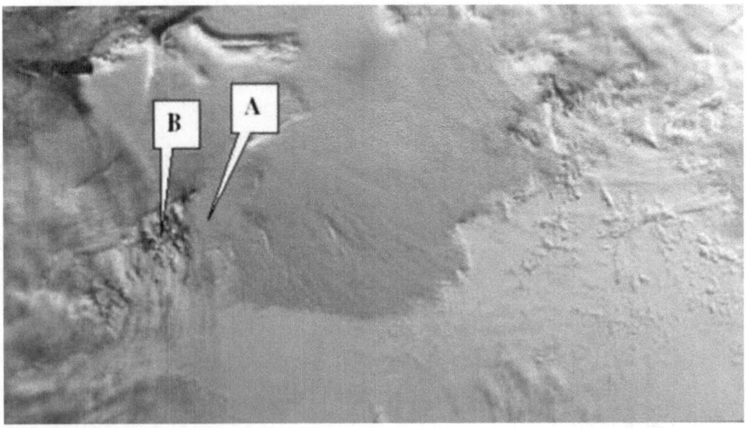

FIGURE 9.15
Resourcesat-1 image of the observation area near Maitri, taken on March 9, 2007. A: uniform and flat glacial ice region where Sankalp was located, B: Schirmarcher Basin of East Antarctica.

radiance due to atmosphere for all bands is from 8% to 15%. Multilayer FCC imagery of Resoursesat-1 B2, B3, and B4 over the test site covering an area of 740 × 740 km and collected on March 9, 2007, is shown in Figure 9.15.

Summary and Conclusions

Ground truth radiometric data collection experiments have been successfully carried out under the 26th Indian Scientific Expedition to Antarctica for vicarious calibration of the AWiFS sensor of Resourcesat-1. The data collection was performed in the fieldwork period from December 1, 2006, to March 17, 2007. During the expedition period, there were 45 cloud-free days. Concurrent/non-concurrent measurements were carried out on all these days.

Among these days, there were 27 days where concurrent measurements were carried out with respect to AWiFS over passes. During the expedition, four types of ground truth radiometers (GTRs) developed at SAC/ISRO, Ahmedabad, were successfully used to collect voluminous radiometric field data. All the instruments worked satisfactorily in the harsh environment of Antarctica.

The measured data is in terms of spectral reflectance/radiance of terrain surfaces and incoming solar radiation in the spectral range from 400 to 3000 nm. The Sankalp area near Maitri (Lat.: 70° 47.914′ S; Long.: 11° 44.277′ E) was used as a natural calibration test site. It was found suitable because of its large and uniform coverage with permanent hard ice. Additionally, the Sankalp area is easily accessible from the Maitri station to undertake measurements. Transportation and other logistic facilities available at Maitri were made use of to conduct the field experiments.

To characterize the test site along with its atmosphere, measurements were carried out near ground as well as at variable ground heights from 200 to 3000 m by mounting the instruments on the helicopter platform. Attempts were made to collect data over the Sankalp and Larsemann Hill areas. The instrument, a narrow band ground truth radiometer, was mounted over the corner-side window of the cockpit. Two other instruments, ratio radiometer modules: voltage type and current type, were mounted outside the helicopter. All instruments were positioned so that they looked vertically downward. To cover the full FOV of the instruments, a snow-covered terrain-surface with an area not less than 4×4 km was ensured. The helicopter was kept stationary during the measurements. The spectral radiance data at different ground heights was collected successfully.

It has been observed that collected ground truth data is consistent. During the months of December and January, the solar zenith angle remains practically constant and consequently the illumination conditions remain constant. Hence, this period is suitable to conduct ground truth experiments. The observed path radiance due to atmosphere for all bands is from 8% to 15%. Data processing and analysis work is being carried out at SAC/ISRO, Ahmedabad. Twenty data sets of the Resourcesat-1 AWiFS sensor have been selected for vicarious calibration analysis. These have been streamlined for procurement. The selection criteria applied were: images should be nearly cloud free. Concurrent ground truth measurements should be available.

The project proposal was made in May 2006 (Manjul and Bhandari, 2006). Activities like project inception, planning, design of field experiments, and design and development of radiometers were successfully completed prior to the expedition. Participation and data collection in the harsh and hostile environment of Antarctica were successfully completed. Presently, post-calibration of instruments, ground truth data analysis, satellite data analysis, interpretation, results and documentation, publications, and so on have been initiated. It has been observed that ground truth data is consistent and useful for further studies.

Future plans consist of processing and analysis of ground truth data and satellite image data and to establish a methodology for vicarious calibration of onboard sensors. Parallel to this, the activity of designing and developing state-of-the-art radiometers, which will be specifically tuned to the exact requirements and to the unique environmental conditions of Antarctica, is also planned. Based on the feedback/outcome results of the activity, it will be decided to plan additional field measurement cycles by participating in future Antarctic expeditions. The possibility of conducting such measurements in the Arctic would also be explored in association with NCPOR.

Acknowledgments

The authors gratefully acknowledge Dr. R. R. Navalgund (Director), Dr. Ram Rattan, (Associate Director), Shri A. S. Kirankumar (Deputy Director), and Smt. Gunbala Navlakha, SEDA, Space Applications Centre (SAC/ISRO), Ahmedabad, for their constant encouragement, guidance, and providing all the necessary support for participation in the expedition. The authors also gratefully acknowledge Dr. P. S. Goel, Secretary, Shri Prakesh Kumar, Joint Secretary, MoES/GOI, New Delhi and Shri Rasik Ravindra, Director, NCAOR, Goa, for their keen interest and for conducting a scientific review of the activity during their visit at Maitri. The authors would also like to express their deep sense of gratitude to Shri T. V. P. Bhaskara Rao, Director (Logistics) and Dr. N. Khare, Program Director (Science), NCAOR, Goa, for providing the logistic support and conducting a scientific review of the project, respectively. The authors also gratefully acknowledge Shri L. Prem Kishore, Shri Ajay Dhar, and Shri Jayapaul D., Leader and Station Commanders, Lt. Col. R. J. Rao, OIC Logistics, Capt. K. Kingston, Station Engineers, and all members of the 25th and 26th Indian Antarctic Expeditions for making available all the logistic support and valuable assistance to make the fieldwork comfortable and successful. Finally, the authors are thankful to Shri M. K. Kalubarme, and Dr. M. P. Oza RESA, SAC, for their helpful counsel.

References

Guenther B., Butler J. and Ardanuy P. 1997. Workshop on strategies for calibration and validation of global change measurement: May 10–12, 1995. NASA Reference Publication #1397.

Manjul S. S. and Bhandari S. M. 2006. Research project proposal on ground truth radiometric/polarimetric data collection experiments under 26th Indian Scientific Expeditions to Antarctica during December 2006 to March 2007. Report No.: SAC/EOSG/SFSD/22/05/2006/27, Space Applications Centre, Ahmedabad.

10

Study of Space Weather on GPS Performance at Low-Latitude Station Bhopal and High-Latitude Station Maitri, Antarctica

A. K. Gwal and S. Choudhary

CONTENTS

Introduction

Extensive research in studying the influence of the propagation media, viz. troposphere, ionosphere, plasmasphere, and magnetosphere, on global positioning system (GPS) signals led to effective use of GPS for atmospheric research, thus contributing to global space weather monitoring. Disturbances on the Sun can produce dramatic effects in the space environment surrounding the Earth. Energetic particle effects become more intense and pose a hazard to astronauts and damage spacecraft electronics, satellite lifetimes are shortened by increased atmospheric drag, and communications and navigation are disrupted by the changing plasma environment. Space weather has become the modern idiom for these effects. The concept of space weather is now widely used in quantitative descriptions of the physical changes in the near-Earth space environment in response to variations in solar radiation, solar plasma ejection, and the electromagnetic status of the interplanetary medium. Even as answers are being sought to questions related to the parameters required for an adequate description of space weather, it is recognized that one of the most important effects of space weather, which has a bearing on operational systems, is that on transionospheric radio wave propagation. Space-weather-related changes in the electron density distribution in the path of a signal due to production or loss processes, electrodynamics of the intervening plasma, and changes in the neutral component that affect the plasma alter the propagation medium and hence its effect on a radio wave signal. In the last decade, the utility of radio wave transmissions from global positioning system satellites in obtaining information about the Earth's ionosphere simultaneously from a global network of stations has been demonstrated. To the lowest order, changes in the total electron content (TEC) along the signal path on transverse spatial scales far more than the Fresnel scale, which is determined by the signal frequency, contribute to only an excess time delay. For an undisturbed ionosphere, where the electron density does not vary significantly over short (<m) scale lengths, this is the only effect that the ionosphere has on GPS signals.

The scattering of radio waves enters the picture when spatial scales of electron-density variations, in a direction transverse to the signal path, are on the order of or smaller than the Fresnel scale. Because of scattering, a diffraction pattern is formed in the plane of the receiver. Movement of irregularities across the signal path results in temporal fluctuations or scintillations in the amplitude and phase of the signal recorded by a receiver. The severity of scintillation effects is determined by the power spectrum of electron-density variations in the propagation medium and the orientation of the signal path. Space weather influences the development of small-scale irregularities in the ionospheric plasma, which may give rise to scintillations on transionospheric very low frequency (VLF), ultra low frequency (UHF), and even higher frequency radio waves. The ionosphere in the equatorial and polar-cap/auroral regions often exhibits such departures from an undisturbed, smooth state. Dual-frequency GPS observations have provided a unique opportunity to simultaneously study the scintillations caused by ionospheric refractive-index variations and the fluctuations in the TEC along the signal path, which directly represent electron density variations associated with scintillation-producing irregularities. Radio wave signals, transmitted from geostationary satellites and recorded by spaced receivers, yield information about the dynamics of ionospheric irregularities, which are affected by space weather. Recent studies of diurnal, seasonal, and magnetic activity-linked variations in TEC and ionospheric scintillations are reviewed here.

Space Weather Effects at High-Latitude Ionosphere

Several unique phenomena affect GPS performance at high latitude, where the satellite constellation changes dramatically due to the inclination of the

GPS satellite orbits. No satellites pass directly overhead as one approaches the North Pole. Instead, all satellites appear closer to the horizon, and at very high latitudes (greater than 60°), satellite visibility increases to 12 or more at low-elevation angles. This produces a reduced geometry as well as a high-potential multipath environment. GPS receiver tracking performance can be degraded during periods of enhanced ionospheric activity, where small-scale scintillation effects are observed in the high-latitude auroral region. During periods of intense scintillation, the availability of carrier phase observations may be limited through loss of signal lock, with a significant impact on precise positioning applications.

Such effects have a larger impact on L2 tracking performance, where codeless and semi-codeless technologies are employed to extract the encrypted L2 signal. GPS data provide adequate information on phase fluctuations associated with middle- and large-scale ionospheric irregularities. Dual-frequency GPS phase measurements enable retrieval of information about variations of the ionospheric total electron content.

Space-weather effects on radio wave signals propagating through the high-latitude ionosphere are a direct result of solar-wind-magnetosphere-ionosphere coupling processes. As far as intermediate-scale irregularities in the ionospheric electron density at high latitudes are concerned, the dayside auroral oval may be the source region and gateway for the transfer of solar-wind energy to the ionosphere. Under active magnetic conditions, the eastward electric field in the dayside auroral oval directs a "tongue of ionization" from the sub-auroral region into the polar cap, which is sliced off into discrete large-scale ionization structures by high plasma flows (Valladares et al., 1994). The role of electric fields in the high-latitude F-region in the generation of sub-kilometer-scale irregularities was examined by Tereshchenko et al. (2000). This was done by studying the correlation between the electric-field direction, obtained from plasma-flow measurements using European Incoherent Scatter Scientific Radar Association (EISCAT), and orientation of the anisotropic irregularities. The orientation was determined from scintillation measurements on satellite signals by a modified version of the method suggested by Tereshchenko et al. (1999). It was found that in nearly all the cases where it was possible to compare the orientation of the irregularities with the F-region plasma flow, the orientation was in the flow direction. This suggests that the instability that creates the density fluctuations were driven by the electric field. Since the pattern of scintillation has a significant dependence on the orientation of the signal path with respect to the anisotropic irregularities, space-weather effects on the electric field will also influence scintillations on trans ionospheric radio waves through this factor.

Ionospheric Effects at Low Latitude

Earth's atmosphere, neutral as well as ionized, is greatly influenced by solar activity. In addition to the regular slowly varying changes with solar activity, there are dramatic events like solar flares and coronal mass ejections that give rise to geomagnetic storms and changes in the upper atmosphere-ionosphere-magnetosphere. The electric fields of magnetospheric origin not only affect the ionosphere at high latitudes but also penetrate low latitudes, causing changes in the ionosphere-thermosphere at low latitudes. The ionosphere at low latitudes is marked by features like equatorial electrojet and the associated equatorial ES, F2 anomaly, and equatorial spread-F. Electrodynamics plays an important role in the low-latitude ionosphere. During daytime counter-electrojet events when electric field reverses, equatorial ES disappears. Equatorial anomaly is also suppressed during the periods of counter-electrojet. The low latitudes are marked with high incidence of spread-F during nighttime. Equatorial spread-F (ESF) is associated with the post-sunset uplift of the F-region. The plasma density structures associated with ESF also give rise to intense scintillations of radio waves propagating through the ionosphere. The features of the low-latitude ionosphere during geomagnetic disturbed conditions are well known. Electric field reversals during daytime give rise to counter-electrojet events and the disappearance of the q type of sporadic E. Equatorial F2 region anomaly is generally weakened or absent on geomagnetically disturbed days. Thus, equatorial electrojet (electric field) strength is an important parameter to model F-region ionization distribution at low latitudes. Equatorial spread-F is inhibited in general except during the June solstices of low solar activity periods, when an increase is noted. However, some geomagnetic storms are associated with intense spread-F. On some occasions, strong scintillations, generated in the early morning hours, extend to daytime hours also. Scintillations get enhanced, inhibited, or unaffected if the maximum excursion in Dst occurs during post-midnight, afternoon, and sunset hours, respectively. Apart from the electric fields, the neutral winds associated with geomagnetic storms affect the low-latitude ionosphere-thermosphere.

The largest contribution to TEC also comes from the F layer of the ionosphere. Hence, most of the modeling efforts have been concerned with this region. Nevertheless, to study the effect of space weather on GPS signals, it is important to model the contribution of the hydrogen-dominated plasma of the high plasmasphere or protonosphere to the slant total electron content (STEC) under different space-weather conditions.

Solar Proton Emission

The solar wind is the supersonic outflow into interplanetary space of plasma from the Sun's corona. It is composed of approximately equal numbers of ions and electrons; the ion component consists predominantly of protons (95%), with a small amount of doubly ionized helium and trace amounts of heavier ions. Embedded in the outflowing solar wind plasma is a weak magnetic field known as the interplanetary magnetic field (IMF). The solar wind varies in density, velocity, temperature, and magnetic field properties with the solar cycle, heliographic latitude, heliocentric distance, and rotational period.

Geomagnetic storms occur when high-speed plasma injected into the solar wind from coronal mass ejections or coronal holes impinges on Earth's geomagnetic field. If the arriving solar wind plasma has a southward magnetic field, energy is coupled efficiently into Earth's magnetosphere and upper atmosphere. Several types of solar events can cause particles with high velocities to be superimposed on this background solar wind. When charged particles ejected from the Sun arrive at the Earth, they can cause perturbations in the geomagnetic field and in the ionosphere; the electron density (number of electrons in a given volume) can vary considerably, both in time and space.

Under some conditions, solar eruptions can also accelerate charged particles to very high energies (protons and heavier particles, such as helium). The enhanced particle density within the Earth's magnetic fields during a geomagnetic storm can also cause degradation of GPS performance. Less energetic particles contribute to a variety of spacecraft surface charging problems, especially during periods of high geomagnetic activity. These events cause major failures to satellite systems, even reducing the position accuracy.

Data and Experimental Setup

Irregularities in the ionosphere due to space weather events caused by solar flares and coronal mass ejection can scatter trans-ionospheric radio signals, producing fluctuations in both amplitude and phase and GPS cycle slips disrupting satellite communications and navigation. We concern our study at equatorial and sub-auroral regions during low solar activity. In the Figure 10.1 shown, a dual frequency GSV 4004A GPS receiver is installed at sub-auroral region Maitri, Antarctica (geomagnetic lat. 62.8° S, geomagnetic long. 52.8° E), and equatorial region Bhopal (geographic lat. 23.2° N, geographic long. 77.6° E), Figure 10.2 shown.

This chapter discusses some space weather effects on GPS signal propagation, total electron content variation, and ionospheric scintillation based on GPS data of January 2005 to March 2005, January 2006 to March 2006, January 2007 to March 2007, and December 2007 to February 2008 at high-latitude region Maitri and January 2005 to December 2005 at low-latitude region Bhopal. We use only those satellite data whose elevation angle is greater than 300. Earth's magnetic field data at the Maitri base station

FIGURE 10.1
GPS receiver installed at the roof of 503 GPS Barkatullah University, Bhopal.

were collected with the digital fluxgate magnetometer. A magnetometer is basically an instrument used for continuous recording of the variations of the geomagnetic elements. At high latitudes, it is a practice to measure the variations X, Y, Z along the geographic north-south, east-west and vertical, rather than H, D, and Z, because the presence of field-aligned currents causes very rapid fluctuations in D. Here we have used the X, Y, Z components to calculate the magnitude of the horizontal component (H) of Earth's magnetic field for our study with minute resolution. The solar wind parameters were obtained from the ACE Satellite Data Centre.

Result and Discussion

Diurnal Variation

Global navigation satellite systems such as GPS and GLONASS offer a unique opportunity to monitor the total electron content of the ionosphere on global scale. Due to the dispersive nature of the ionosphere dual frequency GPS measurements may provide integral information about the ionosphere by computing differential phases of code and carrier phase measurements. In

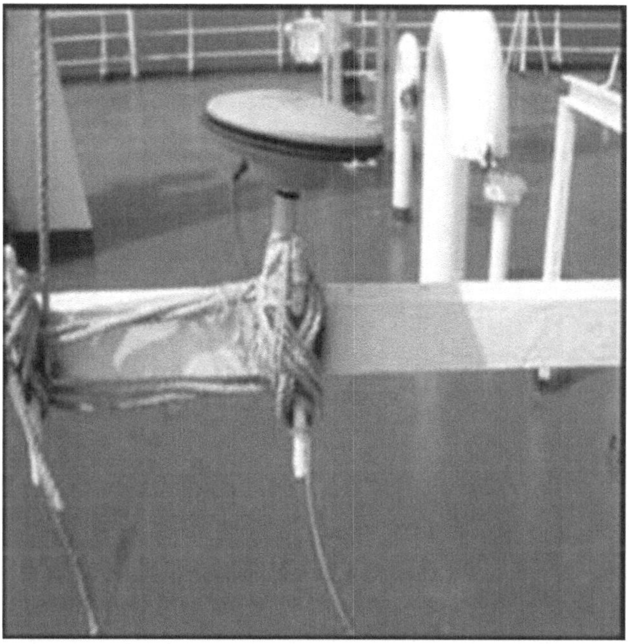

FIGURE 10.2
GPS antenna (NovAtel's model antenna) mounted on the ship, Maitri.

a first-order approximation, this differential phase is proportional to the integral of the electron density along the ray path between the transmitting GNSS satellite and the receiver. After the total electron content along several ray paths has been determined by using a special calibration technique for the ionospheric delay of GPS signals (Sardón et al., 1994), the slant TEC is mapped to the vertical by using a single-layer approximation for the ionosphere at hE = 350 km height.

The geomagnetic storm on 17 January and 21 January 2005 is well documented here. The variation of TEC during ionospheric perturbations is studied by creating differential maps

$$\left[\mathrm{TEC} = \frac{(\mathrm{TEC} - \mathrm{TEC}_{median})}{\mathrm{TEC}_{median}} \times 100\%\right]$$

Both storms were characterized by longitudinal/latitudinal differences in TEC. The 24-hourly plot in Figure 10.3 indicates strong enhancement of DTEC of 112.37% at low-latitude station Bhopal at noon, whereas the same day at Maitri, the high-latitude station shows 108% of DTEC during the daytime. Likewise, Figure 10.4 also indicates the same result of DTEC enhancement at low latitude during the storm event of 21 January 2005. It is observed that on that day, the variation of DTEC rose to 213.03% during the daytime at Bhopal,

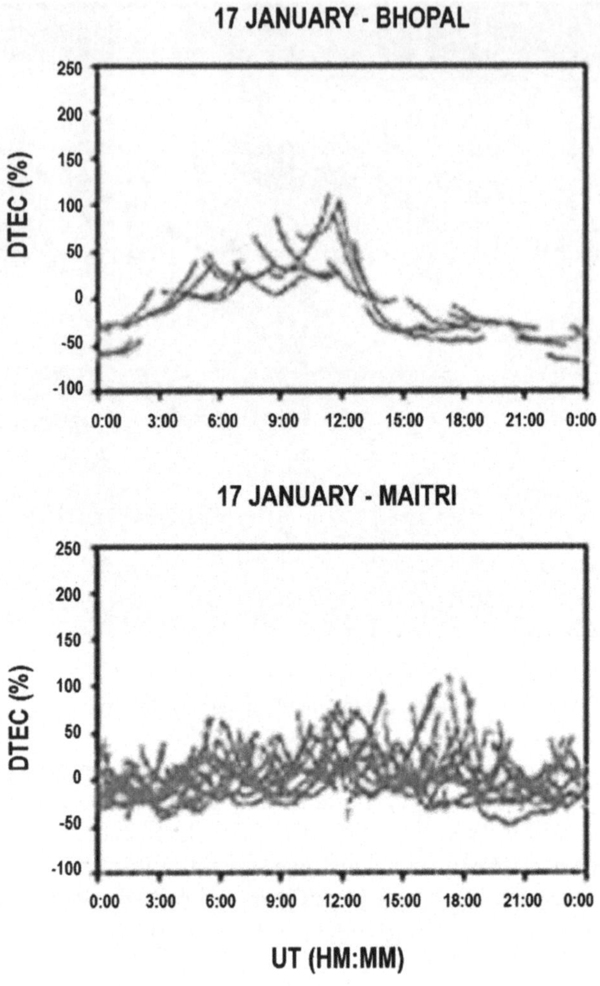

FIGURE 10.3
Diurnal variation of percentage deviations of TEC from monthly medians on 17 January 2005.

whereas in the case of Maitri, the variation of DTEC rose only to 100.26%. It is assumed that equatorward directed neutral winds then caused plasma uplifting in the low latitude region.

Among the possible causes for TEC enhancements over the Australian/Asian sector are fountain effect intensification due to an eastward magnetospheric electric field penetrating to the equatorial ionosphere, thermal expansion caused by the transport of joule and particle heating from high latitudes, and storm-time thermospheric disturbance winds flowing towards lower-latitude regions, transporting ionization upward along the geomagnetic field lines to regions where recombination rates are lower. The effectiveness of thermospheric winds in modifying the ionospheric height is maximum at

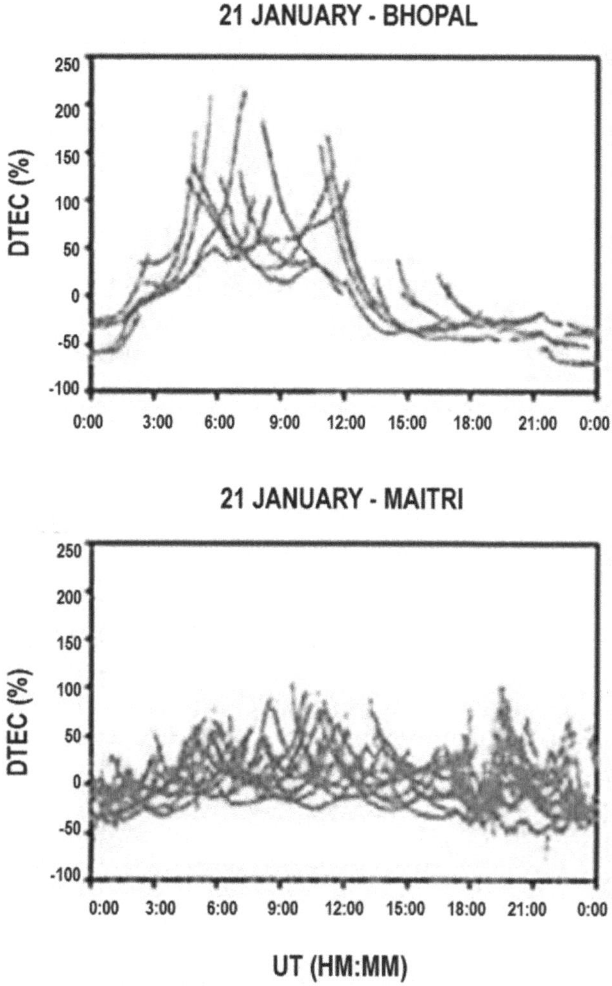

FIGURE 10.4
Diurnal variation of percentage deviations of TEC from monthly medians on 21 January 2005. Anomaly region at Bhopal during the time of January 2005 to December 2005.

middle latitudes and minimum at the geomagnetic equator due to the dip angle, while in equatorial latitudes, ionospheric height modifications are mainly associated with electric field influences (Rishbeth, 1998).

Ionospheric disturbances during magnetic storms are generally explained in terms of thermospheric changes in both neutral air winds and composition, which result in changes of the rates of production and loss of ionization. However, ionospheric day-to-day variability (particularly during storm times) is still not well understood on global scales. Yet, the coupling processes between the magnetosphere and thermosphere become more complicated during magnetic storms. Several investigators have shown that observations of total electron content by GPS can contribute to an understanding of the characteristics of ionospheric perturbation (Jakowski, 1996; Lu et al., 1998; Saito et al., 1998). Lu et al. (1998) have made global ionospheric maps of TEC observations from the worldwide network of GPS receivers to study the upper atmospheric response to a specific major magnetic storm on a simulation basis.

Seasonal Variation

Seasonal variations in the total electron content of the F2 layer in the equatorial anomaly region have been studied by many researchers for solar minimum conditions and for solar maximum conditions in recent years (Aarons et al., 1996; Ho et al., 1996; Kelley et al., 1996; Beach et al., 1997). These studies found that TEC is correlated well with the strength of the equatorial electrojet. The day-to-day behavior of the location of the equatorial anomaly in the columnar electron content was described by Rastogi and Klobuchar (1990). They found that the latitudinal distribution of TEC is related to the strength of the equatorial electrojet. In this section, vertical total electron contents (VTEC) are used to quantify the equatorial.

The month-to-month variation of the day-to-day variability is shown in Figure 10.5. The value of TEC is separately shown for day and night conditions. The variation of TEC is not smooth; the maximum value of daytime TEC is in the months of March and September and minimum in November and December 2005. During the night, the maximum TEC is in the same months of March and September and minimum in November and December 2005.

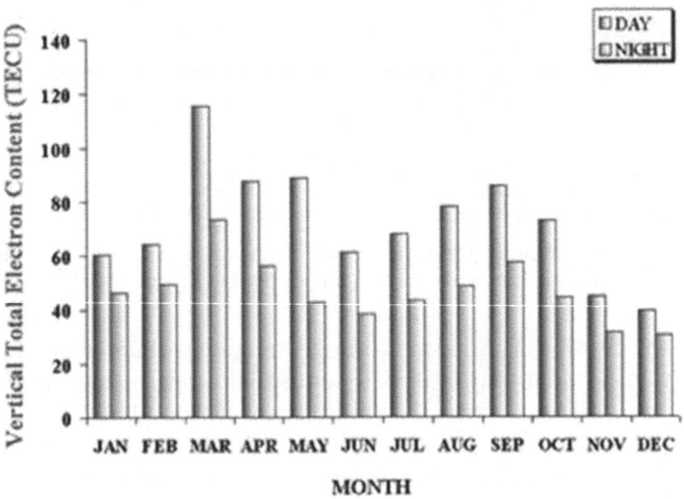

FIGURE 10.5
Month-to-month variability of TEC for daytime and nighttime hours for low-latitude station Bhopal.

It is known that TEC shows marked variation with the geographic location of the observing station. The dominant mechanism controlling the TEC in equatorial and low latitudes is the ExB drifts. Figure 10.6 also depicts the same result of equinoxial maxima during the daytime throughout the year 2005 of solar minima.

The semiannual variation of EIA, is a combined effect of the solar zenith angle and magnetic field geometry. In general, electron populations in the ionosphere are mainly controlled by solar photo-ionization and recombination processes. Localized enhancements/depletions of electrons in the ionosphere are generally caused by electromagnetic forcing. During equinoxes, the subsolar point is around the equator, where the eastward electrojet-associated electric field is often largest. Because of the co-location of the peak photoelectron abundance and the most intense eastward electric field regions, one would expect that the fountain effect would be developed the most. On the other hand, during solstices, photoelectrons at the equator decrease because the subsolar point moves to higher latitudes and the fountain effect is expected to wane.

On the other hand, the qualitative feature of the month-to-month seasonal variations of the monthly average of TEC can also be interpreted based on a simple neutral atmosphere model. It is well known that a meridional component of neutral wind blows from the summer to the winter hemisphere (Bramley and Young, 1968). During the summer solstice, the neutral wind can reduce the crest value by neutral drag as it blows in an opposite direction to the plasma diffusion process originating from the magnetic equator. This results in a low ionization crest value (Stening, 1992). At the equinoxes, the latitude for the midday overhead sun occurs in the region near the magnetic equator. The meridional winds blowing poleward should result in a high ionization crest value. Based on this scenario, a seasonal effect on the crests should be expected, with the crest maximum at the equinoxes and minimum in the summer and winter.

Figure 10.7 shows the result of summer season during the time of January to March of 2006, 2007, and 2008 at high-latitude station Maitri.

It is clearly shown by the figure that the year 2006 has maximum variations in total electron content, 2007 gives the moderate, and 2008 has the minimum variations of TEC. The gradual decrease in the slope of the region with decrease in solar activity indicates that the rate of production of ionization is positively correlated with solar activity.

During the summer season, when Antarctica has continued day for a few months, ionization occurs at a constant rate in the upper atmosphere due to solar radiation and excessive heat. Here a regular measurement of geomagnetic activities in the permanent centers of all the nations is done. More solar radiation reaches the surface of the South Pole than is received at the equator in an equivalent period. To predict the changes in atmosphere in the future, it is necessary to analyze the trend in the past. This continent is ideal for such types of analysis. The ice that has been accumulating for years holds natural secrets in its layers.

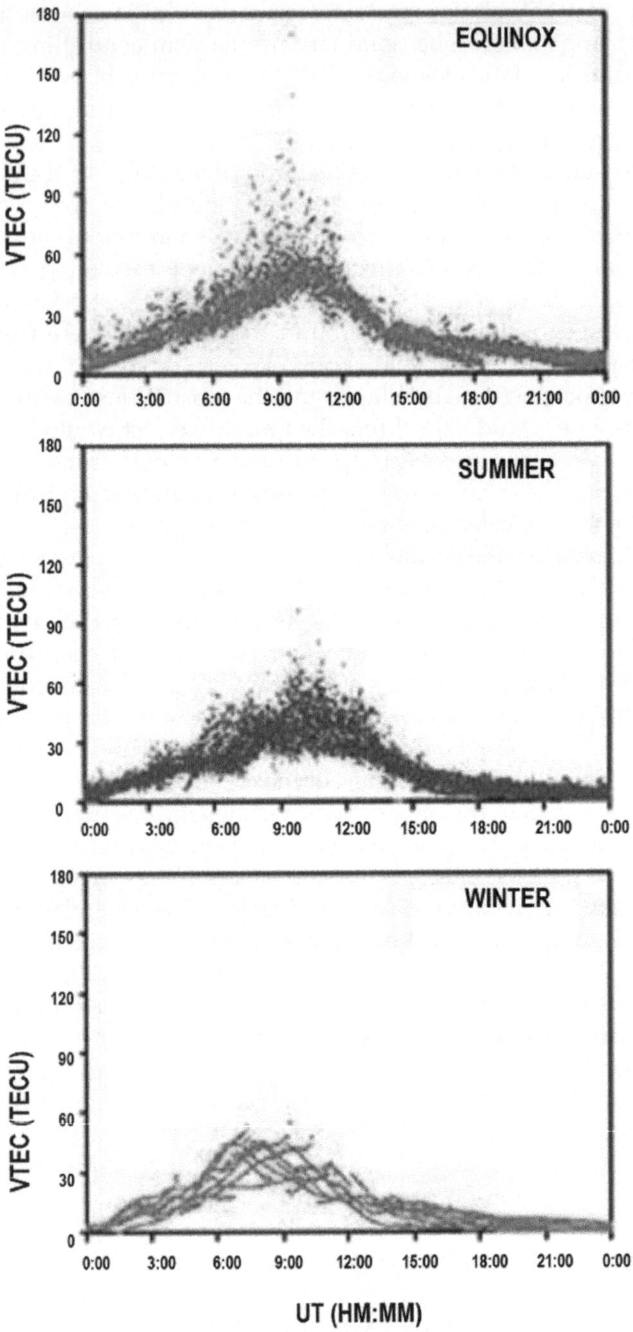

FIGURE 10.6
Seasonal variation of total electron content during January-December 2005 at low-latitude station Bhopal.

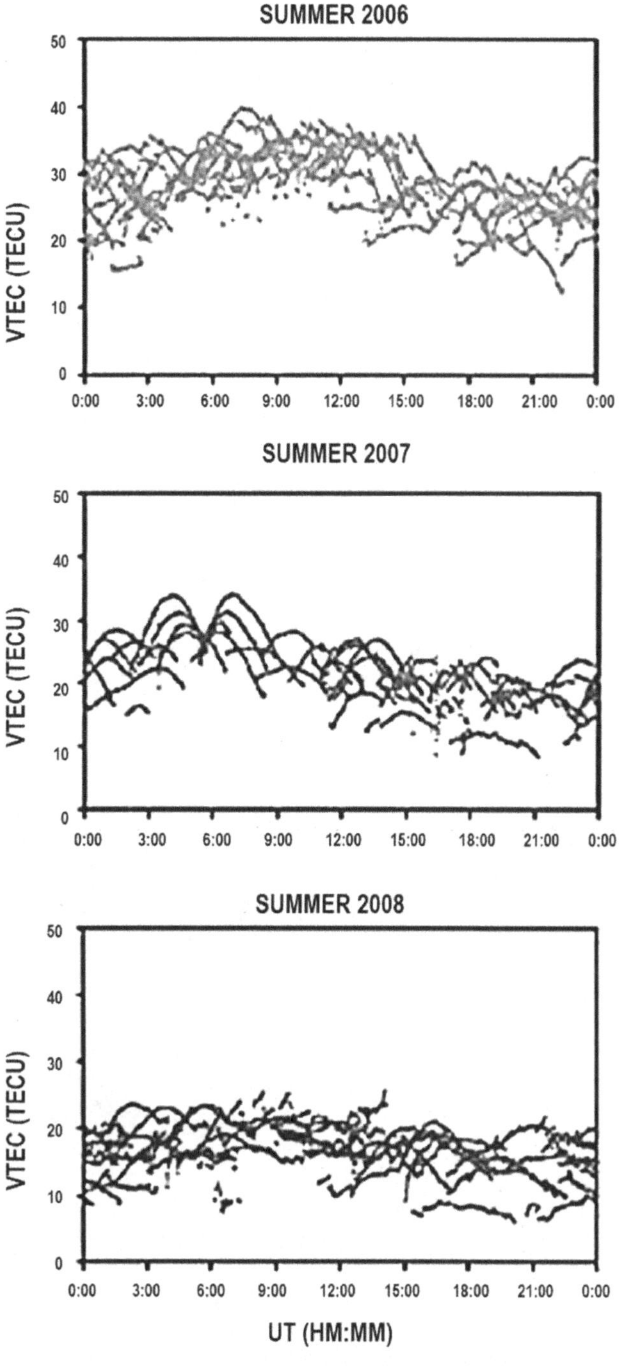

FIGURE 10.7
Variations of total electron content during the summer season at high-latitude station Maitri.

Characteristics of Total Electron Content Variation for Geomagnetic Storms at High-Latitude Region Maitri

Geomagnetic storms are often initiated by the sudden arrival of a high-speed stream of solar wind or high energetic particles from solar flares carrying high particle density and high interplanetary magnetic fields. The storm-time relationship between the perturbation components of TEC and geomagnetic fields was experimentally examined by several investigators in the early days (Nakata, 1966; Mendillo et al., 1969; Titheridge and Buonsanto, 1999). Their results indicate that the TEC increases during the first 24 hr of a storm and then falls to below its quiet-day values with recovery 1 or 2 days later for the typical storm. We examine in this section the relationship between the perturbation components of TEC, which were derived from GPS dual frequency signals, and geomagnetic field variations for the summer storms for the high-latitude region Maitri during 2005 to clarify the low as well as high-latitude characteristics of TEC variations at the time of magnetic storms. The upper panel of Figures 10.8 and 10.9 shows the variation of H-field for the magnetic storms of 17 and 21 January 2005. The perturbation component of GPS-TEC from all observable satellites during the same period is shown in the lower panel of Figures 10.8 and 10.9. We determined 350 km as the altitude of the sub-ionospheric point of GPS slant range and limited analyses to the GPS elevation angles greater than 30°. For this reason, the GPS-TEC in this figure has more than one value at the same time. During the storm of

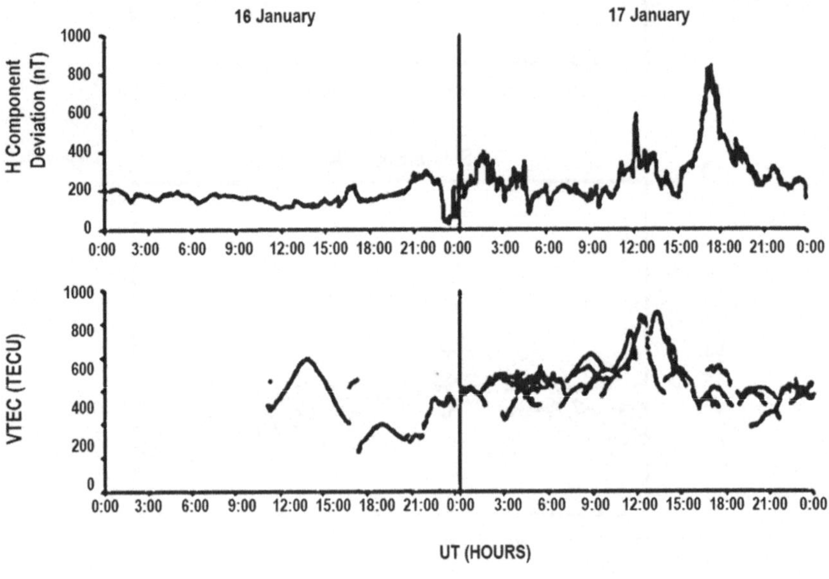

FIGURE 10.8
Variations of (upper panel) magnetic H-component and (lower panel) the perturbation components of TEC associated with the storm event for 17 January 2005 at high-latitude region Maitri.

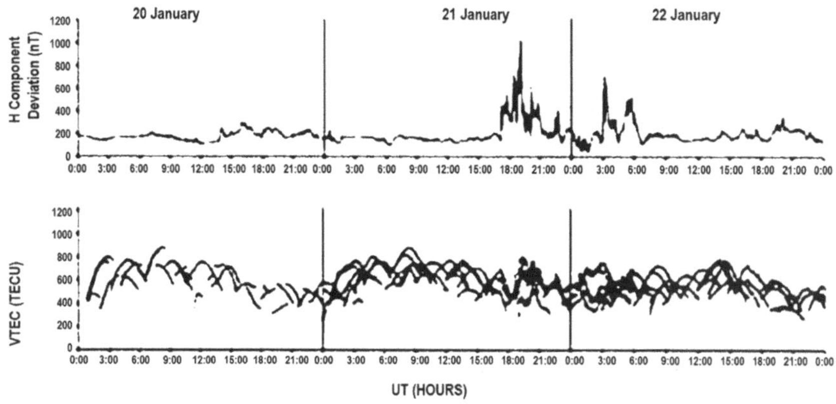

FIGURE 10.9
Variations of (upper panel) magnetic H-component and (lower panel) the perturbation components of TEC associated with the storm event for 21 January 2005 at high-latitude region Maitri.

17 January, sudden storm commencement (SSC) was at 0748 UT, and for the storm of 21 January, it was about 1711 UT.

Figure 10.8 shows that the magnetic field increased with approximately 598 nT; then again, there was another increment in the value of the H-component with a value of approximately 841 nT on 17 January. Accompanying this small-scale magnetic disturbance, the amplitude level of TEC strongly increased to about 53 TECU, but there was no significant variation in TEC at the secondary maxima of the H-component of 17 January. Figure 10.9 illustrates another example showing the storm effect. As seen from the figure, the horizontal component H tends to increase to 1015 nT with the main phase of the storm; simultaneously, the perturbation component of TEC also shows its association with this storm. On the next day, that is, on 22 January when the storm exhibited its maximum excursion, the horizontal component H again showed its secondary maxima of 692 nT and 595 nT during the time before the recovery phase of the storm. The storm affected the upper atmosphere in such a manner that again the perturbation component of TEC was observed during the next day, that is, on 22 January.

The variation in the H component magnitude defines magnetosphere compression or expansion and connectivity under the geomagnetic storm condition. When the interplanetary magnetic field turns northward again, the rate of plasma energization and inward transport slows and the various loss processes that remove plasma from the ring current can begin to restore it to its pre-storm state. Several excellent reviews (Matuura, 1972; Fuller-Rowell et al., 1997; Buonsanto, 1999) concerning ionospheric storms have been published, in which the effects of magnetic storms on TEC are discussed. Ionospheric disturbances during magnetic storms are generally explained in terms of thermospheric changes in both neutral air winds and composition, which result in changes of the rates of production and loss of ionization.

Relationship between Total Electron Content, Scintillation, and Solar Wind at Low and High Latitudes

Solar wind affects the ionosphere by producing changes in the Earth's magnetic field and in particle precipitation. Now we compare our observations with respect to global vertical parameters of the interplanetary magnetic field and solar proton density, temperature, and speed. As the maximum storm effect was observed during the next day or the recovery phase of the storm, similarly, solar indices are also coincident with this time, that is, 22 January 2005. Figure 10.10 has six panels as follows: the first, second, and third panels show the Z, Y, and X components of the interplanetary magnetic field; the

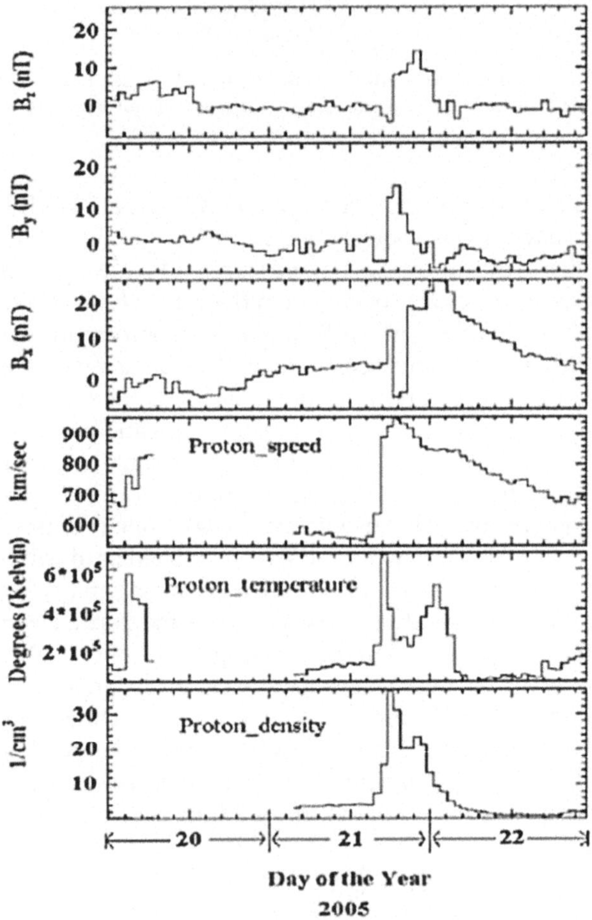

FIGURE 10.10
Interplanetary magnetic field and solar wind plot are panels 1, 2, 3, 4, 5, and 6, which represent Z, Y, and X components of IMF and solar proton speed, temperature, and density. (http://www.srl.caltech.edu/ACE/)

fourth panel shows the solar proton speed; the fifth panel shows the solar proton temperature; and the sixth panel shows the solar proton density. As shown in Figure 10.10, the negative excursion of the X, Y, and Z components of the interplanetary magnetic field was observed during the storm day itself, that is, 21 January. Since the time of sudden storm commencement was at 1711 UT, from this time onward, all the variations started with a southward direction and remained the same up to the next day or the recovery phase of the storm. Figure 10.10 also shows the sudden rise of the solar wind parameters, that is, solar proton speed varies up to 980 (Km/sec), solar proton temperature varies up to 7 × 105 (degrees Kelvin), and solar proton density has the value of 38 (1/cm^3) on the storm day of 21 January 2005.

From Figure 10.11, with this space weather disturbance, we have seen irregularity in total electron content, followed by scintillation activity at both low- and high-latitude stations on 20 January 2005, as this was the solar

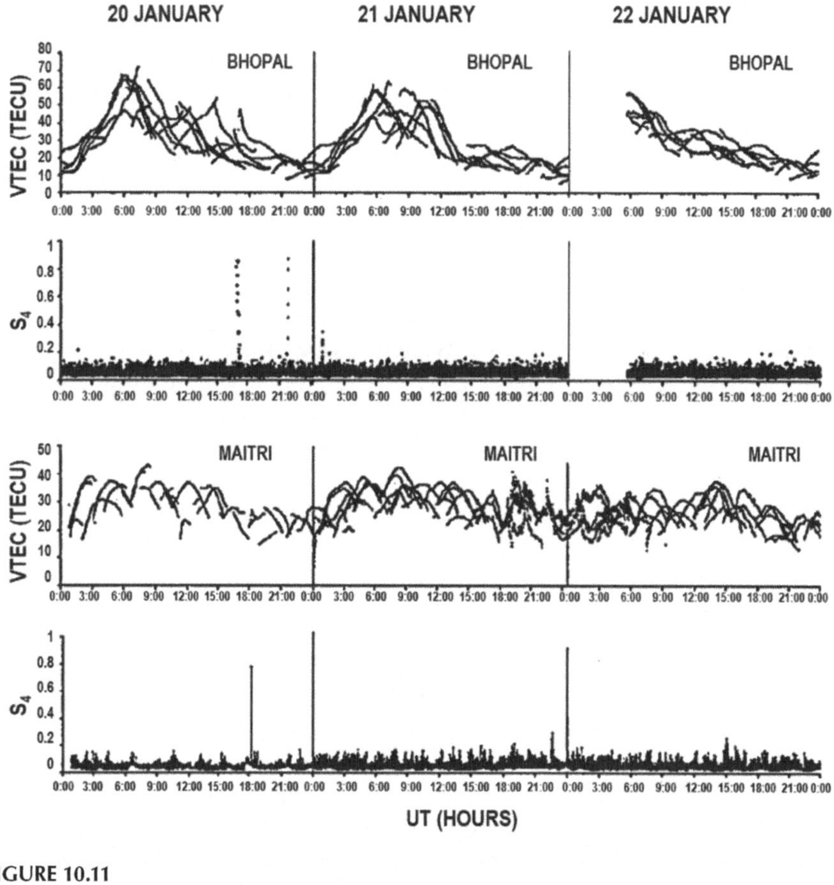

FIGURE 10.11
Variation of total electron content associated with amplitude scintillation during storm event of 21 January 2005 at low and high latitude.

energetic particle (SEP) event that has been called, by some measures, the most intense in 15 years, with a >100 MeV proton intensity comparable to that of the October 1989 event. Second, this event had very hard energy spectra. With the receiver location being in the sub-auroral zone (Maitri), due to solar proton precipitation in the auroral upper atmosphere, TEC shows perturbation with the intense level of scintillation of 0.8. The same irregular behavior is shown by the equatorial anomaly region (Bhopal), as the intense level of scintillation was about 0.9 during pre-midnight hours. During the storm day, 21 January, moderate scintillation activity was observed post-midnight in the low-latitude region Bhopal, and again moderate scintillation activity was observed in the high-latitude region Maitri with TEC perturbation. During this event, these electrons were energized through complex interactions between solar wind and the Earth's magnetic field. On the next day of the storm, when the maximum negative excursion occurred, TEC showed the same irregularity without any significant level of scintillation.

In the ionosphere, the space weather event is accompanied by strong enhancements of nighttime ionization in the polar region. This event is characterized by the geomagnetic or magnetospheric substorm induced due to solar activity, where associated irregularities in electron density lead to scintillation effects; similar activity is reported by Aarons (1982). In Figure 10.11, we see a correlation between TEC and scintillation. Variations of the total electron content can be due to variations of the solar radiation and solar wind. Since the selected time is around solar minimum, the variations of the solar radiation are marginal. Fortunately, the solar wind has recurrent high-speed streams during this time, so the relationship between TEC and solar wind can be well studied.

The irregularities generated on the bottom side of the post-sunset F-region over the magnetic equator by the Rayleigh-Taylor instability mechanism and then rise and align along magnetic field lines propagating to higher latitude (Kelley, 1989). The irregularities, while diffusing along field lines, break into small patches that are observed in the form of scintillations with small patch duration at low latitudes (Singh et al., 1993). Good overviews of scintillations at high and low latitudes were recently given by Basu et al. (2002).

Two main areas of the globe particularly troubled by scintillation are sub-auroral to polar latitudes and a belt surrounding the geomagnetic equator. Scintillations are most severe in the equatorial region, where they often occur after sunset and attain maximum intensity around the peaks of the Appleton anomaly (15° N and S of magnetic equator) (Basu et al., 1988). The electron density irregularities associated with scintillation are geomagnetic, and ionospheric storms generate significant large-scale signatures in the TEC maps (Jakowski et al., 1999). Differential TEC maps generated by subtracting monthly averages from actual TEC maps are helpful to draw conclusions concerning the nature of ionospheric storm processes. So, the corresponding analysis of the storm on 21 January, 2005, indicates both the action of an eastward-directed electric field as well as the action of perturbation-induced meridional winds in the thermosphere. The storm pattern provides deeper insight into storm mechanisms and may be useful for adequate modeling of ionospheric storm behavior in space weather forecast models. The GPS technique provides a unique opportunity to monitor this parameter continuously on global scale in near real time. In addition to solar and geomagnetic activity indices, TEC map-derived ionospheric perturbation indices would be helpful, especially for system operators to estimate the perturbation degree of the ionosphere.

Conclusion

Space-weather effects on transionospheric radio waves fall into two categories: (a) changes in the excess time delay introduced by large-scale variations (>km) in the ionospheric plasma density and (b) amplitude and phase scintillations caused by the appearance of sub-kilometer structures in the ionospheric plasma, usually due to the growth of a plasma instability. The excess time delay is simply proportional to the STEC along the signal path. The F region of the ionosphere is the most important region for transionospheric radio propagation, since the plasma density is the highest in this region. Hence, there has been a greater emphasis on understanding the response of the ionospheric F layer to space-weather variations. Given the complexity of the problem, physics-based models of the ambient ionosphere may provide the insight necessary to understand the observed patterns. The recent trend is to develop global-ionospheric-data assimilation models, which will be driven by appropriate inputs derived from various types of data in real time, to improve the ionospheric specification. The most important inputs are the high-latitude convection electric field, auroral currents and particle precipitation, solar radiation, and tides and gravity waves that propagate from below. The availability of simultaneous GPS data from globally distributed sites makes this an important tool for model validation.

Another challenging task is to find the link between space-weather parameters and the characteristics of ionospheric irregularities, including the very basic question of the prediction of the occurrence of these irregularities, which may have a severe impact on transionospheric radio propagation. During the last few years, observations of ionospheric irregularities of various scale sizes—using radar, scintillation techniques, and airglow images, besides in-situ measurements of density and electric-field fluctuations by satellites—have yielded a large amount of information on the characteristics of irregularities under different conditions. Some crucial parameters are now emerging as the key players in the development of irregularities. Computer simulations of the nonlinear evolution of plasma instabilities that give rise to these ionospheric irregularities may, in the near future, yield those characteristics of the irregularities that are of importance from the point of view of forecasting the outage of communication and navigation systems due to scintillations. An understanding of the pattern of evolution of the irregularities and their dynamics and knowledge of the background ionosphere can be incorporated into an improved regional specification and short-term forecast system.

Acknowledgments

The authors wish to acknowledge National Centre for Antarctic and Ocean Research (NCAOR), Goa, MoES, Govt. of India for the financial support through project.

References

Aarons J. 1982. Global morphology of ionospheric scintillations. *Proceedings of the IEEE*, v. 70, no. 4, pp.

Aarons J., Mendillo M., Yantosca R. and Kudeki E. 1996. GPS phase fluctuations in the equatorial region during the MISETA 1994 campaign. *J. Geophys. Res.*, v. 101, p. 26851.

Basu S., Groves K.M., Basu S.U. and Sultan P.J. 2002. *J Atmos Sol Terres phys*, v. 64, p. 1745.

Basu S., Mackenzie E. and Basu S.U. 1988. *Radio Sci.*, v. 23, p. 363.

Beach T.L., Kelley M.C., Kintner P.M. and Miller C.A. 1997. Total electron content variations due to non classical traveling ionospheric disturbances: Theory and global positioning system observations. *J. Geophys. Res.*, v. 102, p. 7279.

Bramley E.N. and Young M. 1968. Winds and electromagnetic drifts in equatorial F2-region. *J. Atmospheric. Terr. Phys.*, v. 30, pp. 99–111.

Buonsanto M.J. 1999. Ionospheric electron content measured by radio waves from storms—A review. *Space Sci. Rev.*, v. 88, pp. 563–601.

Fuller-Rowell T.J., Codrescu M.V., Roble R.G. and Richmond A.D. 1997. *How does the Thermosphere and Ionosphere React to a Geomagnetic Storm?* Geophysical Monograph, v. 98, pp. 203–225.

Ho C.M., Mannucci A.J., Lindqwister U.J., Pi X. and Tsurutani B.T. 1996. Global ionosphere perturbations by the worldwide GPS network. *Geophys. Res. Lett.*, v. 23, p. 3219.

Jakowski N. 1996. TEC Monitoring by using satellite positioning system. In: *Modern Ionospheric Science*, (Eds. K. Kohl et al.). European Geophysical Society, Katlenburg-Lindau, pp. 371–390.

Jakowski N., Schluter S. and Sardon E. 1999. Total electron content of the ionosphere during the geomagnetic storm. *J. Atmos. Solar-Terr. Phys.*, v. 61, pp. 299–307.

Kelley M.C. 1989. *The Earth's Ionosphere.* Academic, San Diego, CA.

Kelley M.C., Kotsikopoulos D., Beach T., Hysell D. and Musman S. 1996. Simultaneous global positioning system and radar observations of equatorial spread F at Kwajalein. *J Geophys. Res.*, v. 101, p. 2333.

Lu G., Pi X., Richmond A.D. and Roble R.G. 1998. Variations of total electron content during geomagnetic disturbances. *Geophys. Res. Lett.*, v. 25, pp. 253–256.

Matuura N. 1972. Theoretical models of ionospheric storms. *Space Sci. Rev.*, v. 13, pp. 124–189.

Mendillo M., Papagiannis M.D. and Klobuchar J.K. 1969. A seasonal effect in the mid-latitudes lab thickness variations during magnetic disturbances. *J. Atmos. Terr. Phys.*, v. 31(11), pp. 1359–1364.

Nakata Y. 1966. Variation in ionospheric electron content measured by radio waves from Syncom 3. *Radio Sci.*, v. 1, pp. 1145–1148.

Rastogi R.G. and Klobuchar J.A. 1990. Ionospheric electron content within the equatorial F2 layer anomaly belt. *J. Geophys. Res.*, v. 95, p. 19045.

Rishbeth H. 1998. How the thermospheric circulation affects the ionospheric F2-layer. *J. Atmos. Sol. Terr. Phys.*, v. 60, pp. 1385–1402.

Saito A., Fukao S. and Miyazaki S. 1998. High resolution mapping of TEC perturbations with GSI GPS network over Japan. *Geophys. Res. Lett.*, v. 25, pp. 3079–3082.

Sardón E., Rius A. and Zarraoa N. 1994. Estimation of the receiver differential biases and the ionospheric total electron content from global positioning system observations. *Radio Science*, v. 29, pp. 577–586.

Singh R.P., Singh A.K., Singh U.P. and Singh R.N. 1993. *Ind J Rad Spa Phys*, v. 22 pp.

Stening R.J. 1992. Review paper modelling the low latitude F region. *J. Atmospheric. Terr. Phys.*, v. 54, pp. 1387–1412.

Tereshchenko E.D., Khudukon B.Z., Kozlova M.O., Evstafiev O.V., Nygren T., Rietveld M.T. and Brekke A. 2000. Comparison of the orientation of small-scale electron density irregularities and F region plasma flow direction. *Ann. Geophys.*, v. 18, pp. 918–926.

Tereshchenko E.D., Khudukon B.Z., Kozlova M.O. and Nygren T. 1999. Anisotropy of ionospheric irregularities determined from the amplitude of satellite signals at a single receiver. *Ann. Geophys.*, v. 17, pp. 508–518.

Titheridge J.E. and Buonsanto M.J. 1999. A comparison of northern and southern hemisphere TEC storm behaviour. *J. Atmos. Terr. Phys.*, v. 50, pp. 763–780.

Valladares C.E., Basu S., Buchau, J. and Friis-Christensen E. 1994. Experimental evidence for the formation and entry of patches into the polar cap. *Radio Sci.*, v. 29, pp. 167–194.

Glossary

Aeronomy is the meteorological science of the upper region of the Earth's or other planetary atmospheres, which relates to the atmospheric motions, its chemical composition and properties, and the reaction to it from the environment from space.

Aerosol is a suspension of fine solid particles or liquid droplets, in air or another gas. Aerosols can be natural or anthropogenic.

Aerosol optical depth is a measure of the extinction of the solar beam by dust and haze. In other words, particles in the atmosphere (dust, smoke, pollution) can block sunlight by absorbing or by scattering light. An average aerosol optical depth for the U.S. is 0.1 to 0.15.

Albedo is a measure of how much light that hits a surface is reflected without being absorbed. Something that appears white reflects most of the light that hits it and has a high albedo, while something that looks dark absorbs most of the light that hits it, indicating a low albedo.

Antarctic Circumpolar Current: The Antarctic Circumpolar Current (ACC) is an ocean current that flows clockwise from west to east around Antarctica. An alternative name for the ACC is the West Wind Drift. The ACC is the dominant circulation feature of the Southern Ocean and has a mean transport estimated at 100-150 Sverdrup's (Sv, million m^3/s), or possibly even higher, making it the largest ocean current.

Aurora are the result of disturbances in the magnetosphere caused by solar wind. These disturbances are sometimes strong enough to alter the trajectories of charged particles in BOTH solar wind and magnetospheric plasma. These particles, mainly electrons and protons, precipitate into the upper atmosphere (thermosphere/exosphere).

Aurora Australis (also known as the southern lights, and southern polar lights) is the southern hemisphere counterpart to the aurora borealis. In the sky, an aurora Australis takes the shape of a curtain of light, or a sheet, or a diffuse glow; it most often is green, sometimes red, and occasionally other colors too.

Auroral emissions also referred to as polar lights, northern lights (aurora Borealis), southern lights (aurora Australis), is a natural light display in the Earth's sky, predominantly seen in the high-latitude regions (around the Arctic and Antarctic).

Auroral oval an oval-shaped region centered on the earth's magnetic pole in which auroral emissions occur.

Blizzard is a severe snowstorm characterized by strong sustained winds of at least 56 km/h (35 mph) and lasting for a prolonged period—typically three hours or more.

Coriolis force is an inertial or fictitious force that acts on objects that are in motion within a frame of reference that rotates with respect to an inertial frame.

Divergent boundary or divergent plate boundary also known as a constructive boundary or an extensional boundary, is a linear feature that exists between two tectonic plates that are moving away from each other.

El Nino is the warm phase of the El Niño–Southern Oscillation (ENSO) and is associated with a band of warm ocean water that develops in the central and east-central equatorial Pacific (between approximately the International Date Line and 120°W), including the area off the Pacific coast of South America. The ENSO is the cycle of warm and cold sea surface temperature (SST) of the tropical central and eastern Pacific Ocean.

Electrojet an electrojet is an electric current which travel around the E region of the Earth's ionosphere. There are two electrojets: above the magnetic equator (the equatorial electrojet), and near the Northern and Southern Polar Circles (the Auroral Electrojets). Electrojets are Hall currents carried primarily by electrons at altitudes from 100 to 150 km.

Electrodynamics branch of physics that deals with the effects arising from the interactions of electric currents with magnets, with other currents, or with themselves.

Equinox is commonly regarded as the instant of time when the plane (extended indefinitely in all directions) of Earth's equator passes through the center of the Sun.

Extra-tropical cyclone is a storm system that primarily gets its energy from the horizontal temperature contrasts that exist in the atmosphere. Extra-tropical cyclones (also known as mid-latitude or baroclinic storms) are low pressure systems with associated cold fronts, warm fronts, and occluded fronts.

Geomagnetism is the study of the dynamics of the Earth's magnetic field, which is produced in the inner core. The Earth's magnetic field is predominantly a geo-axial dipole, with north and south magnetic poles located near the geographic poles that undergo periodic reversals and excursions.

Ground truth is a term used in various fields to refer to information provided by direct observation (i.e. empirical evidence) as opposed to information provided by inference.

Hadley cell is a global scale tropical atmospheric circulation that features air rising near the Equator, flowing poleward at a height of 10 to 15 kilometers above the earth's surface, descending in the subtropics, and then returning equatorward near the surface.

Indonesian through Flow (ITF) is an ocean current with importance for global climate since it provides a low-latitude pathway for warm, fresh water to move from the Pacific to the Indian Ocean and this serves as the upper branch of the global heat conveyor belt.

Inter Tropical Convergence Zone (ITCZ), is a belt of low pressure which circles the Earth generally near the equator where the trade winds of the Northern and Southern Hemispheres come together. It is characterized by convective activity which generates often vigorous thunderstorms over large areas.

Ionosphere is defined as the layer of the Earth's atmosphere that is ionized by solar and cosmic radiation. It lies 75-1000 km (46–621 miles) above the Earth. (The Earth's radius is 6370 km, so the thickness of the ionosphere is quite tiny compared with the size of Earth.)

Isostatic adjustment refers to the transient (10^2–10^4 years) or long term (> 10^5 years) nonelastic response of the earth's lithosphere to loading and unloading due to erosion, deposition, water loading, desiccation, ice accumulation, and deglaciation.

Katabatic wind is the technical name for a drainage wind, a wind that carries high-density air from a higher elevation down a slope under the force of gravity.

Meridional when the winds cross the latitude lines at a sharp angle, the wind pattern is termed meridional.

Meteorology is the scientific study of the atmosphere that focuses on weather processes and forecasting. Meteorological phenomena are observable weather events which illuminate and are explained by the science of meteorology. Those events are bound by the variables that exist in Earth's atmosphere.

North Atlantic Oscillation is a weather phenomenon in the North Atlantic Ocean of fluctuations in the difference of atmospheric pressure at sea level (SLP) between the Icelandic Low and the Azores High.

Ozonesonde an instrument, carried aloft in a balloon that measures the atmospheric concentration of ozone (and often other gases) and transmits the data by radio.

Planetary boundary layer (PBL), also known as the atmospheric boundary layer (ABL) is the lowest part of the atmosphere and its behavior is directly influenced by its contact with a planetary surface.

Plasmasphere, or inner magnetosphere, is a region of the Earth's magnetosphere consisting of low energy (cool) plasma. It is located above the ionosphere. The outer boundary of the plasmasphere is known as the plasmapause, which is defined by an order of magnitude drop in plasma density.

Polar vortex is a large area of low pressure and cold air surrounding both Earth's poles. It always exists near the poles but weakens in summer and strengthens in winter. The term "vortex" refers to the counter-clockwise flow of air that helps keep the colder air near the Poles.

Protonophore, region in the Earth's upper atmosphere where atomic hydrogen and protons (ionic hydrogen) are the dominant constituents; it can be considered the outermost extension of the ionosphere.

Pyranometer is a type of actinometer used for measuring solar irradiance on a planar surface and it is designed to measure the solar radiation flux density (W/m^2) from the hemisphere above within a wavelength range 0.3 µm to 3 µm.

Radiative forcing or climate forcing is the difference between insolation (sunlight) absorbed by the Earth and energy radiated back to space.

Scintillation is a flash of light produced in a transparent material by the passage of a particle (an electron, an alpha particle, an ion, or a high-energy photon)

Summer solstice the two moments during the year when the path of the Sun in the sky is farthest north in the Northern Hemisphere (June 20 or 21) or farthest south in the Southern Hemisphere (December 21 or 22).

Thermal inversion occurs when a layer of warm air settles over a layer of cooler air that lies near the ground. The warm air holds down the cool air and prevents pollutants from rising and scattering.

Thermohaline circulation is a part of the large-scale ocean circulation that is driven by global density gradients created by surface heat and freshwater fluxes.

Total solar eclipse occurs when the dark silhouette of the Moon completely obscures the intensely bright light of the Sun, allowing the much fainter solar corona to be visible. During any one eclipse, totality occurs at best only in a narrow track on the surface of Earth. This narrow track is called the path of totality.

Troposphere is the lowest layer of Earth's atmosphere and is also where nearly all-weather conditions take place. It contains 75% of the atmosphere's mass and 99% of the total mass of water vapor and aerosols.

Very low frequency or VLF is the ITU designation for radio frequencies (RF) in the range of 3 to 30 kilohertz (kHz), corresponding to wavelengths from 100 to 10 kilometers, respectively.

Zenith angle is the angle between the zenith and the center of the Sun's disc. The solar elevation angle is the altitude of the Sun, the angle between the horizon and the center of the Sun's disc. Since these two angles are complementary, the cosine of either one of them equals the sine of the other.

Zenith sky is the point in the sky vertically straight above the observer. It is the point where the observer's sight-line is perpendicular to the Earth's atmosphere, and hence where the smallest amount of air – termed the airmass – lies along it.

Index